ENVIRONMENTAL AND SUSTAINABLE DEVELOPMENT THROUGH FORESTRY AND OTHER RESOURCES

ENVIRONMENTAL AND SUSTAINABLE DEVELOPMENT THROUGH FORESTRY AND OTHER RESOURCES

Edited by

Arnab Banerjee, PhD

Manoj Kumar Jhariya, PhD

Dhiraj Kumar Yadav, PhD

Abhishek Raj, PhD

AAP | APPLE ACADEMIC PRESS

Apple Academic Press Inc.
4164 Lakeshore Road
Burlington ON L7L 1A4
Canada

Apple Academic Press Inc.
1265 Goldenrod Circle NE
Palm Bay, Florida 32905
USA

International Standard Book Number-13: 978-1-77188-811-0 (Hardcover)
International Standard Book Number-13: 978-0-42927-602-6 (eBook)

Library and Archives Canada Cataloguing in Publication

Title: Environmental and sustainable development through forestry and other resources / edited by Arnab Banerjee, PhD, Manoj Kumar Jhariya, PhD, Dhiraj Kumar Yadav, PhD, Abhishek Raj, PhD.

Names: Banerjee, Arnab (Professor of environmental science), editor. | Jhariya, Manoj Kumar, editor. | Yadav, Dhiraj Kumar, editor. | Raj, Abhishek, editor.

Description: Includes bibliographical references and index.

Identifiers: Canadiana (print) 20190189398 | Canadiana (ebook) 20190189401 | ISBN 9781771888110 (hardcover) | ISBN 9780429276026 (ebook)

Subjects: LCSH: Sustainable forestry. | LCSH: Sustainable development.

Classification: LCC SD387.S87 E58 2020 | DDC 634.9028/6—dc23

CIP data on file with US Library of Congress

Apple Academic Press also publishes its books in a variety of electronic formats. Some content that appears in print may not be available in electronic format. For information about Apple Academic Press products, visit our website at **www.appleacademicpress.com** and the CRC Press website at **www.crcpress.com**

About the Editors

Arnab Banerjee, PhD
Assistant Professor, Department of Environmental Science, Sarguja Vishwavidyalaya, Ambikapur, Chhattisgarh, India

Arnab Banerjee, PhD, is an Assistant Professor in the Department of Environmental Science at Sarguja Vishwavidyalaya (a state university), Ambikapur, Chhattisgarh, India. He completed his MSc degree and PhD (Environmental Science) from Burdwan University, India, and earned his MPhil in Environmental Science from Kalyani University, West Bengal, India. He won the University Gold Medal for securing first-class first position in his MSc examination. He has been awarded a Young Scientist Award for best oral presentation at an international conference held at the University of Burdwan. He was a project fellow under a University Grants Commission-sponsored major research project. He has published 65 research papers in reputed national and international journals. Also to his credit, he has published five books and six book chapters. He is a life member of the Academy of Environmental Biology. He has supervised 23 postgraduate students and is dynamically involved in teaching and research.

Manoj Kumar Jhariya, PhD
Assistant Professor, Department of Farm Forestry, Sarguja University, Ambikapur, Chhattisgarh, India

Manoj Kumar Jhariya, PhD, is an Assistant Professor in the Department of Farm Forestry at Sarguja University, Ambikapur, Chhattisgarh, India, and is the author or co-author of more than 60 research papers in peer-reviewed journals, four books, 20 book chapters, and several extension articles. Dr. Jhariya acquired his BSc (Agriculture), MSc (Forestry), and PhD (Forestry) degrees from Indira Gandhi Krishi Vishwavidyalaya, Raipur, Chhattisgarh, India. He won the University Gold Medal for securing the first-class first position in his PhD examination. He was awarded the Chhattisgarh Young Scientist Award in the year 2013 from the Chhattisgarh Council of Science and Technology. He was also awarded a University

Grants Commission, Rajiv Gandhi National Fellowship, New Delhi, India. He is an editorial board member of several journals. He is also a life member of the Indian Science Congress Association, the Applied and Natural Science Foundation, the Society for Advancement of Human and Nature, the Medicinal and Aromatic Plants Association of India, and the International Society for Development and Sustainability. He has supervised over 40 MSc (Farm Forestry) students and is dynamically involved in teaching (postgraduate) and research.

Dhiraj Kumar Yadav, PhD
Assistant Professor and HOD (In-Charge), Department of Farm Forestry, Sarguja University, Ambikapur, Chhattisgarh, India

Dhiraj Kumar Yadav, PhD, is an Assistant Professor and HOD (In-Charge), Department of Farm Forestry, Sarguja University, Ambikapur, Chhattisgarh, India. He completed his graduation, postgraduation, and PhD in Forestry at Jawaharlal Nehru Agricultural University, Indira Gandhi Agricultural University, and Kumaun University Nainital, respectively. He pursued his MBA in Human Resource Management from Sikkim Manipal University and received a diploma in environmental management. He was awarded a Chhattisgarh Young Scientist Award in 2008 from the Chhattisgarh Council of Science and Technology. He won a merit scholarship during his BSc work. He also worked in several positions at various forestry institutes. He has published many research papers and articles and has published two books and several book chapters. He is also a member of several academic societies of India. He has supervised over 40 MSc (Farm Forestry) students and is dynamically involved in teaching (postgraduate) and research.

Abhishek Raj, PhD
Research Scholar, Department of Forestry, Indira Gandhi Krishi Vishwavidyalaya, Raipur, Chhattisgarh, India

Abhishek Raj, PhD, is a Research Scholar in the Department of Forestry at Indira Gandhi Krishi Vishwavidyalaya, Raipur (Chhattisgarh), India, and is the author or co-author of more than 20 research and review papers in peer-reviewed journals, 10 book chapters, and several extension articles. Mr. Raj acquired his BSc (Forestry) from Sam Higginbottom University of Agriculture, Technology and Sciences, Allahabad, India, and his

MSc (Forestry) and PhD (Forestry) degrees from Indira Gandhi Krishi Vishwavidyalaya, Raipur (Chhattisgarh) India. He was awarded a Young Scientist Award for outstanding contribution in the field of forestry in 2017 from Science and Tech Society for Integrated Rural Improvement (S&T SIRI), India. He has qualified ICAR/ASRB NET (National Eligibility Test) in agroforestry for lectureship. He is an editorial board member of many reputed journals.

Contents

Contributors

Arnab Banerjee
Assistant Professor, Department of Environmental Science, Sarguja University, Ambikapur–497001, Chhattisgarh, India, E-mail: arnabenvsc@yahoo.co.in

Shiboram Banerjee
Research scholar, PG Department of Conservation Biology, Durgapur Government College, Durgapur, West Bengal, India, Mobile: +91-9153083631, E-mail: shiboramb@gmail.com

Pallavi Chattopadhyay
Postgraduate Student, Department of Microbiology, Ranigaunj Girls College, West Bengal; Amity University, Lucknow Campus, Lucknow, Uttar Pradesh, India, Mobile: +91-8116670112, E-mail: pallavi96chatto@gmail.com

Pinaki Chattopadhyay
Undergraduate Student, Department of Botany, Durgapur Government College, Durgapur, West Bengal, India, Mobile: +91-8293807145, E-mail: iampinu7@gmail.com

Anjan Ray Chaudhury
Assistant Professor, Department of Economics, Durgapur Government College, Durgapur, West Bengal, India

Renu Dhupper
Amity Institute of Environment Sciences, Noida, Uttar Pradesh, India

Vishal Gandhi
Central Pollution Control Board, New Delhi, India

Anila George
Assistant Professor, PG Department of Environmental Science, St. John's College, Anchal, Kollam, Kerala, India, Mobile: +91-9496195458, E-mail: george44@gmail.com

Nazir Hussain
Ex-Advisor, Qatar Shell Research, and Technology Center, and Ex-Expert Agriculture, Ministry of Environment, Doha, Qatar; Present Address: Safari Villas 247, Bahria Town, Lahore, Pakistan, Mobile: +923014117927, E-mail: drnazirhussain@gmail.com

Harshita Jain
Amity Institute of Environment Sciences, Noida, Uttar Pradesh, India, Mobile: +91-7838683947, E-mail: harshita13@gmail.com

Manoj Kumar Jhariya
Assistant Professor, Department of Farm Forestry, Sarguja University, Ambikapur – 497001, Chhattisgarh, India, E-mail: manu9589@gmail.com

Geetanjali Kaushik
MGM's Jawaharlal Nehru Engineering College, Aurangabad, Maharashtra, India

Saikat Mondal
Assistant Professor, Department of Zoology, Raghunathpur College, Purulia, West Bengal, Pincode-723133, India, E-mail: sairaniganj@gmail.com

Debnath Palit
Associate Professor and Head, Department of Botany, Durgapur Government College, Durgapur, West Bengal – 713214, India, Mobile: +91-9832175737, E-mail: debnath_palit@yahoo.com

Abhishek Raj
Research Scholar, Department of Forestry, College of Agriculture, Indira Gandhi Krishi Vishwavidyalaya (I.G.K.V.), Raipur – 492012, Chhattisgarh, India, E-mail: ranger0392@gmail.com

Dhiraj Kumar Yadav
Assistant Professor, Department of Farm Forestry, Sarguja University, Ambikapur – 497001, Chhattisgarh, India, E-mail: dheeraj_forestry@yahoo.com

Abbreviations

2MB	2-methyl-1-butanol
AcMNPV	Autographa californica multinucleocapsid nuclear polyhedrosis virus
ACP	acid phosphatase
AMF	arbuscular mycorrhizal fungi
AO	ascorbate oxidase
APHA	American Public Health Association
API	air pollution index
API	anticipated performance index
APOE	apolipoprotein E
APTI	air pollution tolerance index
APX	ascorbate peroxidase
As	arsenic
ASP	activated sludge process
ATP	adenosine triphosphate
B:C ratio	benefit and cost ratio
Ba	barium
BGA	blue-green algae
BHC	benzene hexachloride
BioMEMs	biological microelectromechanical systems
BIS	Bureau of Indian Standards
BNF	biological N_2 fixation
BOD	biochemical oxygen demand
Bt	*bacillus thuringiensis*
Ca	calcium
$CaCO_3$	calcium carbonate
CAT	catalase
Cd	cadmium
CDM	carbon fund project
CFC	chlorofluro carbon
CH_4	methane
CNG	compressed natural gas
CO	carbon monoxide
Co	cobalt

CO_2	carbon dioxide
COD	chemical oxygen demand
COPD	chronic obstructive pulmonary disease
CPCB	Central Pollution Control Board
Cr	chromium
CSP	carbon sequestration potential
Cu	copper
DDT	dichlorodiphenyltrichloroethane
DNA	deoxyribonucleic acid
EC	electrical conductivity
EDTA	ethylenediaminetetraacetic acid
EPA	Environment Protection Agency
ETS	electron transport system
FAO	Food and Agricultural Organization
FC	fecal coliform
Fe	iron
FISH	fluorescent *in situ* hybridization
FNS	food and nutritional security
GDP	gross domestic product
GHGs	greenhouse gases
GM	genetically modified
GoI	Government of India
GPX	glutathione peroxidase
GSA	green screen assay
Gy	gray
H^+	proton
H3	tritium
Hg	mercury
HO_2^-	perhydroxy radical
HP	horsepower
HSO_3^-	bisulfite
ICMR	Indian Council of Medical Research
IPCC	Intergovernmental Panel on Climate Change
ISBR	improved sequential batch reactor
ISPM	inhalable suspended particulate matter
ISS	intensive silvopastoral systems
IUCN	International Union for Conservation of Nature and Natural Resources
IVI	importance value index

JBIC	Japan bank for international cooperation
JICA	Japan International Cooperation Agency
K	potash
KI	potassium iodide
LiP	lignin-peroxidase
LPG	liquefied petroleum gas
MARA	microbial assay for toxicity risk assessment
MCL	maximum concentration level
Mcpe	monthly per capita expenditure
MENA	Middle East and North Africa
Mg ha^{-1}	megagram per hectare
Mg	magnesium
MLD	milliliter per day
MnP	manganese peroxidase
Mo	molybdenum
MoEFCC	Ministry of Environment, Forest, and Climate Change
MPN	most probable number
MPTs	multipurpose tree species
N	nitrogen
NAAQS	National Ambient Air Quality
$NaAsO_2$	sodium arsenite
NaOH	sodium hydroxide
NGOs	Non-Governmental Organizations
Ni	Nichol
NO	nitric oxide
NO_2	nitrogen dioxide
NO_2^{2-}	nitrite
NO_3^{2-}	nitrate
NOx	nitrogen oxides
NPV	nuclear polyhedrosis virus
NRC	National Research Council
NTFPs	nontimber forest products
$OH^{\cdot}$	hydroxyl
$OH_2^{\cdot}$	peroxyl
OM	organic matter
P	phosphorus
PAHs	polycyclic aromatic hydrocarbons
Pb	lead
PBS	poly(butylene succinate)

PCBs	polychlorinated biphenyls
PCL	poly(Å-caprolactone)
PCR	polymerase chain reaction
PE	polyethylene
PERC	Phycospectrum Environmental Research Center
PHA	poly hydroxyl alkanoate
PHB	poly(3-hydroxybutyrate)
PLA	polylactide
PM	particulate matter
PP	polypropylene
PPDO	poly (p-dioxanone)
PUFA	polyunsaturated fatty acid
R&D	research and development
RCF	Raniganj coalfield
ROO·	peroxy radical
ROOH	organic hydroperoxide
ROS	reactive oxygen species
RSPM	respirable suspended particulate matter
RWC	relative water content
SAFAR	System of Air Quality and Weather Forecasting and Research
SBR	sequential batch reactor
SCADA	supervisory and data acquisition system
Se	selenium
SEM	scanning electron microscopy
SO_3^{2-}	sulfite
SOD	superoxide dismutase
SO_X	sulfur oxides
SPCB	State Pollution Control Board
SPM	suspended particulate matter
STP	sewage treatment plant
TC	total coliform
TCE	trichloroethylene
TDS	total dissolved solids
TH	total hardness
TN	total nitrogen
TNT	trinitrotoluene
TP	total phosphorus
TSPM	total suspended particulate matter

TSS	total suspended solids
TV	television
TVOC	total volatile organic compound
UNODC	United Nations Office on Drugs and Crime
USEPA	United States Environmental Protection Agency
VAM	vesicular-arbuscular mycorrhiza
VCRC	Vector Control Research Center
VOC	volatile organic carbon
VOCs	volatile organic compounds
WHO	World Health Organization
WQI	water quality index
Zn	zinc
$ZnSO_4$	zinc sulfate

Acknowledgments

Everything in this world is in the hands of Almighty. It is the supreme divine power that helps us to move forward to our life. Life is an eventful journey in which human beings continuously learn about their surroundings. It is hard work that brings happiness and positive outcomes in life. The present book is the result of tireless effort, positive mental attitude, as well as strong enthusiasm of the group of editors. It is the spiritual bliss and motivation that helped us to reach this platform of completing a book in the field of forestry, environment, and sustainable development.

With a profound and unfading sense of gratitude, we express our deep sense of gratitude to Prof. Rohini Prasad, Honourable Vice-Chancellor, Sarguja University, and Mr. Binod Kumar Ekka, Registrar, Sarguja University, for their continuous support, enthusiasm, and encouragement for helping us in every step of the way to accomplish this venture.

We are grateful to Mr. Ashish Kumar, President, Apple Academic Press, for helping us to accomplish our dream of publishing this book. Further, we acknowledge Dr. Mohammed Wasim Siddiqui, Assistant Professor and Scientist, Bihar Agriculture University, India; Prof. M. N. Hoda, Director Bharati Vidyapeeth's Institute of Computer Application & Management, New Delhi, and Dr. Ritika Wason, Associate Professor and Scientist, Bharati Vidyapeeth's Institute of Computer Application & Management, New Delhi, for their continuous help and support for the execution of the project.

Preface

The dimension of the forestry environment is changing in a rapid way under the era of globalization. As a consequence, incidences of deforestation, desertification, and environmental degradation are putting a huge challenge in front of human civilization. In a modern context, traditional forestry practices have diverged into various modes of sustainable forest management practices. In relation to the environment, newer technologies are designed with eco-friendly approaches to control environmental pollution. Sustainable development is a huge concept that addresses the social, economic, and environmental paradigms of human civilization. Rapid change is taking place across the globe in the field of forestry environment. In the context of the modern world, the rapid growth of human civilization, along with the rapid growth of science and technology, is degrading the world's forests to a considerable extent. Under such a context, addressing sustainability is the biggest challenge for humankind. Some key sustainable approaches include agroforestry, silviculture, biopesticides, biofertilizers, green designing, etc. The new trends and issues toward addressing sustainable development are the central theme of this book.

The book has 12 chapters. Chapter 1 describes the role of forests in the management of resources and the protection of the environment. Chapter 2 looks at different agroforestry systems prevailing in the hills and their ecological role. Different examples are cited in the chapter to enhance the discussion. Chapter 3 deals with the multifunctional roles of leguminous species in relation to agroforestry. Chapter 4 addresses the issue of pastureland for livestock management and mitigation of climate change. Chapter 5 addresses the role of higher plants toward mitigating air pollution. Chapter 6 provides detailed insight about the potential benefits and uses of sludge in order to avoid harm in the environment. Chapter 7 addresses the experimental issue of pit lakes to effect socio-economic upliftment and livelihood security. In Chapter 8, the levels of chlorine and water quality monitoring of a public distribution system were thoroughly discussed. Chapter 9 deals with the various forms of environmental pollution and their health aspects. Chapter 10 addresses issues relating to new innovative approaches toward sustainability and biodiversity conservation. Chapter 11 reflects the various dimensions of biotechnological approaches to facilitate a sustainable world. Chapter 12 reflects the different technological aspects of sewage treatment plants.

This book will be a standard reference work for the disciplines such as forestry, agriculture, ecology, and environmental science and will also be a pathway to addressing new concepts for a sustainable world.

The editors would appreciate receiving comments from readers that may assist in the development of future editions.

—Arnab Banerjee, PhD
Manoj Kumar Jhariya, PhD
Dhiraj Kumar Yadav, PhD
Abhishek Raj, PhD

CHAPTER 1

Forests for Resource Management and Environmental Protection

ABHISHEK RAJ,[1] M. K. JHARIYA,[2] D. K. YADAV,[2] and A. BANERJEE[3]

[1]*Department of Forestry, College of Agriculture, I.G.K.V., Raipur-492012 (C.G.), India, E-mail: ranger0392@gmail.com*

[2]*Department of Farm Forestry, Sarguja University, Ambikapur–497001 (C.G.), India, E-mails: manu9589@gmail.com (M. K. Jhariya); dheeraj_forestry@yahoo.com (D. K. Yadav)*

[3]*Department of Environmental Science, Sarguja University, Ambikapur - 497001 (C.G.), India, E-mail: arnabenvsc@yahoo.co.in*

ABSTRACT

Forests are one of the largest natural resource, the habitat of precious flora and fauna and many other organisms, nurturing soil-inhabiting organism along with maintaining soil quality and health, mitigating climate change and GHGs level in the atmosphere through the activity carbon storage and sequestration into vegetation and soils, respectively. Forests provide both tangible and intangible ecosystem services which help in sustaining the lives on the earth. Tangible (direct) benefits include both timber and non-timber forest products, whereas biodiversity conservation, watershed management, climate change mitigation, soil management, and livelihood security are regarded as intangible and indirect services from the forests. The activities like deforestation, faulty agricultural practices, overexploitation of NTFPs, etc., are destroying the valuable resources and ecosystem health through the release of several greenhouse gases (GHGs) into the atmosphere. In this context, sustainable forest management plays a vital role in productivity and health of forests and related forest products, which recognize an important source of income to the rural farmer and tribal people. Thus, effective policies and its implementation with good governance, scientific-based R&D, and building an effective roadmap for future strategies and scientific oriented programs are the good strategies for forest management.

1.1 INTRODUCTION

Forest has played a very remarkable role in supporting and sustaining livelihoods through providing food and nutrition to the population and make food and nutritional security (FNS). Not to only specific but these systems range from natural forest to plantation and agroforestry systems where people get bulk of tangible and intangible materials for their life and environment sustenance. As we know, forest and trees play an incredible role in well being of human, wild animal, and other organisms with overall proper functioning of ecosystems and maintaining the biodiversity. It is a well-known fact that forest provides both direct and indirect benefits to all organisms, which are the basis of life and provide smooth functioning of the ecosystem and its services to whole biomes. But due to the activity of deforestation, land degradation, diversion of forested land to the non-forested area such as agricultural intensification, development projects like industry, railways lines, power projects, construction of dams and canals, etc. which affects the biodiversity and destroy this treasure at the regional and global level.

Moreover, the forest provides various NTFPs and healthy food and fruits to people which are the basis of healthy life and the source of income that maintain ecosystem health and environmental security (Figure 1.1). Forest stores diverse type of flora and fauna which is life-supporting for humans and very essential for the smooth functioning of ecosystem and its services. Forest stores various type of wild edible plants, medicinal plants, beneficial herbs, timber trees, NTFPs species that provides certain ecosystem services along which is very essential for staple crop production. For example, forest supports some essential pollinators that have significance importance in crop production and tree-based cropping systems throughout the world. NTFPs, which are the primary source from forest, play a vital role in rural economy and strengthening the tribal livelihood by accessing money and providing opportunities of employment to people (Chou, 2018; Harbia et al., 2018; Leßmeister et al., 2018).

Forest and tree-based farming systems offer multifarious advantages over other permanent cropping systems in the term of providing diverse food production with a great adaptability to a wide range of environment and climatic conditions. Forest and other woodland systems in tropical region (particularly moist region) represent a great biodiversity and stores important species of desired importance worldwide (Mace et al., 2005; Lindenmayer, 2009; Gibson et al., 2011). However, some anthropogenic pressure and faulty human activities lead degradation and fragmentation of forest which resulted much loss of biodiversity (Lindenmayer and Franklin, 2002; FAO, 2015a; Newbold et al., 2015; Van Der Plas et al., 2016). Asia, Africa, Central, and South America, have been experiencing continue declining (Newbold et al., 2015; Keenan et al., 2015).

Although, forests offers a remarkable ecosystem services such as climate regulation (through minimizing climate change impact), biomass production (through carbon storage and sequestration), water supply and purification, pollination, and the provision of habitats for forest species (Bauhus et al., 2010; Thompson et al., 2011; Brockerhoff et al., 2013; Decocq et al., 2016; Liang et al., 2016; Mori et al., 2017). There is a great relationship and crystal clear link are exist between biodiversity and most ecosystem services (Gamfeldt et al., 2013). Larger trees with older age have a capacity to create the most benefits as economical return through the better scientific planting technology with care and maintenance (Lindenmayer and Laurance, 2017). Similarly, trees are a very good solution for the problem, which faced by people today for those residing in the both urban and rural region. Likewise, arboreta, and botanical gardens can play an important role in healthy urban forest ecosystem. As per expert suggestions in every dollar of investment on planting, care, and management of trees can result higher return even up to 500% (McPherson et al., 2005, 2016). Large trees have multifarious benefits in the term of habitat creation for other important species (Stagoll et al., 2012; Le Roux et al., 2015; Lindenmayer and Laurance, 2017) and environmental health maintenance through storage and sequestration of carbon from the atmosphere (Stephenson et al., 2014).

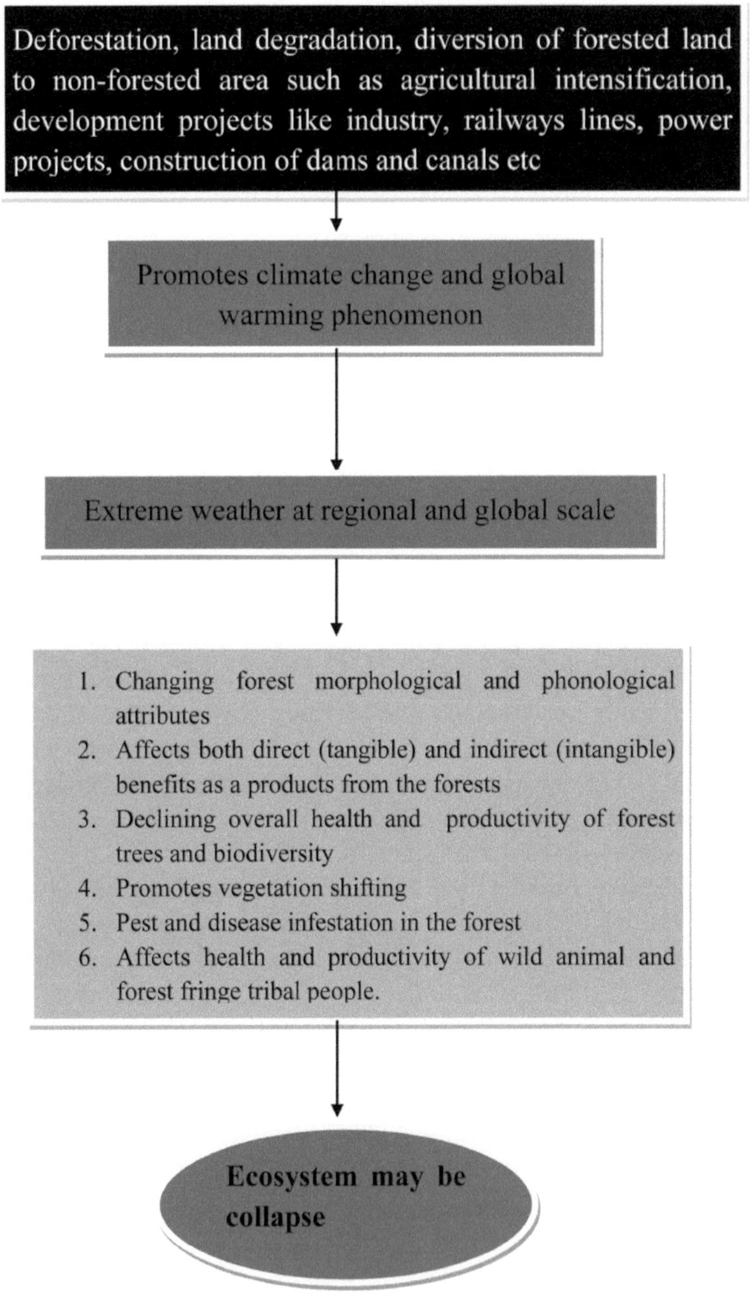

FIGURE 1.1 Climate change impact on forest ecosystem.

This chapter reviews and highlighted importance of forest, its multifarious and uncountable benefits such as tangible (direct) and intangible (indirect) which helps in sustaining the lives on the earth. Forest plays a major role in resources management and its proper utilization through scientific production of both timber and non-timber forest products whereas biodiversity conservation, watershed management, climate change mitigation, soil management and livelihood security are regarded as intangible and indirect services from the forests. This chapter also highlighted about forest role in mitigating climate change and strengthening socio-economic status of rural farmers and forest dwellers.

1.2 FOREST: A GLOBAL CONTEXT

Globally, the forest is the 2nd largest land use terrestrial natural resources after the agriculture which plays a major role socio-economic development of tribal people but due to degradation of 41% of forest cover in the country (India) which affects all biodiversity and environmental health system (MoEF, 2002). Similarly, Indian forest gave shelter to 14% of people's population which directly and indirectly depends on timber and NTFPs for their life (MoEF, 2002; Kumar, 2002). India represents 12 megabiodiverse country of the world with 47,000 (12% of the world flora) and 90,000 (7.28% of the world fauna) species of both flora and animal species (MoEF, 2007).

There are a several definitions of forest reported by various forest and environmental organization. For example, according to FRA (Forest Resource Assessment) (2005) the land having canopy cover of >10% with tree higher than 5 m on >0.5 ha area are regarded as forest and free of agricultural or urban use (FAO, 2006). Recently, the statistics report through FRA I valuable resource in many scientific studies such as study report on forest and agricultural land dynamics (Ausubel et al., 2012), deforestation (Hosonuma et al., 2012), environmental health and its sustainability (Arrow et al., 2012) and the biogeochemical cycle including carbon cycle (Smith et al., 2014). As per FAO (2015b), total forest covers were 3999 million hectares globally (compared to woodland area covered by 1204 million hectares) in the year of 2015 which reflects nearly 31% of worlds land area and contributed 0.6 hectares for every person. Around 44% global forest area share by tropical region followed by 26% (temperate region), 22% (boreal) and 8% (sub-tropical region), respectively (Table 1.1).

TABLE 1.1 Comparative Area of Forest and Other Woodland in the World Forest (FAO, 2015b)

Type	Forest Area (K ha)	Woodland Area (K ha)
Tropical	1,770,156	516,935
Sub-tropical	320,057	399,094
Temperate	684,468	167,255
Boreal	1,224,452	121,187
Total	3,999,134	1,204,471

1.3 FOREST FOR MULTIFARIOUS BENEFITS AND ECOSYSTEM SERVICES

Indeed, the forest provides multifarious benefits, including both tangible (direct) and intangible indirect service in the term of ecosystem services. Tangible services comprise timber, NTFPs, fuelwood, biofuel, etc. while biodiversity conservation, climate change mitigation, watershed management, soil fertility enhancement, and nutrient cycling, etc., are the potential indirect services from forest ecosystem (Figure 1.2).

1.3.1 TANGIBLE BENEFITS

Timber and NTFPs are very important tangible (direct) products derived from the natural and plantation forest and have been used by people and societies from the ancient time around the world. From the ancient time, timber is used for the construction purpose of building, pole, house post, pillar, bridge, etc., which is prerequisite and basic need for the human society for their living and life sustenance. Softwood and hardwood based timber from the forest are used for the construction industry such as brick, steel, timber, sand, cement, and other aggregates. But the timber-based supply chain is unique. As we know, construction materials require ores, rocks, or a soil which is mechanically removed from the earth. While timber is harvested from the forest without disturbing and digging the forest floor and ground. Moreover, any changes in our forests will impact on ecology, environment, and society. Forest can meet the demand of timber for the burgeoning population and makes a positive contribution to ecology, environment, and society along with mitigating climate change and global warming through balancing global carbon budget.

1.3.2 TIMBER AND WOOD-BASED PRODUCTS

Timber is the natural wood products which can directly get through the natural forest. Timber is one of the most important tangible (direct) products. The wood products are more economically profitable globally. For example, the Asian continent contributed 25 to 60% of the global wood-based panel sector in the past two decades (FAOSTAT, 2016). Similarly, plywood production has increased from 11 million m^3 to 104 million m^3 in the duration of fourteen years (2000–2014) (FAOSTAT, 2016). Similarly, China has also become world's largest producer, consumer, and exporter in the context of value-added wood products (Wan et al., 2015; NBSC, 2016) that encourage the European panel and furniture producers to move the production eastward (Wahl, 2008; Poyry, 2013).

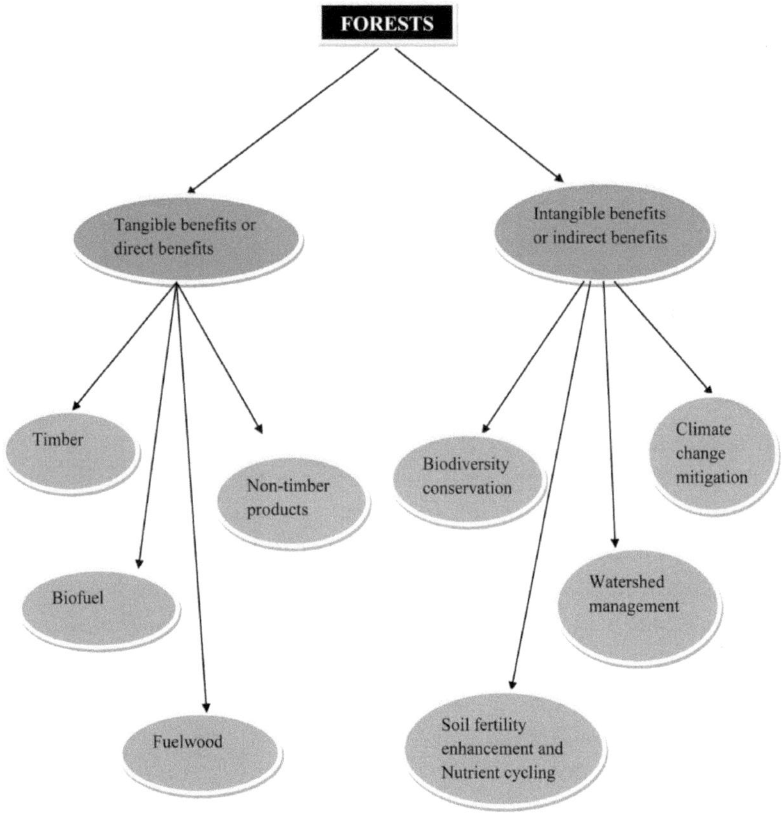

FIGURE 1.2 Multifarious benefits from the forest.

1.3.3 NON-TIMBER PRODUCTS

NTFPs are the plant and animal-based biological resources which can be harvested either from wild and domesticated, i.e., from various sources such as natural forest, farmlands, human-induced plantation, woodland, and TOF (tree outside the forest). These harvested natural and domesticated products are the source of various nutrients and income for forest fringe people and communities throughout the world. Thus, NTFPs are gain attention and attractive source which has the greatest potential to poverty reduction by enhancing socioeconomic upliftments along with biodiversity conservation and ecosystem health maintenance.

NTFPs are defined as those biological resources which are collected either from wild and can be domesticated by rural farmers for both direct consumption and sale for income generation results in socioeconomic upliftments and livelihood generation (Shackleton and Shackleton, 2004). In the broader scene, NTFPs includes a different form of wild and edibles nutritious food material, important medicinal and aromatic plants, horticultural plants, plants fiber, economically valuable fungi, gum, resins, fuelwood material, carvings, etc. (McLain and Jones, 2005). However, NTFPs are harvested regularly from the forest for either subsistence or commercial purpose and help in strengthening the people's livelihood along with nutritional security. Also, NTFPs have valued for its cultural significance (Shackleton and Shackleton, 2004). Although, NTFPs having a variety of products and these products having varying characteristics that leads to varying consumption and utilizing pattern and plays an important role in maintain household livelihood security.

As per Shackleton and Shackleton (2004), the population in remote rural areas exhibits employment generation through extraction, processing, and trading of NTFPs. Similarly, there is a witness in the past decades related to research interest in NTFPs among various development and conservation organizations in the world (Ruiz and Arnold, 1997; Neumann and Hirsch, 2000). The continuous spread of NTFPs has been seen due to its global interest in both harvesting and trade. Therefore, sustainable use NTFPs helps in reducing poverty along with the maintenance of biodiversity.

1.4 INTANGIBLE BENEFITS

1.4.1 SOIL MANAGEMENT

Soil, as a soul of infinite life, and a valuable natural resource and substratum for plant and animals especially forest ecosystem and other woodlands which helps in various ecosystem processes including storing organic matter (OM), its decomposition along with efficient biogeochemical cycle, etc. Soils store several important micro and macronutrients which is quite essential for growth and development of woody perennial trees species in the forest. Alteration of soils may impact on the availability of soil carbon pools that affect both storage and emissions of carbon.

Although, the assessment of soil quality, i.e., both physicochemical and biological factors are prerequisite for better health and productivity of the forest. Therefore, a proper understanding of these characteristics can helps in maintaining the soil quality and related parameter, which is the basis of healthy forest and ecosystem. As we know, forest land represents major terrestrial ecosystem, and the availability of certain microbes helps in decomposing the OM, release essential nutrients and make available to the wooded forest that regulates the overall productivity and ecosystem health (Six et al., 2004; Noguez et al., 2008). In turn, forest regulates soil processes and maintain its quality and health by addition of litter and other wood debris which can decompose by some soil-inhabiting micro-organism and release important nutrients. Moreover, due to varying tree species and forest types, the quality of litter and its decomposition rate will be varied results diverse type of soil properties which influence the occurrence and existence of microbial biomass. Although, due to extreme climatic event and human-induced anthropogenic activities which affect natural regeneration of forest that results in stand-replacing disturbance (Wang and Yang, 2007; Yang et al., 2010).

As per Holden and Treseder (2013) 30% of world land area is represented by forest of tropical, temperate, and boreal forest and these forests are the important source of global terrestrial carbon, helps in mitigating climate change phenomenon by storing and sequestration of atmospheric carbon into both vegetation and soils (Myneni et al., 2001). It has been predicted that naturally regenerated forest stands are generally more efficient in nutrient cycling and related soil quality than artificially managed plantations (Burton et al., 2007). Addition and decomposition of forest floor and leaf litter increase OM status and enhance fertility with maintaining soil health. Thus, there is a linking concept among forest health, soil health, and soil fertility (Jhariya, 2017; Jhariya et al., 2018) (Figure 1.3).

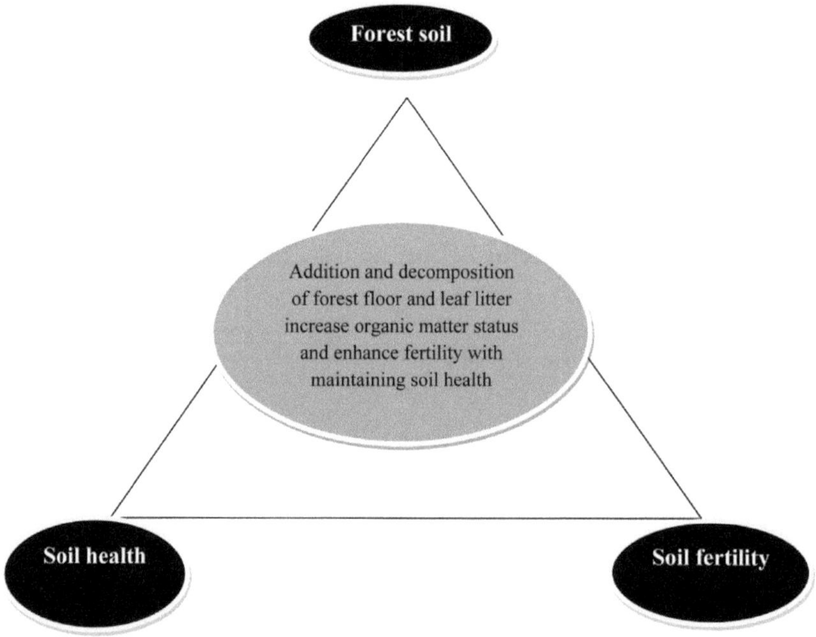

FIGURE 1.3 Forest for soil fertility and health.

1.4.2 WATERSHED MANAGEMENT

The management of watershed is very effective for rehabilitation of soil degradation and related prone area upland eroded areas. Although, various soil and forest parameters are the key measurements which can be assessed in the management and planning of watershed. Therefore, forests are the very important key component for management of the watershed, which is an important integrated approach for the natural resource. Similarly, forest-related measures, i.e., the activity of reforestation which can help in restoring and enhancing the fertility of soil are the integral part of planning and management of any running watershed projects.

1.4.3 NUTRIENT CYCLING

As we know, the availability of essential nutrients (elements or compounds) is very important for metabolism of forest crops and affects overall growth and survival. Soil stores such an important essential micro and macronutrient comprises N, P, K, C, H, O, Ca, and Mg along with small quality of other micronutrients such as B, Mn, Fe, Cu, Zn, and Cl. Although the exchange of nutrients between living and non-living component of the forest ecosystem prerequisite of ecosystem structure and functions (Glossary of Soil Science Terms, 1996). Forest nutrient cycle involves various activities such as nutrient uptake and its storage and transfer in woody perennial vegetation tissue, litter production and its decomposition, nutrients transfer by soil-inhabiting fauna and flora, mineral weathering and atmospheric nutrients input, nutrient leaching through soils and gaseous transfers. Therefore, climatic regime, site quality, and abiotic and biotic communities are the key factors which are controlled nutrient cycling phenomenon in forest ecosystem.

However, the availability of nutrients are also influenced by production of litter and both its quality and quantity. Moreover, elemental and nutrient load of soil and parent material affects the vegetation structure and dry matter dynamics at various landscape units. Wind deposited soils are characterized by fine in texture along with high nutrient supplying activity and can support mixed wooded forest having hardwood in nature rather than conifers and pine forest which is supported by coarse-textured and infertile soil.

1.4.4 BIODIVERSITY MANAGEMENT

Biodiversity represents diversity and variability among living organism at different ecosystem types (terrestrial, aquatic, and other) and shows the ecological complexity along with diversity at genetic, species, and ecosystem level (Gillespie, 1997; Huston and Marland, 2002; Koziell, 2001; UNEP, 2007). Today, forest diverseness is being lost at associate degree baleful rate. Moreover, forest tree species and its ecosystems are threatened globally at alarming rate and are being lost due to deterioration of forest habitat which not only affects flora (woody perennial tree species, herbaceous plants, medicinally important species, and others) but also affects tree dependent wild animals, i.e., fauna. Overall biodiversity gets jeopardized due to overexploitation of these natural resources. Moreover, it was previously discussed that forest ecosystem is the large source of storage and sequestration of carbon, but deforestation activity and related overexploitation of NTFPs can reverse this situation and release billion tonnes of carbon into the atmosphere which can be mitigated by proper scientific management of forests (tropical, temperate, and boreal) and particularly NTFPs which can helps in slowing down the greenhouse gases (GHGs) concentration in the atmosphere (Namayanga, 2002; Gorte, 2009).

There is some example which showing the linking concept among NTFPs, forest sustainability and its ecosystem service, i.e., biodiversity conservation and carbon storage. The plantation and reforestation of *Acacia Senegal* over 17,000 hectares of degraded land in Nigeria (Africa) under the BioCarbonn Fund project of CDM can sequester 1.8 million tonnes of carbon in an upcoming future. Moreover, this plantation system has strong evidence of social benefits through gum production and related income generation activity and profits whereas environmental benefits are related to rehabilitation of degraded land and maintaining biodiversity (NTF-PSI, 2008). Therefore, this system has multifarious and uncountable benefits not only in term of socioeconomic upliftments of people and poor farmers but also mitigating the extreme climatic impact, i.e., climate change by storing excessive carbon into the plantation tree species and soil parts.

1.4.5 CLIMATE CHANGE MITIGATION

The changing climate and extreme weather are the world's most dangerous phenomenon which can affect all biodiversity and related ecosystem structure and its functions. As we know, the activity of deforestation and faulty management of forest results changing the climate. Although, the healthy forest itself keeps a healthy environment and can mitigate climate change phenomenon through absorbing and storing atmospheric carbon into the vegetation parts such as forest and other woody perennial tree species. It is well-known truth that forest and woody vegetation plays a key role in mitigating the climate change and global warming phenomenon and reducing GHGs by storing and sequestration of atmospheric carbon into the vegetation part and helps in enhancing the adaptation capacity of vulnerable communities. Forest ecosystem covers 30% of terrestrial landscape and provides a shelter and habitat for various flora and fauna species along with soil inhabiting fungi, protozoan, invertebrate, vertebrates, and various micro and micro-organism (Yude et al., 2011). Although, forests are known for removing carbon from the atmosphere in both growth and recovery phase after a various disturbance like forest fire, etc., but in some cases, forests have been identified as a potential mechanism to human-induced climate change (Keenan, 2002).

1.5 FOREST FOR ENVIRONMENT MANAGEMENT

Deforestation, land degradation, diversion of forested land to the non-forested area such as agricultural intensification, development projects like industry, railways lines, power projects, construction of dams and canals, etc. promote climate change and global warming phenomenon that resulted in extreme weather at regional and global scale. Changing forest morphological and phonological attributes, degradation of both direct (tangible) and indirect (intangible) benefits as a products from the forests, declining overall health and productivity of forest trees and biodiversity, vegetation shifting, pest, and disease infestation in the forest, deterioration of health and productivity of wild animal and forest fringe tribal people are the major problems that happened due to climate change and global warming. Although, due to continuous emission of several GHGs such as CO_2 (major potent gas), NO_2, CFC, methane, etc., results extreme temperature and uncertain rainfall leads to changing climate. In this situation, forest and tree-based farming system provides nutritious and healthy food which is prerequisite for income and livelihood security along with maintenance of ecosystem health and provision of environmental security (Figure 1.4).

In this context, forest work as a largest sink of carbon as storage and sequestration of carbon and helps in mitigating climate change which is burning issue today (Figure 1.5).

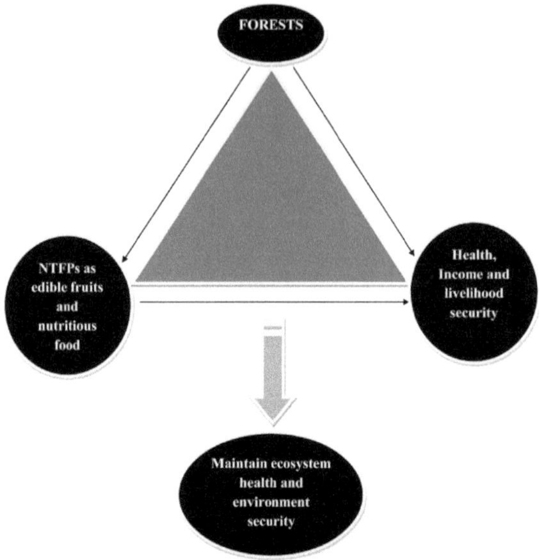

FIGURE 1.4 Linking concepts among forest, food, and income security of people.

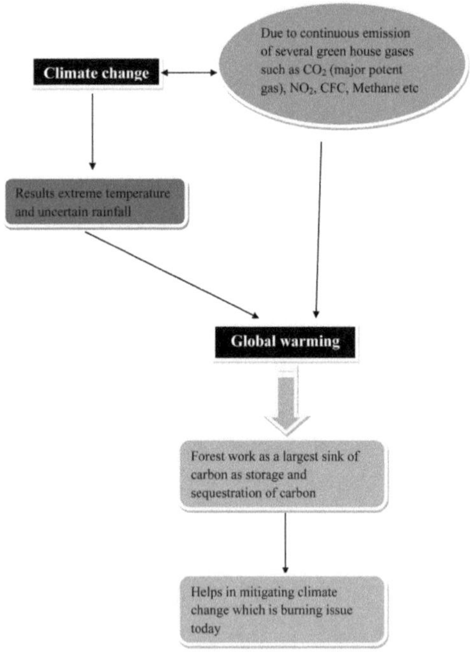

FIGURE 1.5 Forest for climate change mitigation.

1.6 FOREST FOR FOOD AND NUTRITIONAL SECURITY (FNS)

Forest plays an important role in maintaining FNS by providing uncountable non-timber forest products as gum, katha, catechu, sal patta, edible fruits and nutritious fruits. Forests play an inevitable role in the provision of many healthy and nutritious food systems in direct way along with strengthening farmer's income and several other ecosystem services indirectly that makes sustainability in food production and its supply to forest fringe people and poor farmers in the era of climate change and global warming (Figure 1.6).

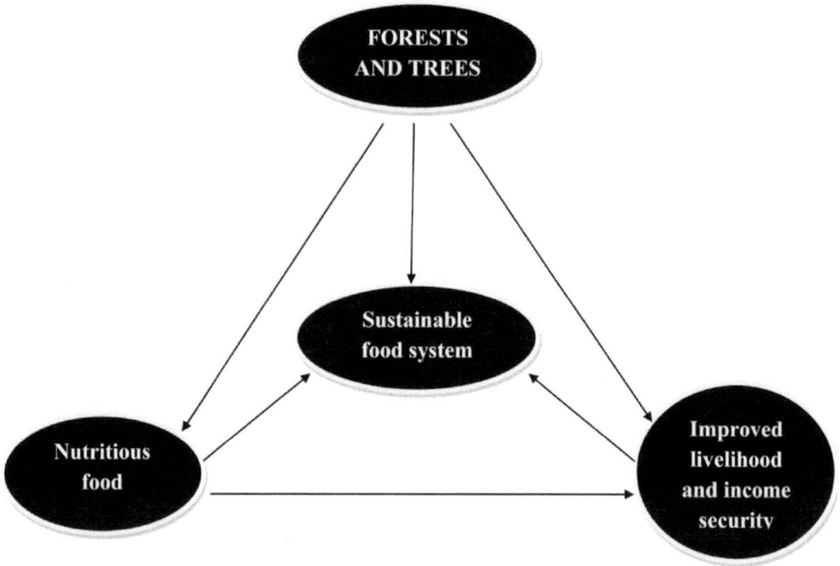

FIGURE 1.6 Forest for sustainable food systems.

Moreover, forest food significantly provides more diverse and nutritious food to people and poor farmers in the forest fringe area which is the source of balance and nutritious diet contacting essential protein, vitamins, and minerals (in leafy vegetables). Fruits and tubers are the essential source of carbohydrate whereas harboring animals are used for protein and fat for people. Forest's food also plays important role during emergency periods of flood, drought, and other extreme climatic situation due to climate change and global warming. Poor farmers and tribal peoples of forest fringe areas are generally rely upon food material that they gathered from the forests as daily and consumes for subsistence purpose and they can sale for their income and livelihood security along with maintaining FNS. Although, forest provides tangible, i.e., direct food provision to people, energy, i.e., fuelwood provision for cooking, income generation and livelihood security and provision of ecosystem services for the better environment which is the four major pillars of FNS and human health well-being (Figure 1.7).

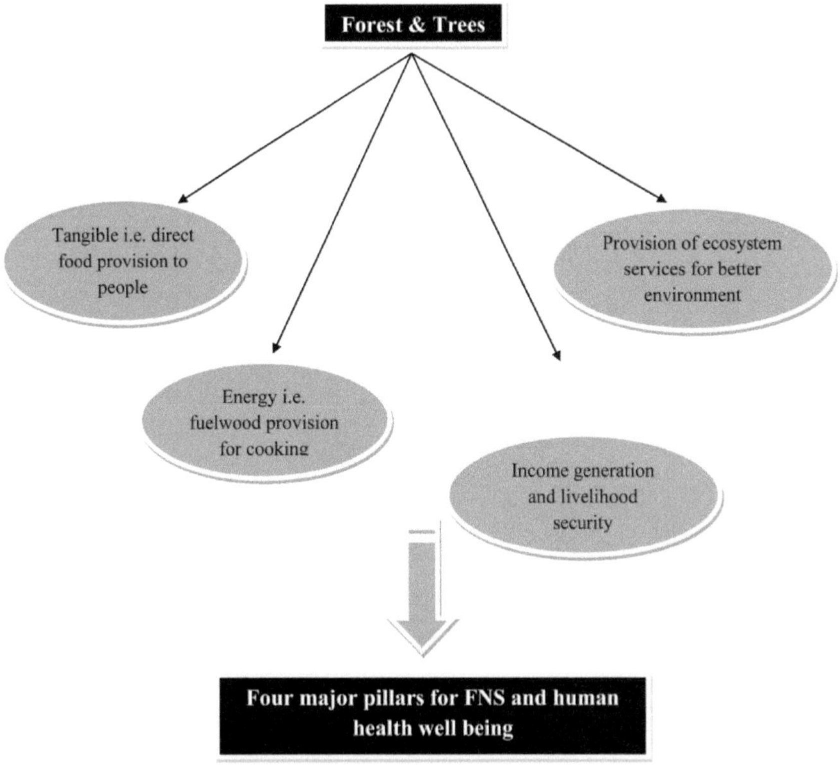

FIGURE 1.7 Forest for FNS, human health and well being.

1.7 FOREST FOR LIVELIHOOD SUSTENANCE

NTFPs and timber products from the forest and tree-based farming system (agroforestry) are the important source for both food supply and income (revenue) generation to local people and governments (Malhotra and Bhattacharya, 2010; Pandey et al., 2011; Jhariya and Raj, 2014; Shackleton et al., 2015). Tree-based incomes offer a considerably more diversified livelihood portfolio given the environmental and economic risks of relying on cash incomes from single commodity crops. NTFPs have been gaining wider recognition in term of contributing household and national economies along with environmental and biodiversity conservation (Arnold and Perez, 2001).

As per the report of MEA (2005), NTFPs, and forest-related services provide up to 96% value and globally recognized as basis of sustainable forestry. Around 275 million people in India are depending on NTFPs for their subsistence and livelihood generation (Pandey et al., 2016). Jones et al. (2004) mentioned sustainable forest management is regarded as the key element in management of NTFPs and sustainable economic development. Therefore, NTFPs are a good strategy which not only helps in raising the income of peoples but also addressing conservation issue and environmental problem, i.e., sort out the problem of climate change and global warming phenomenon (Ruiz-Perez and Byron, 1999).

People from forest fringe areas and those residing in or nearby the forest are conditionally permissible to get some NTFPs for their subsistence and surplus can be sold out in nearby market which helps in earning money. These all activities provide employment opportunities to poor people along with strengthening their socioeconomic conditions and maintain livelihood security. Thus, earned money from the sale of NTFPs supplement agricultural income and contributes in total people household expenses along with providing financial cost of health (Arnold and Perez, 2001).

1.8 NEXUS BETWEEN FOREST AND AGRICULTURE

It is well-known truth that agricultural expansion leads to deforestation which affects food production and nutritional security. Therefore, forest is a very important resource that are degraded and affected by expansion of farming system which not only affects the valuable timber and NTFPs but also disturbs the overall biodiversity and forest ecosystem (Raj et al., 2018).

Although forest and other tree-based farming system are interdependent to each other and play complementary role in overall agricultural production. They can provide multiple benefits as more nutritionally balance diets, fuelwood and can deliver various set of ecosystem services. Similarly, forest degradation will affects the food security by directly as unavailability of forest-based nutritious fruits and food and by indirectly as altering and modifying the important ecological and environmental processes which are important for crop and livestock production.

1.9 FORESTRY RESEARCH & DEVELOPMENT

Deforestation, declining quality timber production, and its health, forest regeneration problem, etc. is a major concern today that requires effective policies and R&D. Effective policies and its implementation with good governance, scientific-based R&D and building effective roadmap for future strategies and scientific oriented programs are the good strategy for forest management (Figure 1.8). Imparting proper education system create basic knowledge and awareness among people and farmers which plays a major role in empowering rural populations and can uplift socioeconomics through delivering tangible and intangible benefits which includes food and health security, sustainable forest management and improving general household nutrition.

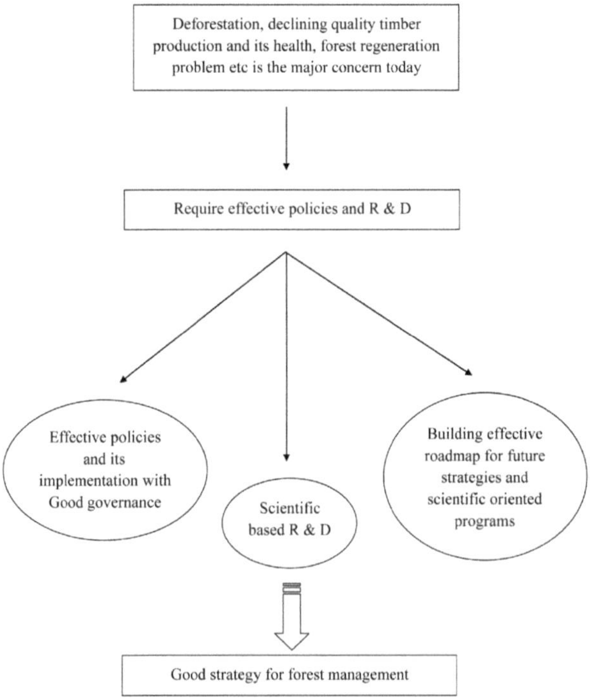

FIGURE 1.8 Policies and R&D for forest management.

1.10 CONCLUSION

The intervention of scientific-based forest management is the key option for quality, and healthy forests which help in managing natural resource and its proper utilization can nurture and protect biodiversity and provides some timber and non-timber forest products along with other ecosystem services. Forest stores carbons (mitigate climate change) into both vegetation and soils and maintain the quality of the soil. Both timber and NTFPs an important source of income to the rural farmer and tribal people. Thus, implementation of some effective policy with good governance well developed effective roadmap with efficient research and development are win-win situation for sustainable forest management and ecosystem health and management.

KEYWORDS

- **biodiversity**
- **ecosystem services**
- **forests**
- **greenhouse gases (GHGs)**
- **nontimber forest products (NTFPs)**
- **watershed**

REFERENCES

Agarwal, M., & Deepak, S. S., (2003). Physiological and biochemical responses of two cultivars of wheat to elevated levels of CO_2 and SO_2, singly and in combination. *Environmental Pollution, 121*, 189–197.

Arnold, M. J. E., & Perez, M. R., (2001). Can non-timber forest products much tropical forest conservation and development objectives? *Ecological Economics*, *39*, 437–447.

Arrow, K. J., Dasgupta, P., Goulder, L. H., Mumford, K. J., & Oleson, K., (2012). Sustainability and the measurement of wealth. *Environ. Dev. Econ.*, *17*, 317–353.

Aslam, M., Saeed, M. S., Sattar, S., Sajad, S., Sajad, M. S., Adnan, M., Iqbal, M., & Muhammad, T. S., (2017). Specific role of proline in heavy metal toxicity in plants. *Int. J. Pure App. Biosci.*, *5*(6), 27–34.

Ausubel, J. H., Wernick, I. K., & Waggoner, P. E., (2012). Peak farmland and the prospect for land sparing. *Popul. Dev. Rev.*, *38*, 217–238.

Baker, E. A., & Hunt, G. M., (1986). Erosion of waxes from leaf surfaces by stimulated acid rain. *New Phytol.*, *1102*, 161–173.

Bauhus, J., Van Der Meer, P., & Kanninen, M., (2010). *Ecosystem Goods and Services from Plantation Forests* (pp. 205–227). Routledge, London. ISBN: 978-1-84971-168-5 hardback.

Brockerhoff, E. G., Jactel, H., Parrotta, J. A., & Ferraz, S. F., (2013). Role of eucalypt and other planted forests in biodiversity conservation and the provision of biodiversity-related ecosystem services. *For Ecol. Manag.*, *301*, 43–50.

Burton, J., Chen, C. R., & Xu, Z. H., (2007). Gross nitrogen transformations in adjacent native and plantation forests of subtropical Australia. *Soil Biol. Biochem.*, *39*, 426–433. doi: 10.1016/j.soilbio.2006.08.011.

Chou, P., (2018). The role of non-timber forest products in creating incentives for forest conservation: A case study of Phnom Prich wildlife sanctuary, Cambodia. *Resources*, *7*(41), 2–16.

Decocq, G., Andrieu, E., Brunet, J., Chabrerie, O., De Frenne, P., De Smedt, P., et al. (2016). Ecosystem services from small forest patches in agricultural landscapes. *Curr. for Rep.*, *2*, 30–44.

FAO, (2006). *Global Forest Resources Assessment 2005*. Rome, Italy.

FAO, (2015a). *Global Forest Resources Assessment 2015*. FAO (Food and Agriculture Organization of the United Nations), Rome.

FAO, (2015b). *Global Forest Resources Assessment 2015*. FAO Forestry Paper No. 1, UN Food and Agriculture Organization, Rome.

FAOSTAT, (2016). *Forestry Production and Trade Database*. Available at: http://faostat3.fao.org/browse/F (Accessed on 4 August 2019).

Gamfeldt, L., Snäll, T., Bagchi, R., Jonsson, M., Gustafsson, L., Kjellander, P., et al. (2013). Higher levels of multiple ecosystem services are found in forests with more tree species. *Nat. Commun.*, *4*, 1340.

Gibson, L., Lee, T. M., Koh, L. P., Brook, B. W., Gardner, T. A., Barlow, J., Peres, C. A., Bradshaw, C. J., Laurance, W. F., Lovejoy, T. E., & Sodhi, N. S., (2011). Primary forests are irreplaceable for sustaining tropical biodiversity. *Nature*, *478*, 378–381.

Gillespie, A., (1997). *Sinks, Biodiversity & Forests: The Implications of the Kyoto Protocol for the Other Primary UNCED Instruments* (pp. 1–26), Waikato University, New Zealand.

Global Forest Resource Assessment, (2010). *Food and Agriculture Organization (FAO): Rome, Italy, 2011*. Available online: http://www.fao.org/forestry/fra2010 (Accessed on 4 August 2019).

Glossary of Soil Science Terms, (1996). *Soil Science Society of America: Madison* (p. 134). WI.

Gorte, R. W., (2009). *Carbon Sequestration in Forests* (pp. 1–23). Congressional Research Service, USA.

Harbia, J., Erbaugh, J. T., Sidiq, M., Haasler, B., & Nurrochmat, D. R., (2018). Making a bridge between livelihoods and forest conservation: Lessons from non timber forest products' utilization in South Sumatera, Indonesia. *For. Policy Econ.*, *94*, 1–10.

Holden, S. R., & Treseder, K. K., (2013). A meta-analysis of soil microbial biomass responses to forest disturbances. *Front Microbiol.*, *4*, 163.

Hosonuma, N., Herold, M., De Sy, D., De Fries, R. S., Brockhaus, M., Verchot, L., Angelsen, A., & Romijn, E., (2012). An assessment of deforestation and forest degradation drivers in developing countries. *Environ. Res. Lett.*, *7*, 044009.

Huston, M. A., & Marland, G., (2002). Carbon management and biodiversity. *Journal of Environmental Management, 00*(2003), 1–10, Elsevier science Ltd., USA.

Jhariya, M. K., & Raj, A., (2014). Human welfare from biodiversity. *Agrobios Newsletter, XIII*(9), 89–91.

Jhariya, M. K., (2017). Influences of forest fire on forest floor and litterfall dynamics in Bhoramdeo Wildlife Sanctuary (C. G.), India. *Journal of Forest and Environmental Science, 33*(4), 330–341.

Jhariya, M. K., Banerjee, A., Yadav, D. K., & Raj, A., (2018). Leguminous trees an innovative tool for soil sustainability. In: Meena, R. S., Das, A., Yadav, G. S., & Lal, R., (eds.), *Legumes for Soil Health and Sustainable Management* (pp. 315–345). Springer, ISBN 978–981–13–0253–4 (eBook), ISBN: 978–981–13–0252–7 (Hardcover). https://doi.org/10.1007/978–981–13–0253–4_10 (Accessed on 4 August 2019).

Jones, E. T., McLain, R. J., & Lynch, K. A., (2004). *The Relationship Between Nontimber Forest Products and Biodiversity in the United States.* National Commission on Science for Sustainable Forestry, Washington, DC, p. 61.

Keenan, R. J., (2002). Historical vegetation dynamics and the carbon cycle: Current requirements and future challenges for quantifying carbon fluxes in Australian terrestrial ecosystems. *Aust. J. Bot., 50*, 533–544.

Keenan, R. J., Reams, G. A., Achard, F., De Freitas, J. V., Grainger, A., & Lindquist, E., (2015). Dynamics of global forest area: Results from the FAO global forest resources assessment. *For Ecol. Manag., 352*, 9–20.

Koziell, I., (2001). *Diversity Not Adversity: Sustaining Livelihoods with Biodiversity* (pp. 1–56). Biodiversity and Livelihoods Group, IIED. Department for International Development.

Kumar, S., (2002). Does participation in common pool resource management help the poor: A social cost benefit analysis of joint forest management in Jharkhand, India. *World Dev., 30*(5), 763–782.

Le Roux, D. S., Ikin, K., Lindenmayer, D. B., Manning, A. D., & Gibbons, P., (2015). Single large or several small? Applying biogeographic principles to tree-level conservation and biodiversity offsets. *Biol. Conserv., 191*, 558–566.

Leßmeister, A., Heubach, K., Lykke, A. M., Thiombiano, A., Wittig, R., & Hahn, K., (2018). The contribution of non-timber forest products (NTFPs) to rural household revenues in two villages in south-eastern Burkina Faso. *Agrofor. Syst., 92*, 139–155.

Liang, J., Crowther, T. W., Picard, N., Wiser, S., Zhou, M., Alberti, G., et al. (2016). Positive biodiversity-productivity relationship predominant in global forests. *Science, 354*, aaf8957. doi: 10.1126/science.aaf8957.

Lindenmayer, D. B., & Franklin, J. F., (2002). *Conserving Forest Biodiversity: A Comprehensive Multiscaled Approach.* Island Press, 1718 Connecticut Avenue NW Suite 300 Washington, DC 20009 USA, 352 pp.

Lindenmayer, D. B., & Laurance, W. F., (2017). The ecology, distribution, conservation and management of large old trees. *Biol. Rev. Camb. Philos. Soc., 92*, 1434–1458.

Lindenmayer, D. B., (2009). Forest wildlife management and conservation. *Ann. NY Acad. Sci., 1162*, 284–310.

Mace, G., Masundire, H., & Baillie, J., (2005). Biodiversity. In: Hassan, R., Scholes, R., & Ash, N., (eds.), *Ecosystems and Human Well-Being: Current State and Trends: Findings of the Condition and Trends Working Group (Chapter 4)* (pp. 77–122). Millennium ecosystem assessment. Island Press, Washington.

Malhotra, K. C., & Bhattacharya, P., (2010). *Forest and Livelihood* (p. 246). CESS Publisher, Hyderabad, India.

McLain, R. J., & Jones, E. T., (2005). *Non Timber Forest Products Management on National Forests in the United States*. General Technical Report PNW-GTR-655. Portland. U.S. Department of Agriculture, Forest Service, Pacific Northwest Research Station. 85 p.

McPherson, E. G., Simpson, J. R., Peper, P. J., Maco, S. E., & Xiao, Q., (2005). Municipal forest benefits and costs in five US cities. *Journal of Forestry*, 411–416.

McPherson, E. G., Van Doorn, N., & De Goede, J., (2016). Structure, function and value of street trees in California, USA. *Urban For. Urban Greening*, *17*, 104–115.

Millennium Ecosystem Assessment-MEA, (2005). *Ecosystems and Human Well-Being: Policy Responses* (Vol. 3, pp. 1–137). Island Press, Washington, DC.

MoEF, (2002). *Sustainable Development—Learning's and Perspectives from India*. Ministry of Environment and Forest (MoEF), Government of India Available: http://envfor.nic.in/divisions/ic/wssd/doc4/main.htm (Accessed on 4 August 2019).

MoEF, (2007). *Interim Country Report—India for United Nations Forum on Forest (UNFF-II)*. Ministry of Environment and Forest (MoEF), Government of India Available: http://envfor.nic.in/nfap/Unff2.pdf (Accessed on 4 August 2019).

Mori, A. S., Lertzman, K. P., & Gustafsson, L., (2017). Biodiversity and ecosystem services in forest ecosystems: A research agenda for applied forest ecology. *J. Appl. Ecol.*, *54*, 12–27.

Myneni, R. B., Dong, J., Tucker, C. J., Kaufmann, R. K., Kauppi, P. E., Liski, J., Zhou, L., Alexeyev, V., & Hughes, M. K., (2001). A large carbon sink in the woody biomass of Northern forests. *Proc. Natl. Acad. Sci.*, *98*(26), 14784–14790.

Namayanga, L. N., (2002). *Estimating Terrestrial Carbon Sequestered in Aboveground Woody Biomass from Remotely Sensed Data*. Serowe, Botswana. MSc Thesis, International Institute for Geo-Information Science and Earth Observation (ITC), Enschede—The Netherlands. pp. 1–55.

National Bureau of Statistics of China (NBSC), (2016). *Statistical Database*. Available at: http://www.stats.gov.cn/english/statisticaldata/ (Accessed on 4 August 2019).

Neumann, R. P., & Hirsch, E., (2000). *Commercialization of Non Timber Forest Products: Review and Analysis of Research*. Center for International Forestry Research, Bogor, Indonesia. ISBN 979-8764-51-X; pp. 1–176.

Newbold, T., Hudson, L. N., Hill, S. L. L., Contu, S., Lysenko, I., Senior, R. A., et al. (2015). Global effects of land use on local terrestrial biodiversity. *Nature*, *520*(11), 45–50.

Noguez, A. M., Escalante, A. E., Forney, L. J., Mendoza, M. N., Rosas, I., Souza, V., & Oliva, F. G., (2008). Soil aggregates in a tropical deciduous forest: Effects on C and N dynamics, and microbial communities as determined by t-FLPs. *Biogeochemistry*, *89*, 209–220.

NTF-PSI, (2008). *Targeted Capacity Building for Avoided Deforestation, Reforestation and Sustainable Land Management in Africa*. Washington DC.

Pan, Y., Birdsey, R. A., Fang, J., Houghton, R., Kauppi, P. E., Kurz, W. A., et al. (2011). A large and persistent carbon sink in the world's forests. *Science*, *333*, 988–993.

Pandey, A. K., Bhargava, P., & Negi, M. S., (2011). Sustainable management of non-timber forest produce through joint forest management. *Indian Forester*, *137*, 105–113.

Pandey, A. K., Tripathi, Y. C., & Kumar, A., (2016). Non timber forest products (NTFPs) for sustained livelihood: Challenges and strategies. *Research Journal of Forestry*, *10*, 1–7.

Poyry, (2013). *The Future of the Wood-Based Panel Industry in Europe*. Markets, industry trends and profitability – Europe 2020. pp. 1–13.

Raj, A., Jhariya, M. K., & Harne, S. S., (2018). Threats to biodiversity and conservation strategies. In: Sood, K. K., & Mahajan, V., (eds.), *Forests, Climate Change and Biodiversity* (pp. 304–320, 381). Kalyani Publisher, India.

Ruiz, P. M., & Arnold, J. E. M., (1997). *Current Issues in Non-Timber Forest Products Research.* Proceedings of the workshop 'research on NTFPs' hot springs, Zimbabwe. ISBN: 979-8764-06-4; pp. 1–264.

Ruiz, P. M., & Byron, N. A., (1999). Methodology to analyze divergent case studies of non timber forest products and their development potential. *Forest Science*, 45.

Shackleton, C. M., Pandey, A. K., & Ticktin, T., (2015). *Ecological Sustainability for Non-Timber Forest Products: Dynamics and Case Studies of Harvesting* (p. 294). Routledge, London, UK. ISBN-13: 978–0415728591.

Shackleton, C., & Shackleton, S., (2004). The importance of non-timber forest products in rural livelihood security and as safety nets: A review of evidence from South Africa. *South African Journal of Science, 100*, 658–664.

Six, J., Bossuyt, H., Degryze, S., & Denef, K., (2004). A history of research on the link between (micro) aggregates, soil biota, and soil organic matter dynamics. *Soil Tillage Res., 79*, 7–31.

Smith, P., Bustamante, M., Ahammad, H., Clark, H., Dong, H., Elsiddig, E. A., et al. (2014). Agriculture, forestry and other land use (AFOLU). In: Edenhofer, O., et al. (eds.), *Climate Change 2014: Mitigation of Climate Change* (pp. 811–922). Contribution of working group III to the fifth assessment report of the IPCC. Cambridge University Press, Cambridge, United Kingdom and New York, NY, USA.

Stagoll, K., Lindenmayer, D. B., Knight, E., Fischer, J., & Manning, A. D., (2012). Large trees are keystone structures in urban parks: Urban keystone structures. *Conservation Letters, 5*, 115–122.

Stephenson, N. L., Das, A. J., Condit, R., Russo, S. E., Baker, P. J., Beckman, N. G., et al. (2014). Rate of tree carbon accumulation increases continuously with tree size. *Nature, 507*, 90–93.

Thompson, I. D., Tylianakis, J. M., Kumar, P., Brockerhoff, E. G., Schellhorn, N. A., Parrotta, J. A., & Nasi, R., (2011). Forest biodiversity and the delivery of ecosystem goods and services: Translating science into policy. *Bio. Sci., 61*, 972–981.

United Nations Environmental Protection-UNEP, (2007). *Biodiversity and Climate Change*. Convention on biological diversity: International day for biological diversity.

Van Der Plas, F., Manning, P., Soliveres, S., Allan, E., Scherer-Lorenzen, M., Verheyen, K., et al. (2016). Biotic homogenization can decrease landscape scale ecosystem multifunctionality. *Proc. Natl. Acad. Sci., 113*, 357–362.

Wahl, A., (2008). *Wood Market Trends in Europe, FP Innovations* (SP–49). Special Publication.

Wan, M., Lähtinen, K., & Toppinen, A., (2015). Strategic transformation in the value-added wood products companies: Case study evidence from China. *Int. J. Emerg. Mark., 10*(2), 224–242.

Wang, C. K., & Yang, J. Y., (2007). Rhizospheric and heterotrophic components of soil respiration in six Chinese temperate forests. *Glob Change Biol., 13*, 123–131.

Yang, K., Zhu, J., Zhang, M., Yan, Q., & Sun, O. J., (2010). Soil microbial biomass carbon and nitrogen in forest ecosystems of Northeast China: A comparison between natural secondary forest and larch plantation. *J. Plant Ecol., 3*, 175–182.

CHAPTER 2

Agroforestry Systems in the Hills and Their Ecosystem Services

ABHISHEK RAJ,[1] M. K. JHARIYA,[2] D. K. YADAV,[2] and A. BANERJEE[3]

[1]*Department of Forestry, College of Agriculture, I.G.K.V., Raipur – 492012 (C.G.), India, E-mail: ranger0392@gmail.com*

[2]*Department of Farm Forestry, Sarguja University, Ambikapur – 497001 (C. G.), India, E-mails: manu9589@gmail.com (M. K. Jhariya); dheeraj_forestry@yahoo.com (D. K. Yadav)*

[3]*Department of Environmental Science, Sarguja University, Ambikapur – 497001 (C.G.), India, E-mail: arnabenvsc@yahoo.co.in*

ABSTRACT

The practices of agroforestry in plain and hill/mountainous region exhibit its wide applicability, possibilities, and scope in agroclimatic zones and represent an ideal system for multifarious benefits and provision of great ecosystem services. Agroforestry is not a new concept; it is working from a time immemorial and today extensively adopted in plain and hills of the tropics. In general, agroforestry provides both tangible and intangible benefits, but it works very efficiently while adopting in a hilly and mountainous region in term of conserving soil and water through minimizing erosion and reduces the loss top organic matter (OM), which is the basic source of essential nutrient for plants and soil-inhabiting organism. Therefore, the provision of ecosystem services is not limited to soil and water conservation and soil health but also provides diverse form of materials (as timber and NTFPs), enhance the overall yield and biomass, store, and sequester atmospheric carbon, helps in mitigating climate change, higher economical returns (basically strengthening the poor farmers income) and maintain food, health, and environment security along with ecosystem stability.

2.1 INTRODUCTION

Agroforestry is not a new farming practice. It has been practiced since the time immemorial and continuing with significant effects on biodiversity and environment through the provision of several direct and indirect benefits with sustainable land and resource utilization. The different model of agroforestry is based on its structure and functions. Structure is based on availability and integration of three elements (tree, crops, and pasture/grasses for livestock's) which forms different models as agrisilviculture, silvipastoral systems, agrisilvipastoral system and other fruits based models such as hortisilviculture, horti-pasture, and agrihortipasture, etc. While well-developed structure of models performs better functions of productions (yield and biomass of timber and NTFPs) and protection (soil and water conservation, enhancing the environmental quality, etc.) (Figure 2.1) (Nair and Garrity, 2012; Jhariya et al., 2015). Although, the intrinsic components of agroforestry are tree, herbs, and pastures for livestock's and its arrangement along with type of management makes several different location specific models which are practices either in plain and hilly regions in the tropics. The common agroforestry practices are agrisilviculture, silvopastoral, agrosilvopastoral, agrihorticultural, hortisilvicultural, hortipastoral, etc.

Agroforestry is the location specific farming practices and its services are modified by existing different elements (trees, field crops and livestock's/pastures), prevailing environmental conditions and socioeconomic factors that influence farmers livelihood improvement. The integration of various components of agroforestry systems and its scientific management provides greater opportunities for sustainable production of materials and incomes generation for farmers and local people (Pandey, 2007). With the provision of diversifying products, agroforestry maintains food security (through nutritious food and fruits), soil fertility improvement (through the addition and decomposition of litter and leaf fall etc) and ensure incomes along with helping in employment generations (Huxley, 1999; Garity, 2004; Nair, 2007; Jose, 2009; Singh and Jhariya, 2016) that are included in the variety of ecosystem services which can fulfill the human needs and build life (Daily, 1997).

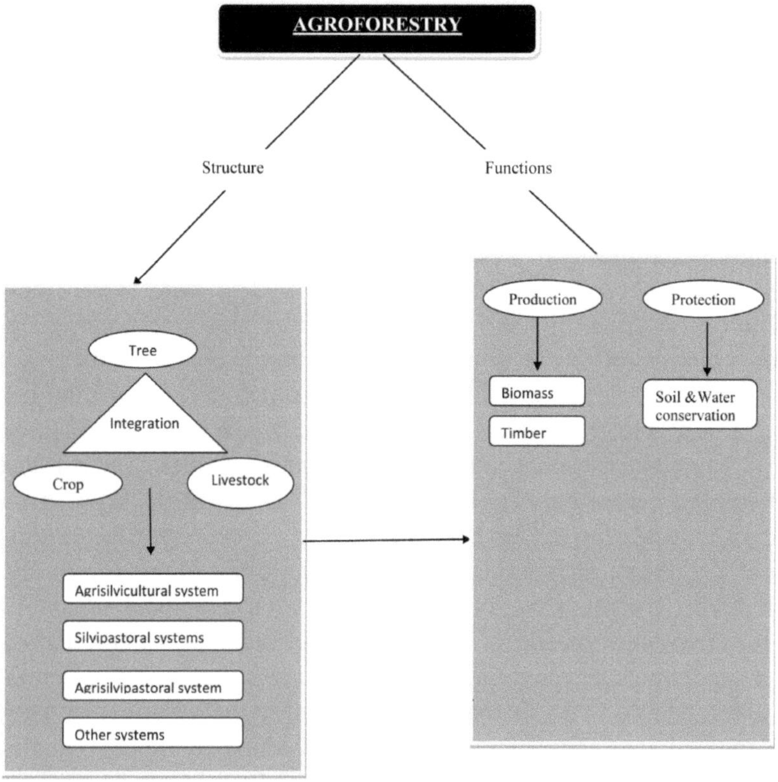

FIGURE 2.1 General agroforestry structure and functions.

Agroforestry is the best solution for adapting and mitigating the burning issue of today, i.e., climate change and global warming which is well recognized by IPCC (Watson et al., 2000). The different models of agroforestry systems are the promising solutions for ensuring food security, minimizing poverty (through employment generation) and enhancing ecosystem resilience for the millions of farmers globally. Well, the type and functions of different agroforestry systems in varying agroclimatic zones are changes as per varying weather regimes (temperature, rainfall, and humidity) along with varying topography, i.e., elevation and soil quality (Combe, 1982; Tejwani, 1987; Nair and Dagar, 1991; Tiwari, 1995; Maikhuri et al., 2000; Pandey, 2007; Singh and Pandey, 2011). The incorporation of woody perennial trees in the agroforestry systems can provide several consumable products (timber and NTFPs) and valuable services to local people and poor farmers. Also, it can fix atmospheric carbon, stabilize, and work as sink by storing in their body parts which helps in minimizing excessive temperature resulting climate change phenomenon and build up biomass (Murthy et al., 2013; Jhariya et al., 2015, 2018; Raj et al., 2018).

The sequestration of carbon into the vegetation and soils generally depend on input materials as dead organic matter (OM) provided by plant parts and changing soil structure and properties along with varying climatic situations (Lal, 2004). In compare to treeless pasture system, tree-based agroforestry systems have potential to store more biomass and carbon in deep soils and a viable system to more carbon to mitigate excessive carbon in the atmosphere (Takimoto et al., 2009). As per Nair et al. (2009) the available carbons in both above ground and in the depth of one meter are 0.29–15.21 Mg C ha^{-1} $year^{-1}$ and 30–300 Mg C ha^{-1}. As per one estimates, carbon sequestration values of agroforestry systems are varied between 12–228 Mg C ha^{-1} with 95 Mg C ha^{-1} of median value (Albrecht et al., 2003). Thus, agroforestry has a greater potential to store and sequester carbon than other monocrop and sole agricultural systems (Nair et al., 2012). Similarly, agroforestry in cool environment of agroecological region contributes more carbon storage potential than other conventional agricultural systems (Bajracharya et al., 2015) and 48.60 t C/ha has been reported in the agroforestry system of Nepal subcontinent (Pandit et al., 2012). Similarly the cool environment of Himachal Pradesh has been characterized for its diverse types of agroforestry systems, floristic diversity, and varying production of biomass and carbon sequestration capacity along with soil and water conservation programme (Toky et al., 1989; Atul and Khosla, 1990; Khosla and Toky, 1996; Thakur, 2002; Thakur et al., 2004; Bijalwan, 2012).

A good research and development of agroforestry have enormous potential to emphasize their structure, diversity, function (as ecosystem services) and climate change mitigation at local, regional, and global level along with economical evaluation and profitability (Beer et al., 2005). Integration of trees in agricultural field crops (say agroforestry system) is the one of the viable farming practices that provides both tangible and intangible products, covers forest areas as given by National Forest Policy, enhance soil and water conservation by deep and tap root systems (particularly in hilly regions), restore soil essential nutrients by reducing the soil erosion, improve fertility status by adding and decompositions organic residues and litter fall from MPTs and leguminous N_2 fixing trees, and recycling the nutrient through efficient close nutrient cycle process (rather than open nutrient cycling process in sole/monocropping agriculture system) (Shrestha, 1995). According to her, incorporation/retaining of perennial trees in agricultural land (as agroforestry system) and afforestation activity can maintain the forest covers in denuded hilly regions that recover soil fertility, health, quality along with ecosystem stability and maintain the food and nutritional security (FNS) for peoples in the hills. This can be achieved through incorporating some MPTs and economically important trees species. In this perspective, Chauhan and Dhyani (1989) mentioned around 80 tree species as per suitability along farm boundaries and home gardens in hilly regions of Meghalaya, India. There are various models/types of agroforestry systems that can be utilized in hill zones such as forest-based, agricultural-based, and horticulture-based farming systems (Singh et al., 1990). This chapter reviews the agroforestry situations, its possibilities and availability in hilly and mountainous regions, floristic diversity and structure and provision of various ecosystem services through its contribution in biomass production, storage, and sequestration of carbon (climate change mitigation), conserving soil and water in steeply slope of hills which helps in restoring fertility and health, providing socioeconomically security along with improving health and living standard of poor farmers by diversifying products and better adaptability in hilly region under the changing climate scenario.

2.2 AGROFORESTRY SITUATIONS

As per one estimates, around 200 m ha area (5%) is considered under agroforestry systems out of 4000 m ha (FAO, 2007). Although, Nair et al. (2009) has reported that around 1023 m ha areas are covered and utilized by agroforestry system. At present, 3.39% of geographical area of the country (111554 km^2) is maintained under agroforestry system (FSI, 2013).

2.2.1 POSSIBILITIES OF AGROFORESTRY IN HILLS

As per Rana (1995), various agroforestry systems such as agrisilviculture, agrisilvipastoral, agrisilvihortipastoral, and sole pasture/grass are adopted and more promising in mid-hill zones of Himachal Pradesh. Similarly, Majumdar (1991) has identified several agroforestry systems such as agricultural, horti-based agriculture, horti-based silvipastoral, grasslands, and wastelands in Nauni, Himachal Pradesh, India. Moreover, as per Kachru (1997) who has conducted an extensive survey and found that some agroforestry practices such as agrisilviculture, agrihorticulture, agrisilvihorticulture, agrihortisilviculture, pasture-based silvicultural, pasture-based hortisilviculture, agrihortisilviculture, and sole pasture are the common suitable for the sub-temperate and sub-humid region of Himachal Pradesh, India. Similarly, six types of agroforestry models such as agrisilviculture, agrisilvihorticulture, pasture-based silviculture, pasture-based silvihorticulture, pastoral-horticulture, and horti-pastoral are practiced and utilized among farmers of Balh valley of Himachal Pradesh.

Another work of Upadhyaya (1997) has been identified 6 agroforestry systems (silviculture, agrisilvihorticulture, pastoral-silvihorticulture, pastoral-hortisilviculture, and silvipasture) in lowlands of Himachal Pradesh which are practiced extensively by farmer's community. Through extensive survey, Bijalwan, (2012) has recognized some adopted tree species of *Grewia optiva, Celtis australis, Melia azedarach* and *Ficus roxburghii* in traditional agroforestry systems such as agrisilviculture, agrihortisilviculture, agrihorticulture, agrisilvipasture, silvipasture, and hortisilviculture, etc. But, the predominant systems were agrisilviculture, agrihortisilviculture, and agrihorticulture which are practiced in between 1000 to 2000 masl of elevations of mid-hill region of Garhwal Himalaya. Further 620, 495 and 705 trees/ha were reported in agrisilviculture, agrihortisilviculture, and agrihorticulture based agroforestry systems, respectively. Also, various agroforestry system such as agrosilvicultural, agrosilvihorticultural, silvopastoral, agrohortisilvicultural, hortiagricultural, and hortisilvicultural systems were recognized by Sood (2006) in Mandi district of Himachal Pradesh.

Thus, extensively adoption of various agroforestry systems in hilly regions are correlated with its diversifying utility in term of provision of several ecosystem services such as enhancement in crop diversification, maximizing biomass productions, ensuring health and food security of hill farmers and people, improving the livelihood through strengthening the farmers' incomes and maintaining ecological security through soil and water conservation and climate change mitigations (Dhakal et al., 2012; Paudel et al., 2017).

2.2.2 *FLORISTIC COMPOSITION AND DIVERSITY*

The quantifications of floristic composition and diversity analysis of agroforestry systems in hilly regions open a great study to analyze structure and major species dominancy that shows the importance of major species and its utilization patterns in the given ecosystems. Various studies have been conducted in this context. For example, Kumar, and Bhatt (2006) has reported *Lannea coromendelica* (IVI 39.80) and *Anogeissus latifolia* (IVI 29.50) were the major dominance tree species at site first and second through extensive survey and studying the structure, composition, and dominancy at tropical foothill of Garhwal Himalaya region. Also, on the both sites most of the species were distributed contagiously than other species characterized by randomly distribution. Although, Sharma et al. (2009) has reported that increasing altitude is negative correlated with density of trees in agroforestry systems in Himalayan temperate forest and concluded that the distribution of species as per increasing altitude are affected by varying climatic and weather patterns. Similarly, Sharma and Sharma (2013) in their investigation at western foothill gaps of central Aravallis reported that response of herbaceous flora varied under different plantations of *Acacia tortilis, Acacia senegal,* and *Eucalyptus camaldulensis* and concluded that many species which showed high value of frequency might not be dominants with respect to IVI. Grasses were common for all the sites of study; however, some species responded differentially, and to some extent, dominance of a species under a particular tree plantation affected the soil properties and floor dynamics.

2.2.3 *CLIMATE CHANGE IMPACT ON HILL-BASED AGROFORESTRY*

Climate change is the results of releasing GHGs (CO_2, NO_2, and methane) into the atmosphere that leads to extreme weather event, i.e., long term changing of weather regimes such as temperature, precipitation, humidity, etc., at local, regional, and global level. The deleterious impacts of climate change are not limited to only yield, biomass, and health of overall agroforestry models but can also affect distribution patterns of certain tree and crop species, i.e., changing distribution pattern and vegetation shifting phenomenon are commonly has been seen due to rising temperature and extreme weather events. The change in phenotype and genotypic characteristics at molecular level are also major concern today due to changing climate (Lynch and Lande, 1993).

Many authors (Parmesan and Yohe, 2003; Walther et al., 2002) have strongly reported that changing climate and weather pattern affects the species distributions pattern in hills. Therefore, the hilly regions are highly vulnerable due to changing climate which accelerates by some anthropogenic activities and natural disaster that exert greater variability of temperature and precipitation. Due to rising temperature and other climatic regimes, the yield and productivity of wheat, barley, mustard, and chickpea were decreased in northern India (Kalra et al., 2008). Similarly, the apple production was gradually increased rather than productivity which was reduced in a great extent (Awasthi et al., 2001).

The rising temperature along with poor chilling events resulted shifting and changing cropping patterns in hilly regions and this can be justified by GoHP (2009) and according to him apple cultivation was shifted towards higher altitude due to due to reduced chilling hours. Due to changing weather, farmers have replaced the apple cultivation by the cultivation of vegetables like tomato and peas in the Kullu and Mandi districts (ENVIS, 2008). Thus, the continuous emissions of several greenhouse gases into the atmosphere are resulting extreme weather event which influence the pattern of rainfall and temperature that affects agroforestry in hilly region by morphological, anatomical, and reproductive changes, species distribution pattern as per altitudinal variation, promotes vegetation shifting, occurrence of insect pest pathogen and fungal attacks, affects distribution of pines and deodar trees at higher altitude due to changing pattern of snowfall and disturbing the health and yield of integrated trees and crops along with reducing overall biomass, productivity, etc. (Figure 2.2).

2.2.4 ECOSYSTEM SERVICES THROUGH HILL AGROFORESTRY

Indeed, an agroforestry system has a promising number of ecosystem services. Ecosystem services through agroforestry in hilly region are categorized into both tangible (yield and biomass production and socioeconomic development through better economics return) and intangible services and benefits (soil health and fertility improvement, soil, and water conservation, close efficient nutrient cycling and climate change mitigation through storage and sequestration of carbon, etc.) which are prerequisite for better environment and ecosystem stability

(Figure 2.3). Major ecosystem services and environmental benefits of agroforestry:

1. Soil fertility maintenance;
2. Soil and water conservation;
3. Biomass production;
4. Climate change mitigation through carbon sequestration;
5. Socioeconomic enhancement, etc.

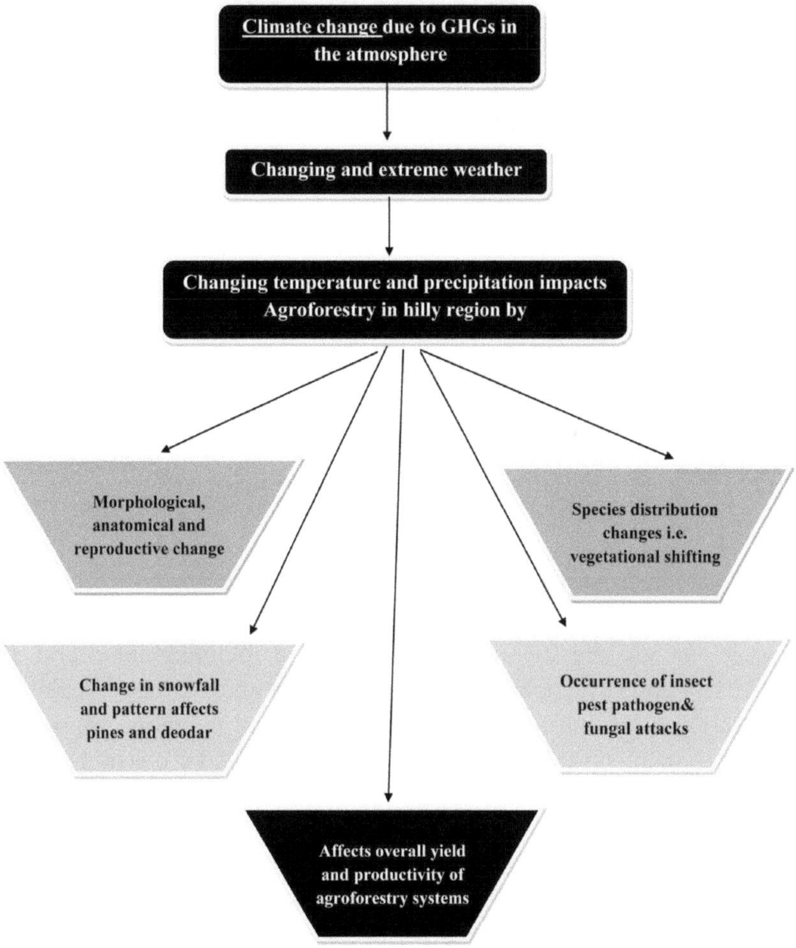

FIGURE 2.2 Climate change impact on agroforestry system in hilly region.

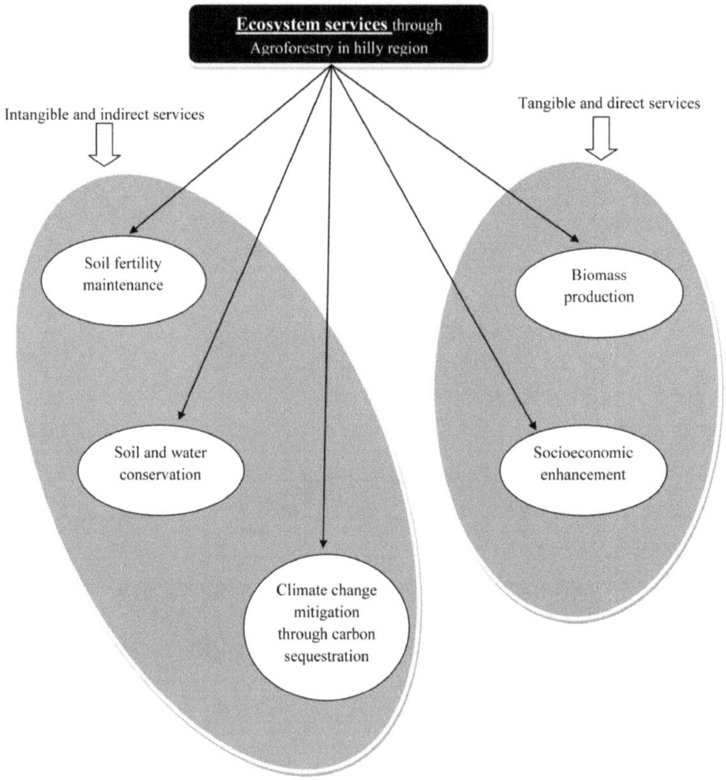

FIGURE 2.3 Ecosystem services through agroforestry system in hilly region.

2.2.5 HILL AGROFORESTRY AND SOIL HEALTH

The role of agroforestry in enhancing and maintaining long-term soil productivity and sustainability has been well documented. Soil health and conservation is an intrinsic concept which can be maintain through the adoption of deep tap-rooted MPTs in agricultural field crops and pastures land that helps in maintaining long-term soil productivity through the addition and decomposition of litter and leaf fall, and other organic residues from incorporating nitrogen and non-nitrogen fixing multipurpose trees in agroforestry of hilly region which can helps in fertility improvement and overall maintain health and quality of soil (Figure 2.4). Thus, integration of some perennials trees and annual crops having extensive root systems can bind and uphold soils mass which helps in protecting soils and minimize the chances of erosion, siltation of rivers, and minimize the fertility reductions. Moreover, the incorporation of trees and crops that are able to biologically fix nitrogen is fairly common in tropical agroforestry systems. Although, non N-fixing trees can also enhance soil physical, chemical, and biological properties by adding significant amount of above and belowground OM and releasing and recycling nutrients in agroforestry systems.

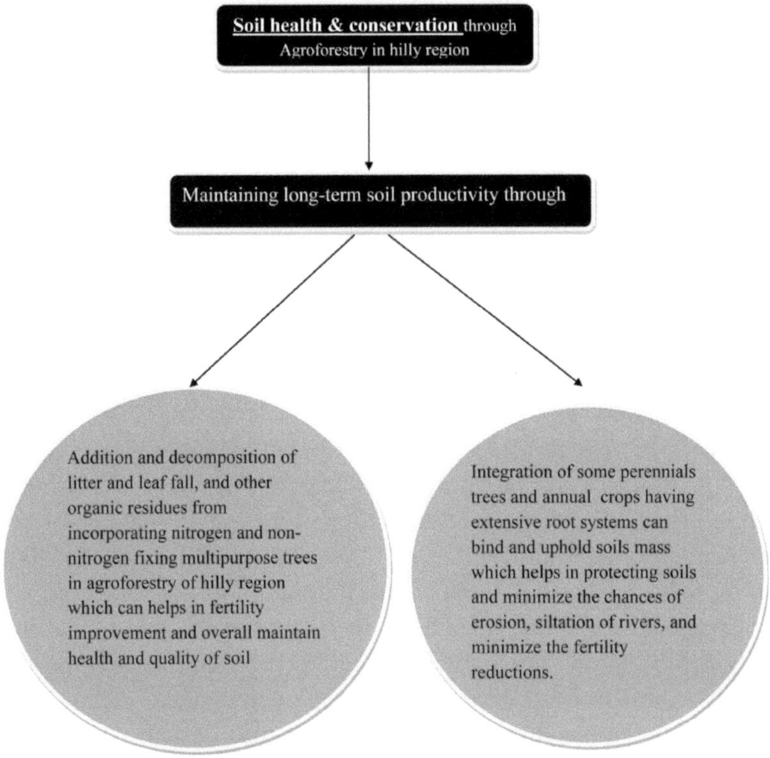

FIGURE 2.4 Soil health improvements through AF in hilly region.

2.3 SOIL CONSERVATION PRACTICES IN HILL AGROFORESTRY

Soil erosion in the hilly area is very common and results wash of OM results declining fertility, siltation of lakes and rivers and thereby creating food insecurity (Khisa et al., 2006). Erosion not only leads to a reduction in soil quality (Zheng, 2005) but also increases the risk of flooding. Extensive soil erosion leads to an increase of areas affected by soil degradation, and soil degradation caused by erosion has long been a serious environmental problem in China. Therefore, application of some control measures is necessary to decrease the amount of soil and water loss and the degree of soil degradation. Afforestation can help to conserve soil on degraded land by reducing soil erosion (Oscar, 2001), increasing soil OM, and improving soil structure (Clark and Clark, 2000), which reduces pressure on forested lands. Cerdà and Doerr (2005) showed that orange plantations in eastern Spain showed increased soil and water losses compared with other land uses such as traditional rainfed olive plantations or scrubland. Huang et al. (2004) reported that the reduction ratio of cumulative runoff in woodland and pastureland gradually decreased from 43% and 52% during the first and second years, respectively, 30% during 8 years for both types, and 12% and 21% during the 14 years, respectively. Integration of Bahia grass into the citrus tree planting or alley cropping can significantly reduce soil loss and protect it (Yang and Zhao, 2014).

2.4 BIOMASS PRODUCTION OF HILL AGROFORESTRY

Various literatures are available on estimation of biomass and yield potential of different agroforestry systems in hilly regions of agro-climatic zones. For example, As per Kumar (1996) both the cropping pattern of maize + lentil and soybean + wheat in agrisilviculture system contributed maximum (1.1, 1.23 and 1.31; and 1.09, 1.22 and 1.29 times higher) biomass than agrihortisilviculture, agrihorticulture, and sole cropping, respectively. Overall biomass and productivity was reported higher under agrisilviculture which is followed by agrihortisilviculture and agrihorticulture whereas least was seen in sole cropping system.

Pasture-based agroforestry plays a viable role in both conservations of soil-water and productions of various edible pasture/grasses for the cattle's which maintains health and sustainability of livestock's in hilly and mountainous regions. In this context, Sarma et al. (1999) was reported the suitability of grasses and pastures with various trees species in term of higher biomass productivity in silvipasture system of Doon valley, Uttarakhand, and according to higher biomass (2929 kg ha^{-1}) was observed in the combination of *Bauhinia purpurea* with grass/pasture rather than the obtained least biomass (2372 kg ha^{-1}) under the combination of *Grewia optiva* + *Eulaliopsis binata* on marginal land.

Similarly, incorporation of MPTs in any abandoned agricultural lands are not only viable in term of providing various timber and NTFPs products but also helps in augmenting the total biomass and productivity of the agroforestry systems and this can be justified by an extensive worked of Maikhuri et al. (2000), according to him the incorporation of local MPTs (*Albizia lebbek, Alnus nepalensis, Boehmeria arugulosa, Celtis australis, Dalbergia sissoo, Ficus glomerata, Grewia optiva, Prunus cerasoides, Pyrus pashia,* and *Sapium sebiferum*) were increased the aboveground tree biomass (3.9 t $ha^{-1}yr^{-1}$) in abandoned agricultural land as compared to degraded forest land comprised less biomass (1.1 t $ha^{-1}yr^{-1}$), respectively at 1200 m altitude in Chamoli district of India. Similarly, as per increasing altitude the rate of biomass production will increase. For example, forest-land use system comprised maximum biomass (371.06 t ha^{-1}) as compared to various agroforestry systems *viz.*, silvipasture, agrisilvipasture, agrihorticulture, agriculture, and grassland systems at the varying higher altitude of temperate zone of northwest Himalayan regions (Sanneh, 2007).

However, various studies were concluded the higher biomass and productivity under pure plantation than mixed and agroforestry systems. For example, sole plantation of poplar (*Populus deltoids*) contributed maximum biomass (188.60 t/ha AGB and 49.04 t/ha BGB of the total 237.70 t ha^{-1} biomass), which is followed by various other agroforestry systems *viz.*, agrisilviculture, silvipasture, pure agriculture and pure grassland systems, respectively (Minj, 2008). The storage of biomass in the woody trees are segregated into various components such as stem, leaf, and braches in agroforestry system which represents biomass storage and production capacity of components and its contribution in overall productivity of agroforestry systems. For example, around 50%, 27%, and 23% of the total biomass production (8314kg ha^{-1}) were contributed by tree components of stem, leaf, and branch in the 4-year *Casuarina* based agroforestry system (Saravanan et al., 2009).

Sharma (2009) studied distribution of biomass in different land uses along altitudinal gradient from 900 to 2100 m of Solan forest division, Himachal Pradesh and revealed that out of 1.07 M t of total biomass in the division, the chir pine land use contributes 36.63%, ban oak 32.78%, other broadleaves 28.94%, deodar 1.15% and culturable land 0.48%. Along altitudinal gradient, it was 44.89% in 1500–1800 m, 42.38% in 1200–1500 m, 10.35% in 1800–2100 m and 2.36% in 900–1200 m elevation. He further reported that Dharampur forest range accounted for 31.60%, Parwanoo 27.38%, Solan 25.77%, Kandaghat 11.22% and Subathu 4.01% of the total biomass. Similarly, Singh, and Lodhiyal (2009) observed total biomass (202.59 t ha^{-1}) of 8-year old *Populus deltoids* agroforestry plantation in tarai region of Central Himalaya of which AGB and BGB contributed 78.68 and 21.32%. Moreover, biomass of any agroforestry and plantation system can vary as per varying structure, composition, and floristic diversity of species in that system. This was justified by the work done of Yadava (2010) and he reported the maximum obtained biomass (57.69 t/ha) under *Populus deltoids* + wheat rather than biomass (56.38 t/ha) of intercropping with lemongrass in different agroforestry systems in Bagawala, Uttarakhand. Similarly, *Dalbergia sissoo* plantation exhibited maximum biomass (9.48 Mg ha^{-1} AGB and 9.48 Mg ha^{-1} BGB) as compared to *Populus deltoids* boundary plantation (1.23 Mg/ha) and *Paulownia salicina* (0.30 Mg/ha), respectively under the biomass production quantification in different tree based land use systems of Central Himalayan tarai region. Moreover, biomass, and production of agroforestry in hill region are depicted in Table 2.1.

TABLE 2.1 Biomass and Production of Agroforestry in Hill Region

Region	Agroforestry System	Production	References
Hill region of Himachal Pradesh, India	Comprised different types of agroforestry systems such as agrisilviculture, agrihortisilviculture, agrihorticulture, and sole cropping	The maximum biomass productivity was observed in agrisilviculture system followed by agrihortisilviculture agrihorticulture and least was seen in the sole cropping system. Also, two species combination such as maize+ lentil and soybean + wheat contributed 1.1, 1.23 and 1.31 and 1.09, 1.22, and 1.29 times higher biomass than agrihortisilviculture, agrihorticulture, and sole cropping systems.	Kumar, 1996
Doon valley of Uttarakhand having Marginal lands	Silvipastoral systems comprised several trees and grasses/ pastures	Intercropping of grass with the tree species of *Bauhinia purpurea* contributed maximum biomass (2929 kg ha^{-1}) as compared to *Grewia optiva* contribution (2372 kg ha^{-1}).	Samra et al., 1999

TABLE 2.1 *(Continued)*

Region	Agroforestry System	Production	References
Wet temperate north-western Himalayas	Comparative studies between forestland and various agroforestry systems (silvipasture, agrisilvipasture, agrihorticulture, agriculture, and grassland systems, respectively)	Maximum biomass (371.06 t ha^{-1}) was seen in forestland followed by agroforestry systems as silvipasture, agrisilvipasture, agrihorticulture, agriculture, and grassland systems, respectively.	Sanneh, 2007
Tarai region in Central Himalaya	*Populus deltoides* (poplar) based agroforestry plantation (8 years old)	Plantation contributed overall biomass of 202.6 t/ha of which 78.7% was aboveground and 21.3% was contributed by belowground biomass.	Singh and Lodhiyal, 2009
Solan, Himachal Pradesh (India)	Agroforestry systems distributed altitudinal zones	Biomass was increased (38.4, 30.8, 65.7, 111.4 and 143.9) as per increasing altitude (600 to 900 m, 900 to 1200 m, 1200 to 1500 m, 1500 to 1800 m and above 1800 m, respectively).	Gupta and Sharma, 2012
Natural vegetation of Himachal Pradesh	Agrihorticultural system comprised *Prunus persica* (Peach) with *withania somnifera* (Indian ginseng), *Ocimum sanctum* (Tulsi), *Andrographis paniculata* (Kalmegh)	Overall biomass was 5.79 t ha^{-1}	Tripathi, 2012
	Agrihorticultural system comprised *Malus domestica* (Apple) with *Zea mays* (Maize) and wheat species	Overall biomass was 108.73 t ha^{-1}	Toppo, 2012
	Grasses/pasture-based systems	Overall biomass was 18.52 t ha^{-1}	Sanneh, 2007
	Panicum maximum (Guinea grass) with *Trifolium alexandrinum* (Berseem) and *Sorghum bicolor* (Sorghum)	Overall biomass was 10.53 t ha^{-1}	Minj, 2008

2.5 CARBON SEQUESTRATION CAPACITY OF HILL AGROFORESTRY

Carbon absorption from atmosphere and its long term retention through carbon sink mediums as terrestrial (vegetations and soils) and oceans is known as carbon sequestration and agroforestry promise this activity in well manner and have greater potential for sequestering and storing carbon than other agricultural field crops and pasture cultivations. Agroforestry is the great substitute to fossil fuel through the provision of fuelwood, charcoal, and other bio-energy sources (energy plantation). Therefore, it works for carbon conservation and carbon substitution (minimizing fossils fuel utilization and it burning by production of fuel and firewood). Many studies have been conducted on quantification of storage and sequestration potential of agroforestry and as per Pandey (2007) the agroforestry has sequestered 12 to 228 Mg ha^{-1} carbon which varies by varying systems in different regions. Therefore, different components of agroforestry systems (woody trees, annual herbaceous crops, grasses/pastures, shrubs, bamboo, palm, and soils) have differential potential to store and fix carbon and maintain efficient carbon and other nutrient cycling along with maintaining ecosystem processes.

Agroforestry in temperate areas have potential to stored 15–198 Mg C ha^{-1} through long term sequestration and in the tropical region of the world, agroforestry practices has regained about 35% of estimated carbon stock due to deforestation compared to pastures and agricultural cropland areas (capacity to regained only 12%). Also, as per one estimates, the yield and productivity of agricultural crops increased (20–40 kg ha^{-1}) due to increasing one ton (Mg) of soil carbon pool of degraded cropland soils (Nair et al., 2008). Due to changing biomass production at considerable level in various regions, the sequestration potential of agroforestry over 96 m ha areas was 25 tonnes carbon in Indian subcontinents (Sathaye and Ravindranath, 1998; Ramnewaj and Dhyani, 2008; Dhyani et al., 2009; Newaj et al., 2013). Also, integrating woody perennial trees (particularly N_2 fixing leguminous trees) in agroforestry systems have a greatest potential to sequestered more carbon compared to sole and mono-cropping system (Kirby and Potvin, 2007).

Therefore, the practices of agroforestry have two benefits, i.e., one for absorbing and retaining atmospheric carbon through the sequestration process to help in mitigation climate change issue and second for providing various timber and non-timber products along with maintain forest cover with reducing pressure on natural forest (Pandey, 2002). Further, he also estimated the carbon sequestration capacity of agroforestry system in the world and according to him around 7 giga tonnes of carbon could be sequestered through agroforestry in between 1995 and 2050 in the worlds. According to Minj (2008) the region of submontane low hills stored maximum biomass carbon and sequestered more carbon which was 2.09 times higher than the mid-hills zones in Western Himalaya. Also, poplar (*Populous deltoids*) sole plantation (1066 trees per hectare of average density) was stored 107.00 t/ha carbon which is 2.68–4.86 and 12.40–22.57 times higher than other perennial plant-based system and annual crops or pure grasslands. Although, aboveground biomass has capacity to store more carbon as compared to belowground biomass this is justified by Chauhan et al. (2009).

Thus, agroforestry is a very good strategy for mitigating climate change and global warming issues by different practices of location specific models in hilly regions which has varying capacity of storing and sequestering carbon from the atmosphere and stabilizes it within vegetations and soils as soil OM which provides essential nutrients to plants and soil inhabiting organisms for sustaining their life on which humans and animals are depends. Agroforestry systems viz. hortipastoral, silvopastoral, agrihorticulture, agrisilviculture, agrihortisilviculture, and one natural grassland system at Nauni were analyzed for their existing carbon stocks to know carbon sequestration potential (CSP). Different agroforestry systems on basis of total carbon stock followed decreasing order: SP > AHS > HP > NG > AS > AH. CSP values with respect to NG taken as base was 2.17, 1.15, 1.08, 0.78 and 0.76 for SP, AHS, HP, AS, and AH, respectively (Verma et al., 2008).

2.6 ECONOMICS OF HILL-BASED AGROFORESTRY

Although, agroforestry has utmost importance at hilly regions due to characteristics of diverse species which provides diversifying and multifarious products for people without harming the environment which is good source of incomes to farmers for their own sustains and sale in nearby markets. Various works have been conducted on economic evaluations of different agroforestry models in hill and mountainous regions of the India subcontinents. For example, as per Dyal et al. (1996) the incorporation of *Leucaena leucocephala*, *Citrus aurantifolia*, *Carrica papaya* and *Curcuma longa* under agrisilvihorticulture gave an annual return of Rs. 17,746 and

15, 092 as compared to Rs. 7,752 and 3,342/ha which obtained from only agriculture-based cropping system (maize + black gram/wheat + mustard/ maize + wheat) under irrigated and rainfed conditions of Shivalik hills, India. Also, due to extensive survey and economical analysis by Kumar (1996) on various adopted agroforestry systems in Himachal Pradesh, he reported that agrihorticulture system yielded the highest net returns followed by agrihortisilviculture and agrisilviculture whereas least return was reported under based sole cropping systems.

Due to increasing population, anthropogenic factors are affecting the hilly regions by more intensive cultivation practices which disturbing the soil and promotes degradation and erosion, affects overall production and incomes of farmers that could be minimize through adopting various sustainable agroforestry and other mixed farming practices and in this context Neupane and Thapa (2001) has reported that adoption of scientifically managed agroforestry system in hills have more B: C ratio and more profitable in term of diversifying products, maintain biodiversity, FNS, and environmental health and security than other sole and conventional agricultural practices. As per Sharma et al. (2007) cardamom (*Amomum subulatum*) based agroforestry systems play a major role in providing ecosystem services and according to his estimates, this systems provides an output to input ratio of more than 13 through the better management practices as compared to other agricultural practices having the value of 1.85. Also, cardamom (*Amomum subulatum*) based agroforestry contributed higher B:C ratio which is 5.7 times more (beneficial for farmers) than other the rainfed agriculture.

Moreover, Sharma et al. (2008) has concluded that over story tree shades affects the yields and productivity of understory crops. They conducted extensive work on this hypothesis and according to him average yields of wheat (18.68q/ha) and gobhi-sarson (10.34q/ha) under kinnow plants were less in comparison to that of wheat (22.34q/ha) and gobhi-sarson (12.00) grown in open conditions. However, agrihorticulture system contributed higher economical returns than sole based cropping systems, i.e., in the open condition. Also, species combination plays a viable role in economical gain (Rs. 56407.55 Rs/ha) and intercropping of gobhi-sarson with Kinnow was observed to be more profitable than intercropping of wheat. Similarly, integration of vegetable crops with apple orchard provided more economical benefits and net profits of Rs 10,23,430/ha/yr as compared to Rs. 9,69,194/ha/yr gained in agrihorticulture in the Kullu district (HP) of temperate north-west Himalaya (Rajput, 2010). Similarly, family size, their numbers and labor availability will affects adoption of agroforestry practices in hilly regions. As per Thangata (1996) integration of tree species in agricultural fields for food crops results in complex type of agroforestry farming practices that need high labor force for managing and sustaining this system which greatly influences the high complex food based agroforestry (home gardens) in the highlands of southwestern Uganda. Gender can also influence the practice and management of agroforestry systems due to capacity to work and their availability in the fields.

2.7 FUTURE PROSPECTS OF HILL AGROFORESTRY

The practices of agroforestry is still gaining recognition either in plain or hills (mountainous) due to varying biomass production and quality food materials as fruits (horticulture based agroforestry system) which is also modified by integrating various multipurpose legumes species and pastures (for livestock's production in silvipasture system and protein bank) in various agroclimatic zones. Hill-based agroforestry systems are recognized potential to minimize soil and water erosion problem along with conservation of lands/topography due to incorporating various N_2 fixing leguminous trees and annual grasses/pasture which hold root systems tightly to protect and reduce the loss of soil essential nutrients. Thus, agroforestry systems are gaining its recognition by its wider applicability in any regions of plain and hill/mountain, and its multifarious benefits through sound ecosystem services to biomes.

2.8 R&D FOR HILL AGROFORESTRY

There is a lack of some potential and scientific research in the context of development and design of location-specific agroforestry models in hilly region and accordingly policy should be made. Hills are very prone to degradations due to some anthropogenic activity (intensive cultivation and deforestation) and natural disaster. In this context, there should be a scientific research and implementing the effective guidelines for the adoption and management of scientifically oriented prioritized agroforestry models (N_2 fixing deep tape root species should be prioritize) as per given slope and topography, i.e., location-specific models which can helps in maintaining overall yields and productivity, minimize the food pressure for burgeoning populations by providing diversifying products and maintain the soil fertility and its conservation in the era of climate change. Governmental programme, NGOs, policymakers, R&D, government, and private institutional agency must works together in this direction and helps in building climate-resilient agroforestry models.

2.9 CONCLUSIONS

The adoption of agroforestry in hilly and mountainous regions have a great opportunities in term of providing greater ecosystem services viz. enhancing yield and overall biomass productivity, improving soil fertility through addition and decomposition of OM, improving nutrient status (particularly nitrogen) through incorporating N_2 fixing leguminous MPTs, conserving soil and water by minimizing erosion and environment and ecosystem stability through mitigation climate change by storing and sequestration of carbon in vegetation and soil pools. Although, good combination of species according to its compatibility and positive relations among woody perennial trees, shrubs, herbaceous crops and pasture/grasses will be great option for making suitability of agroforestry models in any varying climate of hilly region which helps in maintaining FNS, higher economical returns, strengthening livelihood of poor farmers, ecosystem, and environmental stability etc.

KEYWORDS

- **agroclimatic zones**
- **agroforestry**
- **climate change**
- **ecosystem**
- **ecosystem services**
- **environment**
- **hilly region**

REFERENCES

Albrecht, A., & Kandji, S. T., (2003). Carbon sequestration in tropical agroforestry systems. *Agriculture, Ecosystems and Environment*, *99*, 15–27.

Atul, P., & Khosla, P. K., (1990). Classification of traditional agroforestry systems. In: *Proc. IUFRO, 19th World Forestry Congress* (pp. 24–27). Montreal.

Awasthi, R. P., Verma, H. S., Sharma, R. D., Bhardwaj, S. P., & Bhardwaj, S. V., (2001). Causes of low productivity in apple orchards and suggested remedial measures. In: Jindal, K. K., & Gautam, D. R., (eds.), *Productivity of Temperate Fruits* (pp. 1–8).

Bajracharya, R. M., Atreya, K., Raut, N., Shrestha, H. L., Gautam, D. K., & Dahal, N. R., (2015). *Sustainable Diversified Agriculture and Land Management in the Himalaya: Implications for Climate Change Adaptation and Mitigation* (pp. 1–17). Paper presented at the International Conference on Mountains and Climate Change, Karakoram International University, Gilgit, Pakistan.

Beer, J., Ibrahim, M., & Sinclair, F., (2005). The history of future agroforestry research and development: Policy impacts and needs. In: Mery, G., Alfaro, R., Kanninen, M., & Lobovikov, M., (eds.), *Forests in the Global Balance- Changing Paradigms* (Vol. 17, pp. 151–160). IUFRO World Series.

Bijalwan, A., (2012). Land-use and vegetation pattern in traditional agroforestry systems in midhills of Garhwal Himalaya. *Journal of Progressive Agriculture, 3*(2), 6–13.

Cerdà, A., Morera, G. A., & Bodí, M. B., (2009). Soil and water losses from new citrus orchards growing on sloped soils in the western Mediterranean basin. *Earth Surface Process and Landforms, 34*, 1822–1830.

Chauhan, D. S., & Dhyani, S. K., (1989). Important tree species of Meghalaya. *Indian Journal of Hill Farming, 2*, 79–83.

Chauhan, S. K., Gupta, N. R., Yadav, S., & Chauhan, R., (2009). Biomass and carbon allocation in different parts of agroforestry tree species. *Indian Forester, 135*(7), 981–993.

Clark, D. B., & Clark, D. A., (2000). Landscape-scale variation in forest structure and biomass in a tropical rain forest. *Forest Ecology and Management, 137*, 185–198.

Combe, J., (1982). Agroforestry techniques in tropical countries: Potential and limitations. *Agroforestry Systems, 1*, 13–28.

Daily, G. C., (1997). *Nature's Services: Societal Dependence on Natural Ecosystems.* Washington DC, Island Press. 392 p. ISBN: 1-55963-475-8.

Dhakal, A., Cockfield, G., & Maraseni, T. N., (2012). Evolution of agroforestry based farming systems: A case of Dhanusha district Nepal. *Agrofor. Syst., 86*(1), 17–33.

Dhyani, S. K., Newaj, R., & Sharma, A. R., (2009). Agroforestry, its relation with agronomy, challenges and opportunities. *Indian Journal of Agronomy, 54*, 249–266.

Dyal, S. K. N., Grewal, S. S., & Singh, S. C., (1996). An agri-silvi-horticultural system to optimize production and cash returns for Shivalik Foothills. *Indian Journal of Soil Conservation, 24*(2), 150–155.

ENVIS, (2008). *Climate Change: Himachal Pradesh's Perspective* (Vol. II). ENVIS Newsletter. *16*(1), 1–80. http://gbpihed.gov.in/envis/HTML/vol16_1/vol16_1.htm

FAO, (2007). *State of the World's Forest.* www.fao.org (Accessed on 4 August 2019).

FSI (Forest Survey of India), (2013). *The State of Forest Report 2012. Dehradun, India Ministry of Environment and Forests, Government of India.*

Garrty, D. P., (2004). Agroforestry and the achievement of the millennium development goals. *Agroforestry Systems, 61*, 5–17.

GoHP, (2009). *Government of Himachal Pradesh*, Department of Environment, Science, and Technology, State of the Environment Report, Shimla.

Gupta, B., & Sharma, D. P., (2012). Application if GIS in biodiversity-appraisal: Case study of Solan forest division (HP), India. *Journal of Sustainable Forestry*, *31*, 576–588.

Huang, Z. L., Fu, B. J., Chen, L. D., Huang, Y. L., & Wu, X. L., (2004). Comparison of environmental effects on five land use options for cropland conversion program in semiarid Loess hilly and gully area. *Chinese Journal of Soil Water Conservation*, *18*, 37–41.

Huxley, P., (1999). *Tropical Agroforestry* (p. 384). ISBN: 978–0–632–04047–6. Wiley-Blackwell, Oxford, UK.

Jhariya, M. K., Banerjee, A., Yadav, D. K., & Raj, A., (2018). Leguminous trees an innovative tool for soil sustainability. In: Meena, R. S., Das, A., Yadav, G. S., & Lal, R., (eds.), *Legumes for Soil Health and Sustainable Management* (pp. 315–345). Springer, ISBN 978–981–13–0253–4 (eBook), ISBN: 978–981–13–0252–7 (Hardcover). https://doi.org/10.1007/978–981–13–0253–4_10 (Accessed on 4 August 2019).

Jhariya, M. K., Bargali, S. S., & Raj, A., (2015). Possibilities and perspectives of agroforestry in chhattisgarh. In: Miodrag, Z., (ed.), *Precious Forests-Precious Earth* (pp. 237–257, 286). ISBN: 978–953–51–2175–6. InTech, Croatia, Europe, doi: 10.5772/60841.

Jose, S., (2009). Agroforestry for ecosystem services and environmental benefits: Overview. *Agroforestry Systems, 76*, 1–10.

Kachru, S. R., (1997). Diagnostic survey and productivity appraisal of agroforestry systems in sub-temperate and sub-humid region of Himachal Pradesh. MSc Thesis (p. 162). Dr. Y. S. Parmar University of Horticulture and Forestry, Nauni, Solan (H. P.).

Kalra, N., Chakraborty, D., Sharma, A., Rai, H. K., Jolly, M., Chander, S., Kumar, P. R., Bhadraray, S., Barman, D., Mittal, R. B, Lal, M., & Sehgal, M., (2008). Effect of increasing temperature on yield of some winter crops in northwest India. *Current Sci.*, *94*(1), 82–88.

Kanime, N., Kaushal, R., Tewari, S. K., Raverkar, K. P., Chaturvedi, S., & Chaturvedi, O. P., (2013). Biomass production and carbon sequestration in different tree based systems of Central Himalayan region. *Forest, Trees and Livelihoods, 1*(22), 38–50.

Khisa, S. K., Shoaib, J. M., & Khan, N. A., (2006). *Selected Natural Resource Conservation Approaches and Technologies in the Chittagong Hill Tracts of Bangladesh*. Bangladesh conservation approaches and technologies. The Swiss Agency for Development-Intercooperation and Institute of Forestry and Environmental Sciences, Chittagong University.

Khosla, P. K., & Toky, O. P., (1996). Status of agroforestry in the western Himalaya. *Current Researches in Plant Sciences*, pp. 367–377.

Kirby, K. R., & Potvin, C., (2007). Variation in carbon storage among tree species: Implications for the management of a small scale carbon sink project. *Forest Ecology and Management, 246*, 208–221.

Kumar, M., & Bhatt, V., (2006). Plant biodiversity and conservation of forests in foothills of Garhwal Himalaya. *Iyonia*, *11*(2), 43–59.

Kumar, M., (1996). Bio-economic appraisal of agroforestry land use systems. MSc Thesis (p. 96). Dr. Y. S. Parmar University of Horticulture and Forestry, Nauni, Solan (H. P.), India.

Lynch, M., & Lande, R., (1993). Evolution and extinction in response to environmental change. In: Kareiva, P. M., & Kingsolver, J., (eds.), *Biotic Interactions and Global Change* (pp. 234–250). Sinauer Associates Inc., Sunderland, MA, USA.

Maikhuri, R. K., Semwal, R. L., Rao, K. S., Singh, K., & Saxena, K. G., (2000). Growth and ecological impacts of traditional agroforestry tree species in Central Himalaya, India. *Agroforestry Systems, 48*, 257–272.

Majumdar, H. K., (1991). Biomass productivity and nutrient budgeting in different agroforestry system. PhD Thesis (p. 125). Dr. Y. S. Parmar University of Horticulture and Forestry, Nauni, Solan (H. P.), India.

Minj, A. V., (2008). Carbon sequestration potential of agroforestry systems- an evaluation in low and mid hills of western Himalayas. PhD Thesis (p. 124). Dr. Y. S. Parmar University of Horticulture and Forestry, Nauni, Solan, (H.P.), India.

Murthy, I. K, Gupta, M., Tomar, S., Munsi, M., Tiwari, R., Hegde, G. T., & Ravindranath, N. H., (2013). Carbon sequestration potential of agroforestry systems in India. *J. Earth Sci. Climate Change, 4*(1), 1–7.

Nair, P. K. R., & Dagar, J. C., (1991). An approach to developing methodologies for evaluating agroforestry systems in India. *Agroforestry Systems, 16*, 55–81.

Nair, P. K. R., & Garrity, D., (2012). Agroforestry research and development: The way forward. In: Nair, P. K. R., & Garrity, D., (eds.), *Agroforestry—the Future of Global Land Use* (pp. 515–531). Springer, Dordrecht.

Nair, P. K. R., (2007). The coming of age of agroforestry. *Journal of Science and Food Agriculture, 87*, 1613–1619.

Nair, P. K. R., Gordon, A. M., & Mosquera-Losada, M. R., (2008). Agroforestry. In: Jorgensen, S. E., & Fath, B. D., (eds.), *Ecological Engineering. Encyclopedia of Ecology* (Vol. 1, pp. 101–110). Elsevier, Oxford, U.K.

Nair, P. K. R., Kumar, B. M., & Nair, V. D., (2009). Agroforestry as a strategy for carbon sequestration. *Journal of Plant Nutrition and Soil Science, 172*, 10–23.

Nair, P. K. R., Nair, V. D., Kumar, B. M., & Showalter, J. M., (2010). Carbon sequestration in agroforestry systems. *Adv. Agron., 108*, 237–307.

Neupane, R. P., & Thapa, G. B., (2001). Impact of agroforestry intervention on farm income under the subsistence farming system of the middle hills, Nepal. *Agroforestry Systems, 53*, 31–37.

Newaj, R., Chavan, S., & Prasad, R., (2013). Agroforestry as a strategy for climate change adaptation and mitigation. *Indian J. of Agroforestry, 15*(2), 41–48.

Oscar, C., (2001). An analysis of externalities in agroforestry systems in the presence of land degradation. *Ecological Economics, 39*, 131–143.

Pandey, D. N., (2002). Carbon sequestration in agroforestry systems. *Climate Change Policy, 2*, 367–377.

Pandey, D. N., (2007). Multifunctional agroforestry systems in India. *Current Science, 92*(4), 455–463.

Pandit, B. H., Shrestha, K. K., & Bhattarai, S. S., (2014). Sustainable local livelihoods through enhancing agroforestry systems in Nepal. *Journal of Forest and Livelihood, 12*(1), 47–63.

Parmesan, C., & Yohe, G., (2003). A globally coherent fingerprint of climate change impacts across natural systems. *Nature, 421*, 37–42.

Paudel, D., Tiwari, K. R., Bajracharya, R. M., Raut, N., & Sitaula, B. K., (2017). Agroforestry system: An opportunity for carbon sequestration and climate change adaptation in the mid-hills of Nepal. *Oct. Jour. Env. Res., 5*(1), 22–31.

Raj, A., Jhariya, M. K., & Bargali, S. S., (2018). Climate smart agriculture and carbon sequestration. In: Pandey, C. B., Gaur, M. K., & Goyal, R. K., (eds.), *Climate Change and Agroforestry: Adaptation Mitigation and Livelihood Security* (pp. 1–19). ISBN: 9789–386546067. New India Publishing Agency (NIPA), New Delhi, India.

Rajput, B. S., (2010). Bio-economic appraisal and carbon sequestration potential of different land use systems in temperate north-western Himalayas. PhD Thesis (p. 169). Dr. Y. S. Parmar University of Horticulture and Forestry, Nauni, Solan (H.P.), India.

Ramnewaj, & Dhyani, S. K., (2008). Agroforestry for carbon sequestration: Scope and present status. *Indian Journal of Agroforestry, 10*(2), 1–9.

Rana, R. K., (1995). Economic analysis of agroforestry system in mid hill zones of Himachal Pradesh. PhD Thesis. pp. 239. Dr. Y. S. Parmar University of Horticulture and Forestry, Nauni, Solan (H.P.).

Samra, J. S., Vishwanatham, M. K., & Sharma, A. R., (1999). Biomass production of trees and grasses in a silvopasture system on marginal lands of Doon Velly of north-west India. *Agroforestry Systems, 46*, 197–212.

Sanneh, A., (2007). Status of carbon stock under different land use systems in wet temperate North Western Himalaya. MSc Thesis (p. 81). Dr Y. S. Parmar University of Horticulture and Forestry, Nauni, Solan, (H. P.), India.

Saravanan, S., Buvaneshwaran, C., Manivachagam, P., Rajagopal, K., & Gerge, M., (2009). Dry matter production and productivity under *Casuarina equesaetifolia* based agroforestry system. *My Forest, 45*(3), 237–245.

Sathaye, J. A., & Ravindranath, N. H., (1998). Climate change mitigation in the energy and the forestry sectors of developing countries. *Annals of Review on Energy and Environment, 23*, 387–437.

Sharma, B., & Sharma, K., (2013). Influence of various dominant trees on phytosociology of understory herbaceous vegetation. *Recent Research in Science and Technology, 5*(2), 41–45.

Sharma, C. M., Suyal, S., Ghildiyal, S. K., & Gairola, S., (2009). Species richness and diversity along an altitudinal gradient in moist temperate forest of Garhwal Himalaya. *The Journal of American Science, 5*(5), 119–128.

Sharma, D. P., (2009). Biomass distribution in sub-tropical forests of Solan Forest Division (H. P.). *Indian Journal of Ecology, 36*(1), 1–5.

Sharma, K., Thakur, S., Sharma, R., & Kashyap, S. D., (2008). Production and economics of Kinnow cultivation with wheat and gobhi sarson in Himachal Pradesh. *Indian Journal of Soil Conservation, 36*(2), 112–118.

Sharma, R., Jianchu, X., & Sharma, G., (2007). Traditional agroforestry in the eastern Himalayan region: Land management system supporting ecosystem services. *Tropical Ecology, 48*(2), 189–200.

Shrestha, S. S., (1995). Improvements of the hill farming systems of Nepal: An optimization of farm plans and analysis of the impact of new technology using linear programming. MSc Thesis, University of Reading, Department of Agricultural and Food Economics, UK.

Singh, G., Arora, Y. K., Narain, P., & Grewal, S. S., (1990). *Agroforestry Research in India and Other Countries, Surya Publications, 4-B Nashville Road* (p. 189). Dehradun, India.

Singh, N. R., & Jhariya, M. K., (2016). Agroforestry and agrihorticulture for higher income and resource conservation. In: Narain, S., & Rawat, S. K., (eds.), *Innovative Technology for Sustainable Agriculture Development* (pp. 125–145). ISBN: 978–81–7622–375–1. Biotech Books, New Delhi, India.

Singh, P., & Lodhiyal, L. S., (2009). Biomass and carbon allocation in 8-year old poplar (*Populus deltoides* Marsh) plantation in Tarai agroforestry systems of Central Himalaya, India. *New York Science Journal*, *2*(6), 49–53.

Singh, V. S., & Pandey, D. N., (2011). *Multifunctional Agroforestry Systems in India: Science-Based Policy Options* (pp. 1–33). Occasional Paper No. 4/2011 from Climate change and CDM cell, Rajasthan State Pollution Control Board. Jaipur, India.

Sood, K. K., (2006). The influence of household economics and farming aspects on adoption of traditional agroforestry in Western Himalaya. *Mountain Research and Development*, *26*(2), 124–130.

Takimoto, A., Nair, V. D., & Nair, P. K. R., (2009). Contribution of trees to soil carbon sequestration under agroforestry systems in the Western Sahel. *Agrofor. Syst.*, *76*, 11–25.

Tejwani, K. G., (1987). Agroforestry practices and research in India. In: Gholz, H. L., (ed.), *Agroforestry: Realities, Possibilities and Potential* (pp. 109–137). Martinus Nijhoff Publishers. The Netherlands.

Thakur, N. S., (2002). Diagnostic survey and productivity potential of Khair (*Acacia catechu* Wild.) based agroforestry. MSc Thesis (p. 124). Dr. Y. S. Parmar University of Horticulture and Forestry, Nauni, Solan (H. P.).

Thakur, N. S., Gupta, N. K., & Gupta, B., (2004). Phytosociology analysis of woody and nonwoody components under some agroforestry systems in Western Himalaya: Case study. *Indian Journal Agroforestry*, *6*(1), 65–71.

Thangata, P. H., (1996). Resource poor farmers' perception of agroforestry practices: Case study of Musoma. MSc Thesis, University of Edinburgh, Scotland, UK.

Tiwari, D. N., (1995). *Agroforestry Systems* (pp. 31–107). International Book Distributors. Dehradun.

Tokey, O. P., Pradeep, K., & Prem, K., (1989a). Structure and function of traditional agroforestry systems in the western Himalaya-I: Biomass and productivity. *Agroforestry Systems*, *9*, 47–70.

Toppo, S., (2012). Nutrient dynamics under different land use systems in Kullu valley of Himachal Pradesh. MSc Thesis (p. 82). Dr. Y. S. Parmar University of Horticulture and Forestry, Nauni, Solan (H. P.), India.

Tripathi, P., (2012). Effect of organic manures on yield and biomass production of medicinal and aromatic plants under peach based agroforestry system. PhD Thesis (p. 209). Dr. Y. S. Parmar University of Horticulture and Forestry, Nauni, Solan, (H. P.), India.

Upadhyaya, K., (1997). Diagnostic survey of agroforestry system in Balh valley of Himachal Pradesh. MSc Thesis (p. 141). Dr. Y. S. Parmar University of Horticulture and Forestry, Nauni, Solan (H. P.).

Verma, K. S., Kumar, S., & Bhardwaj, D. R., (2008). Soil organic stocks and carbon sequestration potential of agroforestry systems in H. P. Himalayan region of India. *Journal of Tree Science, 27*, 14–27.

Walther, G. R., Post, E., Convey, P., Menzel, A., Parmesan, C., Beebee, T. J. C., Fromentin, J. M., Hoegh-Guldberg, O., & Bairlein, F., (2002). Ecological responses to recent climate change. *Nature, 416*, 389–395.

Watson, R., Noble, I., Bolin, B., Ravindran, N., Verardo, D., & Doken, D., (2000). *Land Use, Land Use Change and Forestry* (pp. 388). Intergovernmental Panel on Climate change. Cambridge University Press, Cambridge, UK.

Yadava, A. K., (2010). Biomass production and carbon sequestration in different agroforestry system in terrain region of Central Himalaya. *Indian Forester, 136*(2), 234–244.

Yang, Z. S., Yang, L. F., & Zhang, B. S., (2010). Soil erosion and its basic characteristics at Karst Rocky-desertified land consolidation area: A case study at Muzhe Village of Xiahou County in Southeast Yunnan, China. *Journal of Mountain Science*, *7*, 55–72.

Zheng, F. L., (2005). Effects of accelerated soil erosion on soil nutrient loss after deforestation on the loess plateau. *Pedosphere*, *15*, 707–710.

CHAPTER 3

Multifunctional Role of Legumes in Agroforestry

ABHISHEK RAJ,[1] M. K. JHARIYA,[2] A. BANERJEE,[3] and D. K. YADAV[2]

[1]*Department of Forestry, College of Agriculture, I.G.K.V., Raipur – 492012 Chhattisgarh, India, E-mail: ranger0392@gmail.com*

[2]*Department of Farm Forestry, Sarguja University, Ambikapur – 497001 Chhattisgarh, India, E-mails: manu9589@gmail.com (M. K. Jhariya); dheeraj_forestry@yahoo.com (D. K. Yadav)*

[3]*Department of Environmental Science, Sarguja University, Ambikapur–497001, Chhattisgarh, India, E-mail: arnabenvsc@yahoo.co.in*

ABSTRACT

Plantation and integration of leguminous trees/crops are good choices of agriculturist and forester due to its multipurpose in nature as providing fuelwood, fodder (for livestock's), fixing nitrogen (essential nutrient for plants) into the soil and combating climate change/global warming phenomenon through enhancing carbon storage and sequestration capacity of both soil and trees (as leguminous in nature). Although, herbaceous legumes having greater diversity than tree legumes but if once it exists, then it is more profitable to farmers and provides greater ecosystem services (both tangible and intangible). The integration of legumes in various agroforestry practices are good choice for both farmer's livelihood security and environmental health concern. Improving overall productivity, close, and efficient nutrient cycling, food, and health security, and climate security through carbon sequestration are the major promising services provided by legume-based AF systems. Moreover, legume-based AF maintains/add optimum nitrogen into the soils, particularly nitrogen-deficient soil of any regions with the help of biological N_2 fixation (BNF). However, there is a lack of scientific research and development in the context of location-based specific agroforestry model and type of legumes that can be incorporated according to models in the given areas. Research should be explored in term of the suitability of leguminous trees/shrubs in the given agroforestry models, its impact of soil health and recycling of nitrogens and the provision of ecosystem services for better environment and biodiversity.

3.1 INTRODUCTION

Thanks to legume-based agroforestry which gives one step more benefits and services to the ecosystem as compared to simply agroforestry systems. Because of the nitrogen-fixing ability and multipurpose in nature of legume trees, it can provides timber and fuelwood, mulch for soil and water conservations, provide nutrient-rich high quality pastures/fodders for feeding purpose of livestock's, make availability of nutritious fruits and food for human consumptions and ameliorate microclimates with greater ecosystems stability (Gutteridge, 1994; Jhariya et al., 2018). Moreover, legume-based agroforestry systems have utmost important in the era of changing climate due to emission of various GHGs through both natural and human activities as changing the pattern of land utilization, faulty land conversion, intensive agricultural practices, shifting cultivation, flooded rice cultivation and various industrial development.

In this context, legume-based agroforestry systems have enormous potential to capture atmospheric carbon and fix into the vegetation as woody biomass, capability to store and retain into the deeper soil profile along with the promotion of growth and development through higher sequestration capacity of carbon (Takimoto et al., 2008; Jensen et al., 2011; Verchot et al., 2011; Tonucci et al., 2011; Montagnini et al., 2012; Jhariya et al., 2015). Further, leguminous trees and shrubs in any farming/cropping systems have greatest potential to reduce purchasing bill of chemical fertilizers for soil and higher yield through enhancing the nitrogen and other nutrients in soil through N_2 fixing ability of these legumes which is very promising in the direction of higher yield and economy of farmers along with greater ecosystem stability (Ribeiro-Barros et al., 2018).

This chapter reviews the scope, importance, and possibilities legumes in agroforestry systems and its productive potential, role-playing for efficient nutrient cycling, soil health and fertility enhancement, food and nutritional security (FNS), climate change mitigation through storage and sequestration of atmospheric carbon and related various ecosystem services that help in making sustainability in ecosystem and environmental health.

3.2 LEGUMES: AN OVERVIEW

As we aware about the legumes which having peculiar characteristic of nitrogen-fixing ability (as BNF) and multipurpose in nature, represented by overall 750 number of genus along with 19,000 species worldwide (Stevens, 2001; Bargali, 2016; Jhariya et al., 2018) of which 1152 species under 179 genera are reported in India (Husain and Kapoor, 1990; Sanjappa, 1991), comprises three subfamily viz., Caesalpinioideae, Papilionoideae (largest subfamily having 14,000 species under 476 genera; Lewis et al., 2003) Mimosoideae and having varying diversity of life forms, species, growth characteristics, morphological parameters (height, colors, etc.) and life cycle (Rao and Husain, 1993; Bargali, 2016; LPWG, 2017).

Fabaceae/Leguminoceae regarded as 3rd largest family of plants and distributed in tropical, temperate, Mediterranean, Savannah biomes in the world (Schrire et al., 2005). Legume has conspicuous characteristics of nitrogen-fixing ability with the help of symbiotic bacteria *Rhizobium* (exist in the root nodules of plants) and due to this characteristics legumes can be incorporated in any type of degraded land, plantation areas and both traditional and modern agroforestry systems. Due to N_2 fixing multi-purpose in nature, it has capacity to enhance nitrogen content in the soil which can improve fertility status of soil, mitigate climate change through enhancing the carbon sequestration, proving various tangible products in short duration as timber, fuelwood, fodder, and foods for livestock's and peoples that helps to reduce hunger and maintain both FNS along with economic benefits to the rural people (Jhariya et al., 2018).

3.3 LEGUMES: UNCOUNTABLE BENEFITS

No doubt, legume provides uncountable benefits through both tangible as food, fuel, and fodder and intangible way as improving the soil quality through N_2 fixation and minimizing the low nitrogen availability constraint in soil, ameliorating capability of degraded and wasteland, enrich, and protect the wild animals, enhance the carbon sequestration capacity to mitigate changing climate situations nowadays (Bargali and Bargali, 2009; Jhariya et al., 2018). Degradation of cultivated and forest land due to higher input of chemical fertilizer is the major problem today's and its consequences can be observed in term of spreading eutrophication, enhancing climate change due to ozone depletion and loss of biodiversity (Erisman et al., 2011; Serpil, 2012; Rahman and Zhang, 2018; Raj et al., 2018). This can't only degrade land and soil health but also affects overall biomass productivity and health of the ecosystems. In this context, integration of leguminous trees/crops improve the health and quality of soil due to less disturbance of soil and abundant availability of nitrogen through nitrogen-fixing ability (as BNF) and enhance the biomass productivity and profitability to poor farmers by enhancing the incomes and reducing the money cost for fertilizer application (Crews and Peoples, 2004; Rosenstock et al., 2014).

FNS, soil health security, humans, and livestock's health security along with environmental security are the precious benefits of leguminous trees/ crops while incorporating in the agroforestry systems. Thus, improve soil physicochemical properties and soil health security, extensive root systems to make more efficient systems, provision of nutritious food to make FNS, minimizing the possibility of crop failure due to providing essential nutrients and nurturing the crops, availability of fuelwood, fodders, green leaves for humans and animals consumptions, mitigating climate change and global warming through enhancing carbon sequestration, ameliorating nutrient-deficient soil, i.e., minimize land degradation and improve livelihood security through promoting economic benefits to poor farmers, etc. are the uncountable benefits while incorporating nitrogen-fixing leguminous trees in agroforestry systems (Figure 3.1).

3.4 TREE LEGUMES VS. HERBS LEGUMES

As we know, leguminous is a very diverse family having a variety of life forms as herbaceous and perennial trees and are nitrogen-fixing multipurpose in nature. Majorly, highly protein-rich source of herbaceous legume plants in cultivated forms comprises *Arachis hypogeal* (peanut), *Phaseolus vulgaris* (French bean), *Glycine max* (soybean), *Pisum sativum* (field pea), etc. whereas major legumes trees and shrubs of tropics comes under the subfamily of Mimosoideae and Caesalpinioideae (Hong and Bhatnagar, 2007). However, leguminous trees and shrubs were not gaining equal importance and assessed as legume herbaceous crops but some countries like Asia, Africa, Australia, and America have assessed and implemented these trees and shrubs legumes successfully (Hasniati and Shelton, 2005;

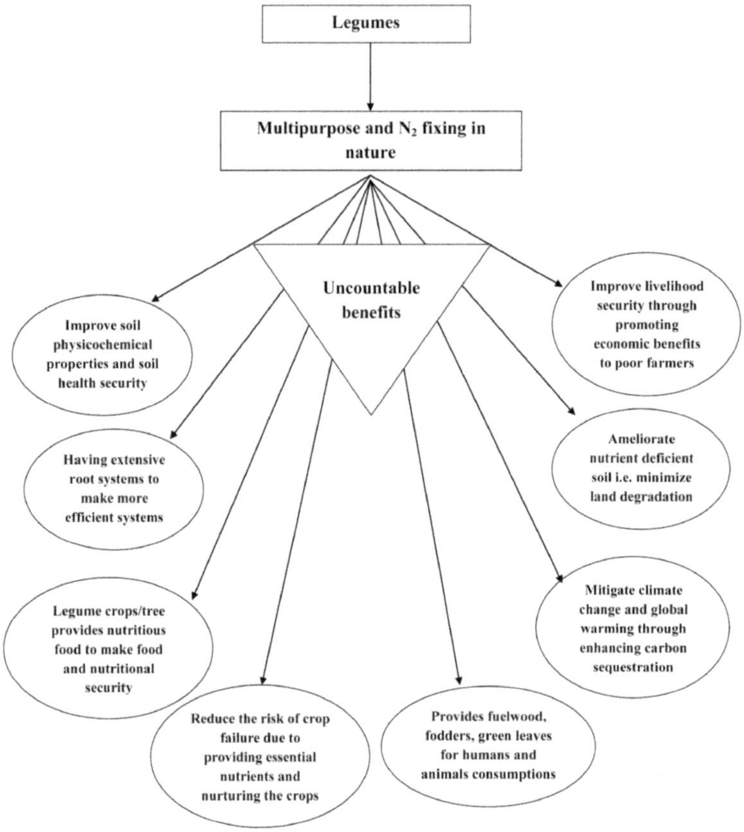

FIGURE 3.1 Legumes: uncountable benefits.

Wambugu et al., 2011; Virgona et al., 2012; Franzel et al., 2014; Xavier et al., 2014; Apolinário et al., 2015; Hernández-Muciño et al., 2015).

Legume trees species such as *Leucaena leucocephala* (Subabul), *Calliandra calothyrsus*, *Gliricidia sepium* (quickstick), *Sesbania grandiflora* (Agati), and *Acacia* species are gaining extreme importance and successfully adopted by farmers in various countries like Australia, Asia, Southeast Asia, Indonesia, and Africa, respectively (Palme et al., 1994; Gutteridge, 1994; Middleton et al., 1995; Wickens et al., 1995; Stewart, 1996; Moog et al., 1998). These legumes have very broad and wide adaptation in every climatic situation as arid, semiarid, and humid areas which can potentially enhance the scope and potential of these legumes with their utilization ability by peoples in agroclimatic zones.

Further, tree legumes are more persistent in nature than herbaceous legumes, once they established. Also, integration of legume trees along with livestocks/pasture cultivations is more viable for sustainable intensification and gaining wider importance with showing potential role in developmental paradigm (Tedeschi et al., 2015). Nitrogen fixation capacity by tree legumes in different regions is depicted in Figure 3.2.

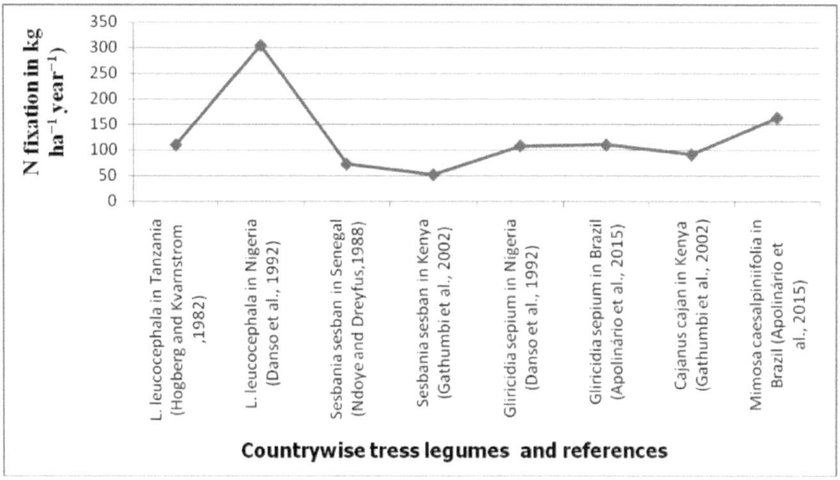

FIGURE 3.2 Nitrogen fixation capacity by tree legumes in different regions.

3.5 POSSIBILITIES OF LEGUMINOUS PLANTS IN AGROFORESTRY

Since a time immemorial, the practices of agroforestry spread continuously through the tropics which do compromise between agricultural and forestry sector by integrating into single piece of land/area with fullest utilization of natural resources in sustainable manner without depleting and degradation of environment. Integration of three elements such as woody perennial trees, herbaceous crops and livestock's/pastures are the basic characteristics of agroforestry systems and type and level of combinations in any locality represent different models which are now prevailing in agroclimatic zones.

For example, trees, and shrubs legume such as Acacia species are gaining an importance due to provision of protein-rich pasture, and satisfying forage requirement to the livestock's in dry regions particularly in the Sahel and North Africa (Baumer, 1992; Halli et al., 2018). Similarly, incorporation of legumes in agroforestry makes the availability of protein-rich food as forage and pasture for livestock's and cattle and such type of practice has been seen in the Zimbabwe where legume pastures in agroforestry practices are gaining importance and promising the availability of forage to cattle's along with enhancing overall farm productivity (Campbell et al., 1991). Also, Leucaena and grass-based agroforestry system in Australia are more profitable for livestock's and farmers in term of forage availability and farm productivity (Larsen et al., 1998).

The practices of agroforestry in any locality can ability to improve soil physicochemical property and fertility through enhancement in the beneficial soil microorganism through deposition and decomposition of soil organic matter (OM) (Raj et al., 2016; Singh and Jhariya, 2016). Integration of legume trees/crops in agroforestry can boost overall biomass production, improve soil fertility and health, and enhance availability of essential nutrients for plants and maximized production and profitability to the farmers. In this context, addition of MPTs such as *Acacia nilotica* in traditional agroforestry systems of Chhattisgarh (India) are more ecologically acceptable and economically viable that can utilized natural resources in the fullest potential with enhancing the soil fertility through addition of nitrogen-rich leaf as litter-fall and fixing ability of multipurpose Acacia tree along with the contribution in enhancing the economics of farmers through gum production (Das et al., 2014; Raj, 2015a). Moreover, this *Acacia* based model can protect our environment by conserving biodiversity (Raj, 2015b) in tropical climate. Thus, there is a tremendous possibility of integration of leguminous trees/crops in the agroforestry systems in the tropics. Agroforestry having both components, i.e., forest trees and crops either may be leguminous or not and can exist along with livestock's/pasture. Incorporation of leguminous trees can promotes efficient nitrogen cycling and provides nutrients to plant, improve soil health and quality through close and efficient nutrient cycling process, enhance overall biomass and productivity of agroforestry models, maintain food security along with human health and environmental security and provision of fodder (for livestock's), fuelwood, and timber, etc. due to its MPTs in nature, etc. (Figure 3.3).

3.6 LINKING CONCEPTS BETWEEN LEGUMES IN AGROFORESTRY AND SOIL HEALTH

Soil degradation is the major challenge today due to unstoppable additions of inorganic/chemical fertilizers that results soil nutrient losses leads to erosion and promotes pollutions (Zentner et al., 2004). Leguminous trees play an enviable role in this problematic soil by conserving, storing the nutrients (majorly nitrogen) through fixing into the soil, which can enhance soil fertility and whole agroforestry productivity (Figure 3.4). Soil carbon sequestration is also promoted by legume-based agroforestry systems. Thus, nitrogen-deficient soil can be ameliorated through the integration of legumes in any agroforestry models in agroclimatic zones.

The integration of legumes in agroforestry not only positive in yield and productivity of crops and perennial trees but it also helps in enhancing the soil physicochemical properties through addition and decomposition of OM results accumulation of nutrient in soil which in turn consume by plants. This can be justified by research of Collier et al. (2018) on legume-based agroforestry system having *Dipteryx alata* tree species and its effects on yield of maize plant (intercropped) and overall effects on soil physicochemical properties. Under the practices of alley cropping, incorporation of legumes species as the hedgerow on the bunds of any field has two benefits; first as reduce the contamination and destruction of species from the outside and seconds as this model can provide very essential nutrients for soil and nutrient-rich feeding material for the cattle's. For example various leguminous species like *Indigofera tinctoria, Cajanus cajan, Desmodium rensonii, Tephrosia candida*, etc., have the ability to add on average 5 tonnes of fresh leaves per hectare per year which is quite interesting subject for the further study in the context of yield and soil health sustainability through legumes in agroforestry system (Das and Ghosh, 2012).

As per Bhatt and Bujarbaruah (2006) integration of nitrogen-fixing legumes in alley cropping viable in term of nutrient addition in the soil as nitrogen (20 to 80 kg), Phosphorus (2 to 14 kg) and Potassium (8 to 38 kg), respectively through the tending operation, i.e., pruning activity which can helps in lowering the acidity of soil, improves fertility and conserve the soil moisture by acting as a mulch (Kang et al., 1990). A numbers of studies were conducted on species to species nature for their contribution in addition of nutrients (N) into the soil. For example, in compared to *Dalbergia* and *Albizia* legume tree species, *Leucaena* green manuring has significant effects on the higher yield of *Ricinis communis* plant along with higher addition of nitrogen as 75 kg/ha (AICRPDA, 1984). Similarly as per Palled et al. (2000), addition of nitrogen through Leucaena species varies in between 35 to 41 kg per hectare and helps in maintaining soil fertility along with improving productivity of overall agroforestry model. Thus, nitrogen fixation rates (0.1–700 kg N/ha/yr) can vary as per varying NFTs species, prevailing climatic conditions, nature of properties of soil and its type (Vanlauwe et al., 2006; Adams et al., 2010; Ajayi et al., 2011). Increasing the availability and mobility of phosphorus due to mycorrhizal associations in the soil is also another beneficial service provides by nitrogen-fixing trees in agroforestry systems (Houlton et al., 2008).

3.7 NEXUS BETWEEN LEGUMINOUS TREES IN AGROFORESTRY AND FOOD SECURITY

The term food security represents both the quantity and quality (in term of nutrition value) of food material that are available to people in a continuous and sustainable way. Food security is not a single concept that can be achieved through addressing the both nutritional and environmental challenges of today (Garnett, 2014). As we know, food productions in current day are not enough to satisfying the burgeoning population and provide very less nutrition and calories to the needy people (Foresight, 2011; Lobell et al., 2011).

In this context, the practices of legume-based intercropping can ensure the productivity along with nutritious food materials which help in building the health status of people and ecosystems. Thanks to the leguminous family which provides various tangible benefits to poor farmers in term of varying food, edible gum, medicines, and fruits while integration into the various agroforestry models which can meet out the problem of hunger and supply sufficient quantity and quality of feed materials to both animals and humans in every climatic zones (Rao and Husain, 1993; Bargali, 2016). This can maintain both FNS along with human health. Thus, legume-based agroforestry systems are the more profitable in enhancement of ecological diversity and socioeconomic benefits along with maintaining FNS for the farmers (Gliessman, 2015).

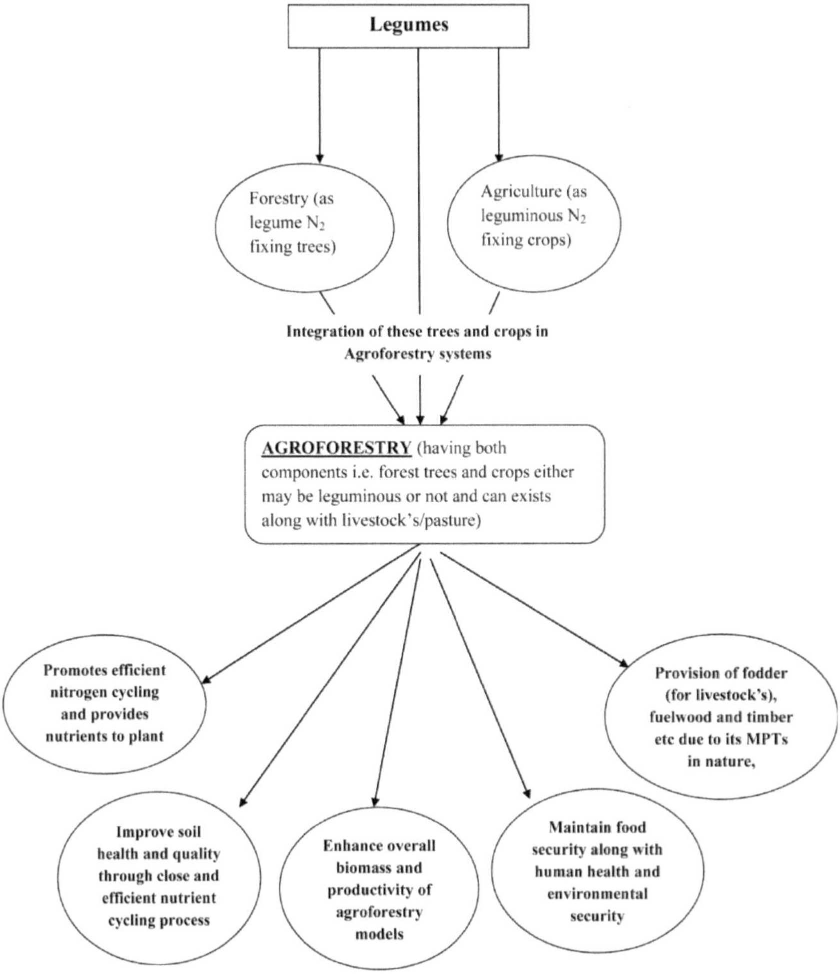

FIGURE 3.3 Possibilities of legumes in agroforestry systems.

3.8 LEGUMINOUS TREES FOR NUTRIENT CYCLING IN AGROFORESTRY

As we know, a single elements (either trees and crops) are unable to minimize the loss of nutrients through leaching (and other means) in the soil, therefore, agroforestry (combination of these two elements) has greatest potential to resolve the losses through efficient close nutrient cycling

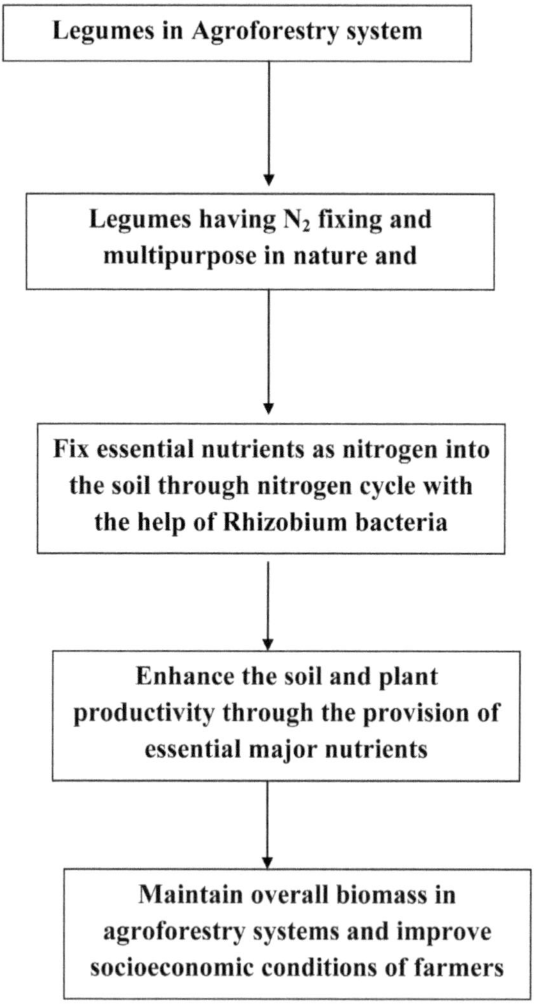

FIGURE 3.4 Legumes for soil fertility enhancement in agroforestry system.

process and guarantee for the better soil and environment health. Close nutrient cycling System is exist in agroforestry systems which help in minimizing the nutrients loss due to extensive deep root systems in leguminous tree and make availability of essential nutrients to both agricultural crops and woody perennials trees as a source of food. This hypothesis is supported by various scientist and researchers globally (Sanchez, 1995; Young, 1997). Beyond the benefits of efficient nutrient cycling (N cycle), leguminous trees having greater potential to enhance the soil fertility through inherit capacity of fixing nitrogen which can modify soil properties in term of higher availability of nutrients and water to plants in the agroforestry practices (Cooper et al., 1996; Young, 1997; Sanchez et al., 1997; Buresh and Tian, 1998; Rao et al., 1998).

Leguminous trees can fix atmospheric nitrogen with the help of *Rhizobium* (bacteria) which exist in the root nodules. Integration of short-rotation woody perennial trees and shrubs such as *Sesbania sesban* species (legumes in nature) with herbaceous crops are benefited for both farmers (through enhancing crop yields and biomass productivity) and environment (through efficient nutrient cycling and soil fertility enhancement). However, deep, and extensive root systems of leguminous trees in agroforestry have the capacity to remove nutrients from the deeper layer of soil, which is minimizing the loss of nutrient through the leaching process. In contrast, nutrient loss is more prominent through leaching due less extensive root systems of agricultural crops. Therefore, this loss can be mitigating by planting of multipurpose NFTs along with agricultural crops (Figure 3.5). However, type, and nature of species and available soil nitrogen status can affects the N_2 fixation rate of legume trees in agroforestry systems prevailed in any locality and this rate can varies from 5 to 300 kilogram N/ha/year (Vanlauwe and Giller, 2006; Ajayi et al., 2011).

3.9 ECOSYSTEM SERVICES PROVIDED BY LEGUME BASED AGROFORESTRY

Legume based agroforestry systems particularly provides greater ecosystem services in term of both tangible and intangible benefits, i.e., providing timber, fuelwood, protein-rich fodder (for livestock's), green leaves, NTFPs, and environmental protection services as maximizing biodiversity, retaining nitrogen content in soil and protecting soil by reducing erosion and compaction, etc. (Harmand et al., 2007; Tscharntke et al., 2011). OM deposition and its decomposition are still gaining wide importance for addition of mineral nitrogen and other basic essential nutrients into the soil under legume-based agroforestry system and hence improve soil quality and health which is prerequisite for the better ecosystem services (Aranguren et al., 1982; Glover and Beer, 1986). Also, legume-based silvopastoral systems perform a variety of ecosystem services as biological nitrogen fixation, enhancing soil fertility through increasing SOM which in turn increase soil carbon sequestration value, makes availability of pastures for cattle's, providing shade to livestock's

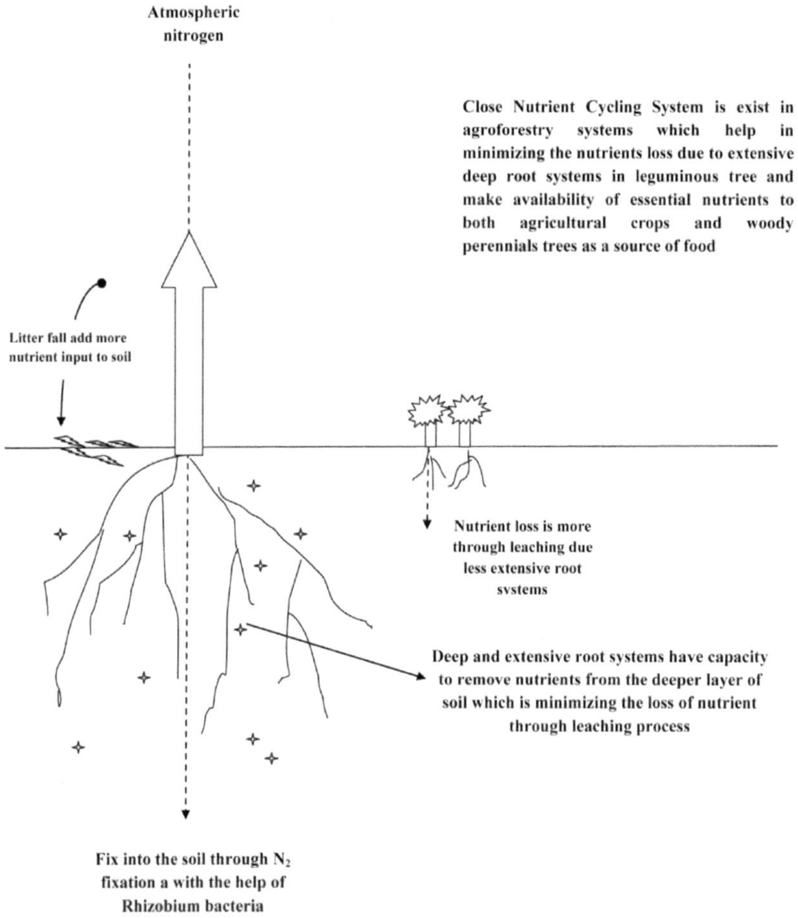

FIGURE 3.5 Legumes for close nutrient cycling in agroforestry systems.

in harsh climatic situations, conservation of biodiversity and helpful in climate change mitigation through reducing GHGs emissions to the atmosphere (Dubeux et al., 2015).

3.10 PRODUCTION POTENTIAL OF LEGUME BASED AGROFORESTRY

Legume based agroforestry system has tendency to build up more biomass productivity by increasing nitrogen availability in soil rather than other non fertilized farmers practices. For example, incorporation of *Gliricidia sepium* with the maize crop resulted higher yield by 42% as compared to other fields having non-fertilized method (Sileshi et al., 2012). Similarly, crop production can also increased by integrating leguminous woody trees in the fallow periods between the crops and which is justified by the work of Hall et al. (2005) and according to him the yields of sorghum was enhanced by 55% in consecutive two seasons which following the *Gliricidia* as leguminous trees species as compared to traditional based fallows system. However, there are some reports were published which can contradict the concept of beneficial impact of legume-based agroforestry system in term of yield and biomass enhancement. For example, coffee-based agroforestry has lower yield and biomass than the monoculture system (Tully and Lawrence, 2011).

Similarly, cereal-legume based intercropping is more viable in term of both higher productivity and profitability for resource-poor farmers (Naudin et al., 2010). Moreover, integration of leguminous trees increased the yield of maize as doubled in compared to traditional based practices and make the availability of nitrogen (60 kg/ha/yr through BNF) into the soil which helps in enhancing soil fertility along with minimizing (up to 75%) the farmers bill to purchase chemical nitrogen fertilizers for application in soil. Thus legume-based AFs are ecologically viable and economically profitable to poor farmers/people. Therefore, various studies have been concluded that annual intercropping are more productive and profitable as compared to sole cropping based production systems (Bedoussac et al., 2015; Zhu et al., 2016; Kermah et al., 2017) in any agroclimatic zones.

3.11 HEALTH SECURITY THROUGH LEGUME BASED AGROFORESTRY SYSTEMS

Agroforestry itself sustains billions of human's population, and livestock's globally through providing various tangible (direct) products to organisms and intangible benefits as efficient nutrient cycling, conservation of soil and water, mitigation of extreme climate through carbon sequestration, etc. But the integration of legumes in agroforestry can add more benefits to biodiversity in term of fixing nitrogen into the soils which are the primary sources of nutrient for the plants metabolism and absorbing nitrogen can reduce excessive emissions of nitrous oxide (N_2O) in the atmosphere results changing climate and global warming phenomenon.

Legume based agroforestry systems provides protein-rich leaves (as protein bank) and fodders as feed for livestock's cattle and has ability to build soil fertility and productivity through fixation of nitrogen which is component of amino acid and protein which is the building block of living cell and tissues and prerequisite for the good health of plants which in turn consume by livestock's and human population results healthy people and ecosystems (Figure 3.6). We can say legume-based agroforestry gives better security to environment and organism residing within it through climate security and FNS that results ecosystems health security.

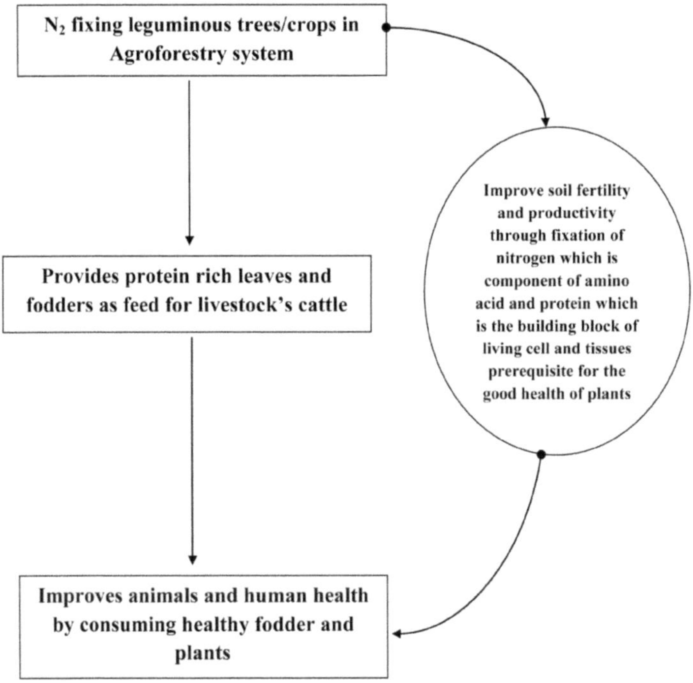

FIGURE 3.6 Legumes for livestock's and human health in agroforestry.

3.12 CARBON SEQUESTRATION POTENTIAL (CSP) IN LEGUME-BASED AGROFORESTRY

In general, the practices of sole cropping system (in contrary to legume-based intercropping system) require higher doses of chemical fertilizers and pesticides which are the major contributors of GHGs emission into the atmosphere results global warming and climate change phenomenon. This can be minimized through integration of legume-based agroforestry systems which can not only fix atmospheric nitrogen into the soil but can supply these nutrients to plants in available form (Carlsson and Huss-Danell, 2014) that helps to farmers to save money for purchasing nitrogenous chemical fertilizers and reduce the dependency on higher application of synthetic fertilizers. However, legume-based agroforestry also contributes somehow in greenhouse gases (GHGs) production, i.e., nitrous oxide emission through the process of nitrification and denitrification through microorganism (Hergoualch et al., 2007; Verchot et al., 2008) and as per one estimate, about 3 to 4 times more emission through legume-based agroforestry systems than other natural forest and cropland due to rapid nitrogen cycling process (Baggs et al., 2006; Verchot et al., 2006). Legume based agroforestry system has enormous potential to fix nitrogen into the soil and combat climate change problem through enhancing storage and sequestration capacity of carbon in the tropics. Incorporation of legume trees in agroforestry has the greatest potential to minimize continuous emissions of CO_2 and N_2O in the atmosphere. However, legume trees itself absorbs and fix atmospheric carbon in their parts and whole agroforestry systems can potentially sequester atmospheric carbon into both tree (as legumes) and soils. This will helps to mitigate climate change and global warming through reducing emission of GHGs and maintain climate security and agroecosystem health (Figure 3.7). For example, legumes trees like Leucaena and Acacia have great potential to carbon into the soil that helps to enhance soil organic carbon pools and fertility. Also, as per Cadisch et al. (1998) the higher rate of OM build-up can be possible through integrating N_2 fixing tree species in the tropical soil ecosystem. Legume tree-like *Leucaena leucocephala* can sequester more carbon with the help of integration of *Dalbergia sissoo* rather than sole plantation that system can mitigate global warming and combat climate change phenomenon (Sheikh et al., 2015). Similarly, the integration of some leguminous tree species into the eucalyptus based plantation system was benefited in term of higher storage and sequestration of carbon (Kaye et al., 2000). Thus, incorporation of nitrogen-fixing leguminous trees in the forest vegetation are more sustainable in term of higher carbon sequestration as one gram nitrogen is equivalent to the fixation of 12–15 gram carbon (Binkley and Menyailo, 2005).

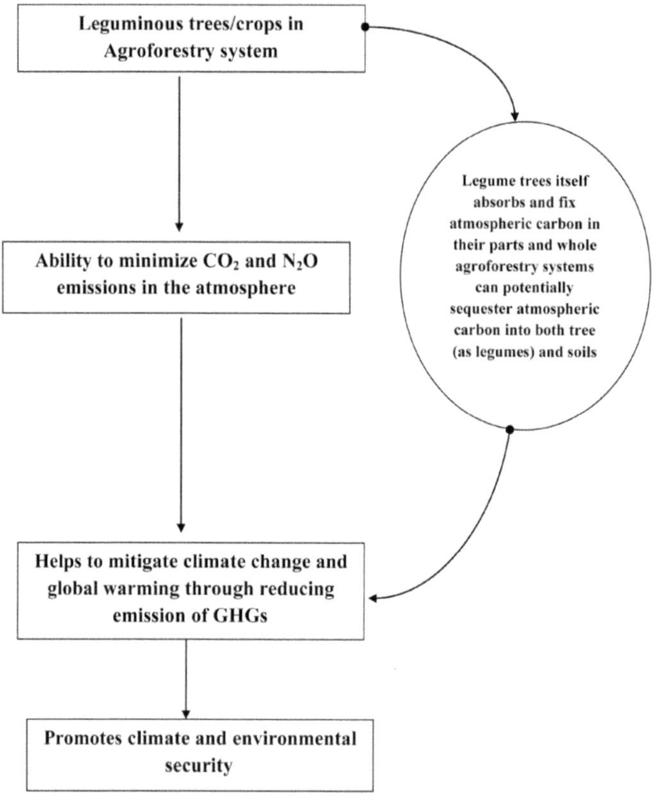

FIGURE 3.7 Legumes for climate change mitigation in agroforestry.

3.13 FUTURE ROADMAP FOR R&D IN LEGUME-BASED AGROFORESTRY

Deficient nitrogen in the soil of developing countries including Indian subcontinent has appeared loss in soil fertility and biomass productivity. Such deficiency can be minimized by the process of biological nitrogen fixation through incorporating leguminous tree in different agroforestry models. In this context, there is a need to explore more potential legumes species through appropriate scientific research and development in the field of production, protection, and profits to farmers in varying ecological zones of the world. Also, border application of this leguminous species is becoming thrust field of research and development in modern science and technology.

3.14 CONCLUSIONS

Indeed, legumes play undeniable role in both fixation of nitrogen and enhancing the storage and sequestration of carbon into the soil treasure which can enhance overall productivity under the changing climate phenomenon. Carbon fixation by legumes can enhance biomass accumulation that leads to build-up of soil OMs and improves the fertility. Therefore, legume-based agroforestry systems can be adopted wherever possible and a boon for the reclamation of wasteland and degraded land. Moreover, due to advanced photosynthetic characteristics of leguminous trees, it is a very good option to combat climate change under elevated carbon dioxide.

KEYWORDS

- **agroforestry**
- **biodiversity**
- **biological N_2 fixation (BNF)**
- **ecosystem services**
- **environment**
- **legumes**
- **productivity**

REFERENCES

Adams, M., Simon, J., & Fautsch, P., (2010). Woody legumes: A review from the South. *Tree Physiology, 30*, 1072–1082. doi: 10.1093/treephys/tpq061.

AICRPDA, (1984). *Annual Reports, 1979–80 to 1983 to 1983–1984*. All India Coordinated Research Project for Dryland Agriculture, ICAR, Hyderabad.

Ajayi, O. C., Place, F., Akinnifesi, F. K., & Sileshi, G. W., (2011). Agricultural success from Africa: The case of fertilizer tree systems in southern Africa (Malawi, Tanzania, Mozambique, Zambia and Zimbabwe). *Int. J. Agric. Sustain., 9*, 129–136.

Apolinário, V. X. O., Dubeux, Jr. J. C. B., Lira, M. A., Ferreira, R. L. C., Mello, A. C. L., Santos, M. V. F., Sampaio, E. V. S. B., & Muir, J. P., (2015). Tree legumes provide marketable wood and add nitrogen in warm-climate silvopasture systems. *Agronomy Journal, 107*, 1915–1921.

Aranguren, J., Escalante, G., & Herrera, R., (1982). Nitrogen cycle of tropical perennial crops under shade trees. I. Coffee. *Plant Soil, 67*, 247–258.

Baggs, E. M., Chebii, J., & Ndufa, J. K., (2006). A short-term investigation of trace gas emissions following tillage and no-tillage of agroforestry residues in western Kenya. *Soil Tillage Res., 90*, 69–76.

Bargali, K., & Bargali, S. S., (2009). *Acacia nilotica*: A multipurpose leguminous plant. *Nature and Science, 7*(4), 11–19.

Bargali, K., (2016). Leguminous plants of Kumaun Himalaya: Diversity, distribution, threats and management. In: Chand, R., & Leimgruber, W., (eds.), *Globalization and Marginalization in Mountain Regions, Perspectives on Geographical Marginality* (Vol. 1, pp. 199–204). doi: 10.1007/978–3–319–32649–8–15.

Baumer, M., (1992). Trees as browse to support animal production. In: Speedy, A., & Pugliese, P. L., (eds.), *Legume Trees and Other Fodder Trees as Protein Sources for Livestock* (pp. 1–10). Proceedings of an FAO Expert Consultation, Kuala Lumpur, Malaysia. Rome, FAO.

Bedoussac, L., Journet, E. P., Hauggaard-Nielsen, H., Naudin, C., Corre-Hellou, G., Jensen, E. S., Prieur, L., & Justes, E., (2015). Ecological principles underlying the increase of productivity achieved by 47 cereal-grains legume intercrops in organic farming: A review. *Agronomy for Sustainable Development, 35*(3), 911–935. doi: 10.1007/s13593–014–0277–7.

Bhatt, B. P., & Bujarbaruah, K. M., (2006). *Agroforestry in North East India: Opportunities and Challenges, 2005.* ICAR Research Complex for NEH Region, Umiam, Meghalaya. pp. 603.

Binkley, D., & Menyailo, O., (2005). Tree species effects on soils: Implication for global change. *NATO Science Series, IV: Earth and Environmental Sciences, 55*, 155–164.

Buresh, R. J., & Tian, G., (1998). Soil improvement by trees in sub-Saharan Africa. *Agroforestry Systems, 38*, 51–76.

Cadisch, G., Oliveira, O. C., De Cantarutti, R., Carvalho, E., & Urquiaga, S., (1998). The role of legume quality in soil carbon dynamics in savannah ecosystems. In: Bergstrom, L., & Kirchmann, H., (eds.), *Carbon and Nutrient Dynamics in Natural and Agricultural Tropical Ecosystems.* Wallingford, UK, CAB International. pp. 47–70.

Campbell, B. M., Clarke, J. M., & Gumbo, D. J., (1991). Traditional agroforestry practices in Zimbabwe. *Agroforestry Systems, 14*, 99–111.

Carlsson, G., & Huss-Danell, K., (2014). Does nitrogen transfer between plants confound 15N-based quantifications of N_2 fixation? *Plant and Soil, 374*(1–2), 345–358. doi: 10.1007/s11104–013–1802–1.

Collier, L. S., Arruda, E. M., Campos, L. F. C., & Nunes, J. N. V., (2018). Soil chemical attributes and corn productivity grown on legume stubble in agroforestry systems. *Rev. Caatinga, Mossoró, 31*(2), 279–289.

Cooper, P. J. M., Leakey, R. R. B., Rao, M. R., & Reynolds, L., (1996). Agroforestry and the mitigation of land degradation in the humid and sub-humid tropics of Africa. *Experimental Agriculture, 32*, 235–290.

Crews, T. E., & Peoples, M. B., (2004). Legume versus fertilizer sources of nitrogen: Ecological tradeoffs and human needs. *Agriculture, Ecosystems & Environment, 102*, 279–297.

Danso, S. K. A., Bowen, G. D., & Sanginga, N., (1992). Biological nitrogen fixation in agroecosystems. *Plant and Soil, 141*, 177–196.

Das, A., & Ghosh, P. K., (2012). Role of legumes in sustainable agriculture and food security: An Indian perspective. *Outlook on Agriculture*, *41*(4), 1–6.

Das, I., Katiyar, P., & Raj, A., (2014). Effects of temperature and relative humidity on ethephon induced gum exudation in *Acacia nilotica. Asian Journal of Multidisciplinary Studies, 2*(10), 114–116.

Dubeux, Jr., J. C. B., Muir, J. P., Nair, P. K. R., Sollenberger, L. E., Silva, H. M. S., & Mello, A. C. L., (2015). The advantages and challenges of integrating tree legumes into pastoral systems. In: Evangelista, A. R., et al. (eds.), *Proceedings of the 1st International Conference on Forages in Warm Climates* (pp. 141–164). Universidade Federal de Lavras, Lavras, MG, Brazil.

Erisman, J. W., Galloway, J. N., Seitzinger, S. P., Bleeker, A., & Butterbach-Bahl, K., (2011). Reactive nitrogen in the environment and its effect on climate change. *Curr. Opin. Environ. Sustain.*, *3*, 281–290.

Foresight, (2011). *The Future of Food and Farming: Challenges and Choices for Global Sustainability*. Final Project Report. The Government Office for Science, London.

Franzel, S., Carsan, S., Lukuyu, B., Sinja, J., & Wambugu, C., (2014). Fodder trees for improving livestock productivity and smallholder livelihoods in Africa. *Current Opinion in Environmental Sustainability*, *6*, 98–103.

Garnett, T., (2014). Three perspectives on sustainable food security: Efficiency, demand restraint, food system transformation. What role for life cycle assessment? *Journal of Cleaner Production*, *73*, 10–18. doi: 10.1016/j.jclepro.2013.07.045.

Gathumbi, S. M., Cadisch, G., & Giller, K. E., (2002). 15N natural abundance as a tool for assessing N2-fixation of herbaceous, shrub and tree legumes in improved fallows. *Soil Biology and Biochemistry*, *34*, 1059–1071.

Gliessman, S. R., (2015). *Agroecology: The Ecology of Sustainable Food Systems* (No. 3). CRC Press. ISBN 9781439895610; pp. 406.

Glover, N., & Beer, J., (1986). Nutrient cycling in to traditional Central American agroforestry systems. *Agroforestry Systems*, *4*, 77–87.

Gutteridge, R. C., (1994). The perennial *Sesbania* species. In: Gutteridge, R. C., & Shelton, H. M., (eds.), *Forage Tree Legumes in Tropical Agriculture* (pp. 49–64). Wallingford, UK, CAB International.

Hall, N. M., Kaya, B., Dick, J., Skiba, U., Niang, A., & Tabo, R., (2005). Effect of improved fallow on crop productivity, soil fertility and climate forcing gas emissions in semi-arid conditions. *Biol. Fert. Soils*, *42*, 224–230.

Halli, H. M., Rathore, S. S., Manjunatha, N., & Wasnik, V. K., (2018). Advances in agronomic management for ensuring fodder security in semi arid zones of India: A review. *International Journal of Current Microbiology and Applied Sciences*, *7*(2), 1912–1921.

Harmand, J. M. A., Vila, H., Dambrine, E., Skiba, U., Miguel, S., Renderos, R. V., Oliver, R., Jimenez, F., & Beer, J., (2007). Nitrogen dynamics and soil nitrate retention in a *Coffea arabica—Eucalyptus deglupta* agroforestry system in Southern Costa Rica. *Biogeochemistry*, *85*, 125–139.

Hasniati, D., & Shelton, H. M., (2005). *Sesbania grandiflora*: A successful tree legume in Lombok, Indonesia. In: *XX International Grassland Congress: Offered Papers, Dublin, Ireland* (p. 324). Wageningen Academic Publishers, Wageningen, The Netherlands.

Hergoualch, K., Skiba, U., Harmand, J. M., & Oliver, R., (2007). Processes responsible for the nitrous oxide emission from a Costa Rican Andosol under a coffee agroforestry plantation. *Biol. Fert. Soils*, *43*, 787–795.

Hernández-Muciño, D., Sosa-Montes, E., & Ceccon, E., (2015). *Leucaena macrophylla*: An ecosystem services provider? *Agroforestry Systems*, *89*, 163–174.

Hogberg, P., & Kvarnstrom, M., (1982). Nitrogen fixation by the woody legume *Leucaena leucocephala* in Tanzania. *Plant and Soil*, *66*, 21–28.

Hong, Y., & Bhatnagar, S., (2007). Tropical tree legumes. In: Pua, E. C., & Davey, M., (eds.), *Transgenic Crops V. Biotechnology in Agriculture and Forestry* (p. 60). Springer, Berlin, Heidelberg.

Houlton, B. Z., Wang, Y. P., Vitousek, P. M., & Field, C. B., (2008). A unifying framework for dinitrogen fixation in the terrestrial biosphere. *Nature*, *454*, 327-330. doi: 10.1038/nature07028.

Husain, T., & Kapoor, S. L., (1990). *Enumeration of Legumes in India (Indigenous and Introduced)*. Lucknow, India, National Botanical Research Institute. pp. 61.

Jensen, E. S., Peoples, M. B., Boddey, R. M., Gresshoff, P. M., Hauggaardnielsen, H., Alves, B. J. R., & Morrison, M. J., (2011). Legumes for mitigation of climate change and the provision of feedstock for biofuels and biorefineries: A review. *Agron. Sustain. Develop.*, *32*, 329–367.

Jhariya, M. K., Banerjee, A., Yadav, D. K., & Raj, A., (2018). Leguminous trees an innovative tool for soil sustainability. In: Meena, R. S., Das, A., Yadav, G. S., & Lal, R., (eds.), *Legumes for Soil Health and Sustainable Management* (pp. 315–345). Springer, ISBN 978–981–13–0253–4 (eBook), ISBN: 978–981–13–0252–7 (Hardcover). https://doi.org/10.1007/978–981–13–0253–4_10 (Accessed on 4 August 2019).

Jhariya, M. K., Bargali, S. S., & Raj, A., (2015). Possibilities and Perspectives of Agroforestry in Chhattisgarh. In: Miodrag, Z., (ed.), *Precious Forests-Precious Earth* (pp. 237–257, 286). ISBN: 978–953–51–2175–6, InTech, Croatia, Europe, doi: 10.5772/60841.

Kang, B. T., Reynolds, L., & Atarrah, A. N., (1990). 'Alley farming.' *Advances in Agronomy*, *43*, 315–359.

Kaye, J. P., Resh, S. C., Kaye, M. W., & Chimner, R. A., (2000). Nutrient and carbon dynamics in a replacement series of *Eucalyptus* and *Albizia* trees. *Ecology*, *81*(12), 3267–3273.

Kermah, M., Franke, A. C., Adjei-Nsiah, S., Ahiabor, B. D. K., Abaidoo, R. C., & Giller, K. E., (2017). Maize-grain legume intercropping for enhanced resource use efficiency and crop productivity in the Guinea savanna of northern Ghana. *Field Crops Res.*, *213*, 38–50.

Larsen, P. H., Middleton, C. H., Bolam, M. J., & Chamberlain, J., (1998). *Leucaena* in large-scale grazing systems: Challenges for development. In: Shelton, H. M., Gutteridge, R. C., Mullen, B. F., & Bray, R. A., (eds.), *Leucaena – Adaptation, Quality and Farming Systems* (pp. 324–330). Proceedings of workshop, Hanoi, Viet Nam. ACIAR Proceedings No. 86. Canberra, Australia, Australian Centre for International Agricultural Research.

Lewis, G. P., Schrire, B. D., Mackinder, B. A., & Lock, J. M., (2003). *Legumes of the World*. Royal Botanic Gardens, Kew.

Lobell, D. B., Schlenker, W., & Costa-Roberts, J., (2011). Climate trends and global crop production since 1980. *Science*, *333*(6042), 616–620. doi: 10.1126/science.1204531.

LPWG (The Legume Phylogeny Working Group), (2017). A new subfamily classification of the Leguminosae based on a taxonomically comprehensive phylogeny. *Taxon, 66*, 44–77. doi: 10.12705/661.3.

Middleton, C. H., Jones, R. J., Shelton, H. M., Petty, S. R., & Wildin, J. H., (1995). Leucaena in northern Australia. In: Shelton, H. M., Piggin, C. M., & Brewbaker, J. L., (eds.), *Leucaena – Opportunities and Limitations* (pp. 214–221). Proceedings of workshop, Bogor, Indonesia. ACIAR Proceedings No. 57. Canberra, Australia, Australian Centre for International Agricultural Research. (Also available at: www.aciar.gov.au/proc57/proc57.htm (Accessed on 4 August 2019)).

Montagnini, F., & Nair, P. K. R., (2012). Carbon sequestration: An underexploited environmental benefit of agroforestry systems. *Agroforestry Systems, 61*, 281–295.

Moog, F. A., Bezkorowajnyj, P., & Nitis, I. M., (1998). *Leucaena* in smallholder farming systems in Asia: Challenges for development. In: Shelton, H. M., Gutteridge, R. C., Mullen, B. F., & Bray, R. A., (eds.), *Leucaena – Adaptation, Quality and Farming Systems* (pp. 303–310). Proceedings of workshop, Hanoi, Viet Nam. ACIAR Proceedings No. 86. Canberra, Australia, Australian Centre for International Agricultural Research.

Naudin, C., Corre-Hellou, G., Pineau, S., Crozat, Y., & Jeuffroy, M. H., (2010). The effect of various dynamics of N availability on winter pea–wheat intercrops: Crop growth, N partitioning and symbiotic N_2 fixation. *Field Crops Research, 119*(1), 2–11. doi: 10.1016/j.fcr.2010.06.002.

Ndoye, I., & Dreyfus, B., (1988). N_2 fixation by *Sesbania rostrate* and *Sesbania sesban* estimated using 15N and total N difference methods. *Soil Biology and Biochemistry, 20*, 209–213.

Palled, Y. B., Desai, B. K., & Prabhakar, A. S., (2000). Integrated nutrient management in alley cropped maize (*Zea mays*)–groundnut (*Arachis hypogea*) system with subabul (*Lucaena leucocephela*). *Indian Journal of Agronomy, 45*(3), 520–525.

Palmer, B., Macqueen, D. J., & Gutteridge, R. C., (1994). *Calliandra calothyrsus* – a multipurpose tree legume for humid locations. In: Gutteridge, R. C., & Shelton, H. M., (eds.), *Forage Tree Legumes in Tropical Agriculture* (pp. 65–74). Wallingford, UK, CAB International.

Rahman, K. M. A., & Zhang, D., (2018). Effects of fertilizer broadcasting on the excessive use of inorganic fertilizers and environmental sustainability. *Sustainability, 10*(759), 1–15.

Raj, A., (2015a). *Evaluation of Gummosis Potential Using Various Concentration of Ethephon*, MSc Thesis (p. 89). I. G. K. V., Raipur, Chhattisgarh.

Raj, A., (2015b). Gum exudation in *Acacia nilotica*: Effects of temperature and relative humidity. In: *Proceedings of the National Expo on Assemblage of Innovative Ideas/Work of Post Graduate Agricultural Research Scholars* (p. 151). Agricultural College and Research Institute, Madurai (Tamil Nadu).

Raj, A., Jhariya, M. K., & Bargali, S. S., (2016). Bund based agroforestry using eucalyptus species: A Review. *Current Agriculture Research Journal, 4*(2), 148–158.

Raj, A., Jhariya, M. K., & Harne, S. S., (2018). Threats to biodiversity and conservation strategies. In: Sood, K. K., & Mahajan, V., (eds.), *Forests, Climate Change and Biodiversity* (pp. 304–320, 381). Kalyani Publisher, India.

Rao, M. R., Nair, P. K. R., & Ong, C. K., (1998). Biophysical interactions in tropical agroforestry systems. *Agroforestry Systems, 38*, 3–50.

Rao, R. R., & Husain, T., (1993). Himalayan legumes: Diversity and conservation. In: Dhar, U., (ed.), *Himalayan Biodiversity: Conservation Strategies* (pp. 253–266). Gyanodaya Prakashan, Nainital.

Ribeiro-Barros, A. I., Silva, M. J., Isabel, M., Ramalho, J. C., Máguas-Hanson, C., & Ribeiro, N. S., (2018). *The Potential of Tree and Shrub Legumes in Agroforestry Systems*, Nitrogen in Agriculture. Intech Open, doi: 10.5772/intechopen.69995. Available from: https://www.intechopen.com/books/nitrogen-in-agriculture-updates/the-potential-of-tree-and-shrub-legumes-in-agroforestry-systems (Accessed on 4 August 2019).

Rosenstock, T. S., Tully, K. L., Arias-Navarro, C., Neufeldt, H., Butterbach-Bahl, K., & Verchot, L. V., (2014). Agroforestry with N2-fixing trees: Sustainable development's friend or foe? *Current Opinion in Environmental Sustainability*, *6*, 15–21.

Sanchez, P. A., (1995). Science in agroforestry. *Agroforestry Systems*, *3*, 5–55.

Sanchez, P. A., Shepherd, K. D., Soule, M. J., Place, F. M., Buresh, R. J., Izac, A. N., Kwesiga, F. R., et al. (1997). Soil fertility replenishment in Africa: An investment in natural resource capital. In: Buresh, R. J., Sanchez, P. A., & Calhoun, F., (eds.), *Replenishing Soil Fertility in Africa* (pp. 1–47). SSSA Special Publication Number 51, SSSA, Madison, WI, USA.

Sanjappa, M., (1991). *Legumes of India*. Dehradun, Bishen Singh Mahandra Pal Singh. pp. 338.

Schrire, B. D., Lavin, M., & Lewis, G. P., (2005). Global distribution patterns of the Leguminosae: Insights from recent phylogenies. *Biologiske Skrifter*, *55*, 375–422.

Serpil, S., (2012). Investigation of effect of chemical fertilizers on environment. *APCBEE Proc.*, *1*, 287–292.

Sheikh, M. A., Kumar, M., & Todaria, N. P., (2015). Carbon sequestration potential of nitrogen fixing tree stands. *Forestry Studies Metsanduslikud Uurimused*, *62*, 39–47.

Sileshi, G. W, Debusho, L. K., & Akinnifesi, F. K., (2012). Can integration of legume trees increase yield stability in rainfed maize cropping systems in Southern Africa? *Agron. J.*, *104*, 1392–1398.

Singh, N. R., & Jhariya, M. K., (2016). Agroforestry and agrihorticulture for higher income and resource conservation. In: Sarju, N., & Sudhir, K. R., (eds.), *Innovative Technology for Sustainable Agriculture Development* (pp. 125–145). ISBN: 978–81–7622–375–1. Biotech Books, New Delhi, India.

Stevens, P. F., (2001). *Angiosperm Phylogeny*. Website version 9. University of Missouri, St Louis, and Missouri Botanical Garden.

Stewart, J. L., (1996). Utilization. In: Stewart, J. L., Allison, G. E., & Simons, A. J., (eds.), *Gliricidia Sepium – Genetic Resources for Farmers* (pp. 33–48). Tropical Forestry Papers No. 33. Oxford, UK, Oxford Forestry Institute.

Takimoto, A., Nair, P. K. R., & Nair, V. D., (2008). Carbon stock and sequestration potential of traditional and improved agroforestry systems in the West African Sahel. *Agric. Ecosyst. Environ.*, *125*, 159–166.

Tedeschi, L. O., Muir, J. P., Riley, D. G., & Fox, D. G., (2015). The role of ruminant animals in sustainable livestock intensification programs. *International Journal of Sustainable Development & World Ecology*, *22*, 452–465.

Tonucci, R. G., Nair, P. K. R., Nair, V. D., Garcia, R., & Bernardino, F. S., (2011). Soil carbon storage in silvopasture and related land-use systems in the Brazilian cerrado. *J. Environ. Quality*, *40*, 833–841.

Tscharntke, T., Clough, Y., Bhagwat, S., Buchori, D., Faust, H., Hertel, D., Holscher, D., Juhrbandt, J., Kessler, M., & Perfecto, I., (2011). Multifunctional shade-tree management in tropical agroforestry landscapes-a review. *J. Appl. Ecol.*, *48*, 619–629.

Tully, K. L., & Lawrence, D., (2011). Closing the loop: Nutrient balances in organic and conventional coffee agroforests. *J. Sustain. Agric.*, *35*, 671–695.

Vanlauwe, B., & Giller, K. E., (2006). Popular myths around soil fertility management in sub-Saharan Africa. *Agricultural Ecosystems and Environment*, *116*, 34–46. doi: https://doi.org/10.1016/j.agee.2006.03.016 (Accessed on 4 August 2019).

Verchot, L. V., Brienzajunior, S., Deoliveira, V., Mutegi, J., Cattanio, J. H., & Davidson, E. A., (2008). Fluxes of CH_4, CO_2, NO, and N_2O in an improved fallow agroforestry system in eastern Amazonia. *Agric. Ecosyst. Environ.*, *126*, 113–121.

Verchot, L. V., Dutaur, L., Shepherd, K. D., & Albrecht, A., (2011). Organic matter stabilization in soil aggregates: Understanding the biogeochemical mechanisms that determine the fate of carbon inputs in soils. *Geoderma*, *16*, 182–193.

Verchot, L. V., Hutabarat, L., Hairiah, K., & Van Noordwijk, M., (2006). Nitrogen availability and soil N_2O emissions following conversion of forests to coffee in southern Sumatra. *Global Biogeochem. Cycles*, *20*, 1–12.

Virgona, J. M., Harris, C., Kemp, S., Evans, J., & Salmon, R., (2012). Australian legume research – synthesis and future directions. *Crop & Pasture Science*, *63*, 918–926.

Wambugu, C., Place, F., & Franzel, S., (2011). Research, development and scaling-up the adoption of fodder shrub innovations in East Africa. *International Journal of Agricultural Sustainability*, *9*, 100–109.

Wickens, G. E., Sief El Din, A. G., Sita, G., & Nahal, I., (1995). *Role of Acacia Species in the Rural Economy of Dry Africa and the Near East*. FAO Conservation Guide No. 27. Rome, FAO. ISBN 92-5-103651-9; http://www.fao.org/3/V5360E/v5360e00.htm

Xavier, D. F., Lédo, F. J. S., Paciullo, D. S. C., Urquiaga, S., Alves, B. J. R., & Boddey, R. M., (2014). Nitrogen cycling in a Brachiaria based silvopastoral system in the Atlantic forest region of Minas Gerais, Brazil. *Nutrient Cycling in Agroecosystems*, *99*, 45–62.

Young, A., (1997). *Agroforestry for Soil Management*. CAB International, Wallingford, UK. ISBN:9780851991894; pp. 320.

Zentner, R. P., Campbell, C. A., Biederebeck, V. O., Selles, F., Lemke, R., Jefferson, P. G., & Gan, Y., (2004). Long-term assessment of management of an annual legume green manure crop for fallow replacement in the brown soil zone. *Canadian Journal of Plant Science*, *83*, 475–482.

Zhu, J., Werf, W., Vos, J., Anten, N. P. R., Putten, P. E. L., & Evers, J. B., (2016). High productivity of wheat intercropped with maize is associated with plant architectural responses. *Annals of Applied Biology*, *168*(3), 357–372. doi: 10.1111/aab.12268.

CHAPTER 4

Multifunctional Pastures for Livestock Management and Climate Change Mitigation

ABHISHEK RAJ,[1] M. K. JHARIYA,[2] D. K. YADAV,[2] and A. BANERJEE[3]

[1]*Department of Forestry, College of Agriculture, I.G.K.V., Raipur – 492012, Chhattisgarh, India, E-mail: ranger0392@gmail.com*

[2]*Department of Farm Forestry, Sarguja University, Ambikapur–497001, Chhattisgarh, India, E-mails: manu9589@gmail.com (M. K. Jhariya); dheeraj_forestry@yahoo.com (D. K. Yadav)*

[3]*Department of Environmental Science, Sarguja University, Ambikapur–497001, Chhattisgarh, India, E-mail: arnabenvsc@yahoo.co.in*

ABSTRACT

Indeed, pasture/grass (particularly N_2 fixing) based farming systems (such as silvopastoral system) plays a viable role in enhancing biomass, productivity, improving soil fertility and health, conserving soil and water by minimizing erosion, improving socio-economic condition of farmers by diversifying products and mitigating the deleterious impacts of climate change through storage and sequestration of carbon into both vegetation and soils, respectively. Integrating the MPTs in cropland and pastureland are the good options for providing both tangible and intangible benefits, which is prerequisite for better environment and ecosystem services. Good quality pasture and grasses are better feed materials for livestock's that maintains their health and productivity, which affects their milk and other consumable products and can be a good economic source to poor farmers. Therefore, in this context, proper, and good grazing management under better scientific practices are necessitates the healthy environment and nurturing the biodiversity. However, due to some faulty farming practices and activities along with livestock's production and activities such as manure management and enteric fermentations are contributed GHGs emission into the atmosphere but it can be reduced at certain level through the better management and scientific practices of whole pasture-based farming systems comprising nitrogen-fixing trees (MPTs), good pastures/ grasses and livestock's which not only managed the degraded wasteland but also provide various products for people, helps in gaining greater economic return, maintains cattle's health along with the maintenance of food and environmental security.

4.1 INTRODUCTION

Integration of pastures/grasses (Poaceae family) in agroforestry system having woody perennial trees and herbaceous crops can influences response of tree and overall pasture-based agroforestry systems (silvipasture, agrisilvipasture, hortipasture, etc.). Tree presence in pasture-based agroforestry can provides some environmental services as enhancing the biomass and provides shade and protections to cattle's and livestock's along with stress reduction and health and environmental security (Cuartas et al., 2014). Incorporation and livestock's play a crucial role in both structure and management of pasture-based farming system (silvopastoral system) and healthy pasture is prerequisite for feeding grazing animals which can fulfill the criteria of protein bank where woody perennial trees and shrubs (particularly N_2 fixing in nature) can provide protein-rich leaves to hunger animals in a lean period, i.e., when there is no availability of any feeding materials.

Dry matter production of different silvipasture systems in the world is depicted in Table 4.1. Although, animals play the greatest role in agricultural economy through providing various consumable and non-consumable products and their good health and productivity can strengthen the socioeconomic status farmer's. Grazing management involves various practices, i.e., management based on carrying capacity of pastureland and incorporation of preferred good quality pasture's and crops are the basis of healthy cattle's and livestock's animals (both ruminant and non-ruminant animals). As we know, GHGs emissions due some anthropogenic and natural factors are affecting the biodiversity and ecosystem structure and functions (Raj et al., 2018). However, emission of CO_2 due to anthropogenic fuelwood consumption is the largest contributors which is increased at a rate of 3.4% per year in the duration of eight years (between 2000 to

2008) as compared to 1.0% in the year of 1990 (Le Quere et al., 2009). Moreover, faulty practices of land-use changes such as lands conversion into farms and deforestation activity are the 2nd largest source of CO_2 emission which can be mitigated by the process of bio-sequestration, i.e., through planting trees, crops, and afforestation program, etc. Besides carbon dioxide, methane, and nitrous oxides are two another GHGs that contributed in global warming with the potential of 25 and 298 kg CO_2e per kg, respectively (Solomon et al., 2007). Similarly, the process of enteric methane production due to ruminant contributed highest rank in GHGs emission which is followed by methane production from manure practices while the production of nitrous oxide is commonly seen in soils system. However, these all GHGs emissions are less contributed by non-ruminant livestock systems rather than ruminants (Hristov et al., 2013).

This chapter reviews about importance of pasture-based farming system, scope, and applicability of intensive silvopastoral systems (ISS), livestock's situation and its importance in farming practices. Livestock's contribution in climate change, impacts, and its mitigation strategy through various pasture-based farming models and its both storage and sequestration capacity which can mitigate the extreme climatic impacts on biodiversity and maintains ecosystem health.

TABLE 4.1 Dry Matter Production of Silvipasture Systems in the World

Region	Systems	Dry Matter Production	References
Nothofagus antarctica (ñire) forest of silvopastoral systems, Patagonia	Silvopastoral systems	1094 kg/ha	Fertig et al., 2009
	Silvopastoral systems	2005 kg/ha	Bahamonde et al., 2012
	Silvopastoral systems	140–3760 kg/ha	Peri et al., 2016
	Silvopastoral systems	1200 kg/ha	Hansen et al., 2008
Chubut province in Patagonian Argentina	Orchardgrass (*Dactylis glomerata*) – Festuca sp.- red clover (*Trifolium pratense* pastures	2178 kg/ha	Hansen et al., 2013
Kerala, India	Models having hybrid Napier grass + Mulberry and Calliandra tree species	31.49 Mg/ha	Raj et al., 2016

TABLE 4.1 *(Continued)*

Region	Systems	Dry Matter Production	References
Kerala, India	Models having hybrid Napier grass + Mulberry tree species	48 Mg/ha	Varsha et al., 2017
	Sole cropping of hybrid Napier grass	51.20 Mg/ha	
Uttar Pradesh, India	Silvopastoral system in a natural grassland species of *Albizia procera, Eucalyptus tereticornis, Albizia lebbeck, Embilica officinalis* and *Dalbergia sissoo*	Stored 8.6, 6.92, 6.52, 6.25 and 5.41 t/ha/yr of biomass	Rai et al., 2009

4.2 INTENSIVE SILVIPASTORAL SYSTEMS

Intensively cultivation of pastures/grasses for the cattle production under silvopastoral systems and is gaining an importance due to its wider applicability in any cultivated and degraded land makes sustainable land utilization (Nair et al., 2009). Rotational grazing with both occupation and resting periods (12 to 24 hours and 40 to 50 day) are basic management practices for the smooth functions of ISS (Calle et al., 2012). As compared to treeless farming systems, the biodiversity will be more in the ISS. This is justified by Fajardo et al. (2009) and according to him there was 5 times more bird species in silvopastoral systems than sole pasture system. Diversity of ant and dung beetle are 30% and 2 times higher in silvopastoral system than sole pasture system (Rivera et al., 2008; Giraldo et al., 2010) and this increment in species diversity are helpful for controlling the plant pest and cattle's parasites along with good indicator for the sustainable land management.

As per Murgueitio et al. (2013), ISS have more diverse species, all they have positive interaction among them, close, and efficient nutrient cycling, more biomass and dry matter production, higher availability of crude protein per hectare, higher availability of milk and meat, etc.

Also, animal grazing in ISS were reported for higher intake of dry matter by percentage body weight, crude protein, fat, and calcium than grazing in sole pasture systems, respectively (Cuartas et al., 2014). Also, the amount of meat production were high and increased (74 to 1.060 and 0.45 to 1.060 kg year^{-1} in the Colombia and Mexico) by shifting sole pasture cropping to improved ISS (Mahecha et al., 2011; Solorio-Sánchez et al., 2011). Therefore, reduction in methane emission along with higher production of milk and meat are related to production of higher quality of fodder under well-managed ISS. Well, presence of trees in ISS can promote the higher storage and sequestration of atmospheric carbon per hectare which is the important process for mitigating climate change and global warming phenomenon. As per one estimates the sequestration potential of carbon from the atmosphere are ranges from 1.5 to 6.55Mg/ha/yr (Kumar et al., 1998; Montagnini et al., 2013). Addition of trees and grasses in silvopastoral systems gives a greater environmental services as minimizing the evaporation losses by enhancing the soil humidity which is good resilience to the drought (Broom et al., 2013).

4.3 LIVESTOCKS SITUATIONS IN FARMING SYSTEMS

Indeed, the livestock population represents the integral part of farming systems and extended overall map of the country. Globally, Indian subcontinents represent largest (520.6 million) position in the livestock population and accounts 16.1, 57.9, 16.7 and 5.7% shares in the population of cattle, buffaloes, goats, and sheep (FAOSTAT). Similarly, the figure of cattle, buffaloes, goat, sheep, pigs, and poultry birds in India were 185, 98, 124, 61, 14 and 489 millions that reported by Livestock Census which conducted in the year 2003. However, the prioritized species of both large and small ruminant species are considered and retained in farming field for both mechanization of agricultural operations and production of various consumable and edible products as milk, meat, eggs, etc. (Birthal and Taneja, 2006). Animals of both large and small ruminant species and their preference for different trees and crops in the world is depicted in Table 4.2.

TABLE 4.2 Animal's Preference for Different Trees and Crops in the World

S. No.	Tree and Crops	Regions	Animal's Preference
1	*Tectona grandis* (Teak)	Myanmar	Cattles and Buffaloes*
2	*Cocos nucifera* (Coconut)	Asia, Sri Lanka, Indonesia, Philippines	Goat, Sheep, Cattles, and Buffaloes
3	*Elaeisguineensis* (Oil palm trees)	India, Thailand, Costa Rica	Goat and Sheep**
4	*Heveabrasiliensis* (Para rubber tree)	Indonesia, Philippines, Brazil, Thailand	Goat and Sheep
5	*Theobroma cacao*(Cocoa tree)	Indonesia, Nigeria, and Malaysia	Goat and Sheep
6	*Anacardium occidentale* (Cashew tree)	India and Vietnam	Goat and Sheep
7	Various citrus trees	India, Thailand, Vietnam, and the Philippines	Goat and Sheep

Note: Goat and Sheep** (small ruminants) and Cattles and Buffaloes* (Large ruminants)

Moreover, different livestock's plays a great role in maintaining the structure and functions of different farming situations and their distribution across different agro-ecological regions are depicted in are given in Figure 4.1 where the highest percentage of livestock's population (cattle, buffalo, sheep, goat, pig, and poultry) are supported by rainfed region followed by irrigated region and least population was shown in arid region, respectively.

4.4 IMPACTS OF CLIMATE ON LIVESTOCK'S HEALTH AND PRODUCTIVITY

The overall health and performance of livestock's, i.e., their growth, milk, and meat productions and reproduction are affected by the prevailing climate in both direct and indirect way. In this context, an interlinking concept of climate change and its impacts livestock's has been developed and shown in Figure 4.2. Moreover, livestock's are homoeothermic in nature, i.e., they have a capacity to regulate and

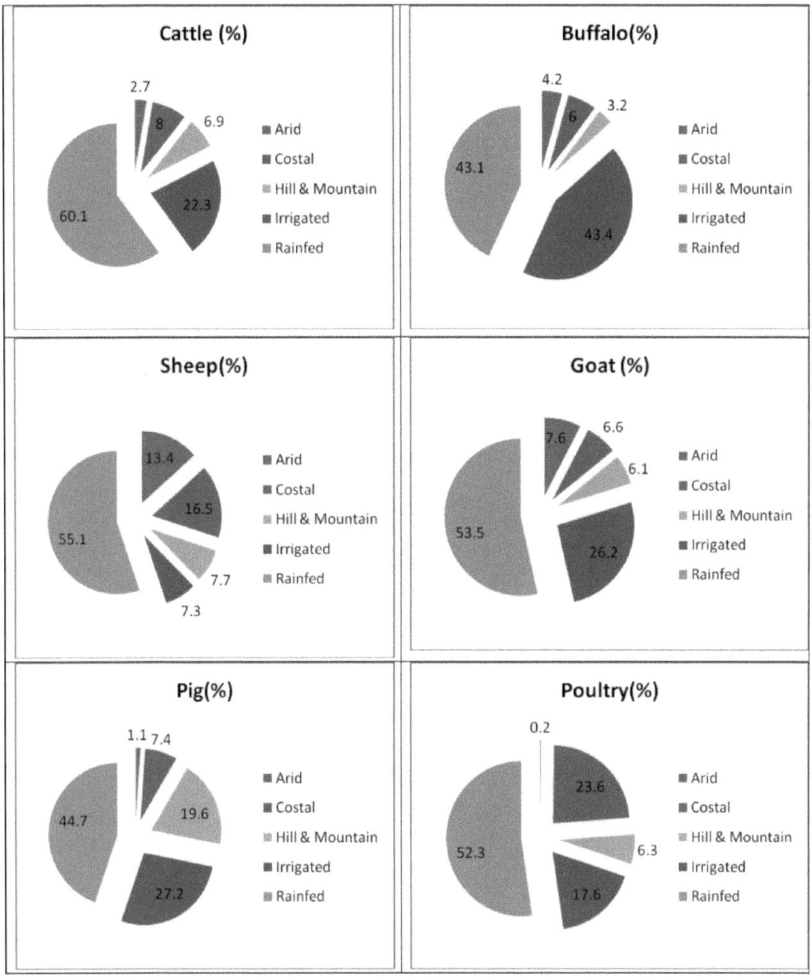

FIGURE 4.1 **(See color insert.)** Distribution of major livestock's across different agro-ecological regions in the year 2003.

maintain their body temperature as per changing the climatic and weather regimes to stay healthy and productive. Hahn (1999) has reported that the temperature ranges of 5 to 15°C are the comfort for livestock's good health and performance rather than 5 to 25°C which is unfavorable and severely affects the physiological process of cattle's (McDowell, 1972).

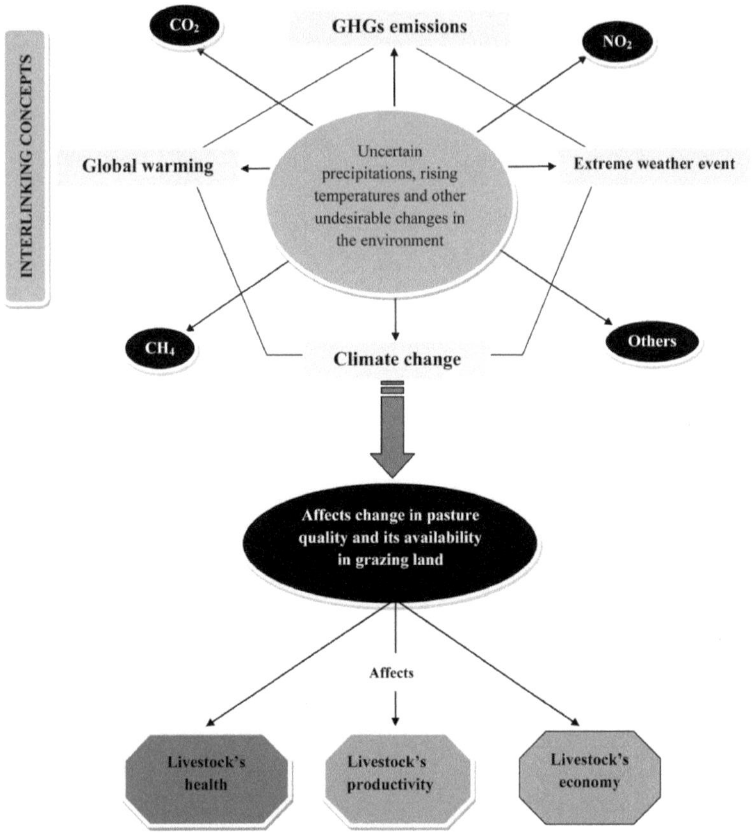

FIGURE 4.2 Interlinking concepts of climate change and its impacts livestock.

However, dairy cattle's have a lower limit of critical temperature than other livestock's species (Wathes et al., 1983). Rising temperature and heat that induce some metabolic changes in livestock's leads to reduce feed intake and decline their productivity (NRC, 2001). Temperature variations affect the average yield and productivity of daily milk production of the crossbred animals in the hot–humid eastern part of the country (Kale and Basu, 1993). As per Lal et al. (1987) the total intake of dry matter and milk production in cows (more adoptable in tropic) were affected by continuous rise in temperature in Haryana, India. Similarly, rising temperature and heat stress have deleterious impacts on productivity and reproduction of both Sahiwal cows and buffaloes (Kaur and Arora, 1982; Mandal et al., 2002b; Nagda, 2005).

The higher temperature limits for maximum milk production was recorded 27°C in India rather than 25°C in other country (Dutt et al., 1992). However, changing climate exert the negative impact on Indian agriculture which in turn affects the health and productivity along with overall livestock's economy (Dinar et al., 1998). Due to less areas (3.4%) of India covered by permanent pasture and grazing land, animals feed less and poor quality of fodder and grasses which affects the livestock's productions and their economy (GOI, 2002). Climate change promotes several livestock's disease as buffalo fly and sheep blowfly (Sutherst et al., 1996). Change in weather, i.e., extreme weather situations resulted cattle's food and mouth diseases of 52 and 84% in Andhra Pradesh and Maharashtra, respectively (Ramarao, 1988). Diseases outbreaks were also correlated with animal's movement and their climatic dependency (Sharma et al., 1991). As per one report, the outbreak and infestation of clinical mastitis in dairy animals and some cattle ticks such as *Boophilus microplus, Hyalom maanatolicum,* and *Haemaphysalis bispinosa* were due to hot and humid climate (Singh et al., 1996, 2000; Kumar et al., 2004; Basu and Bandhyopadhyay, 2004).

4.5 GRAZING MANAGEMENT FOR HEALTHY PASTURE AND LIVESTOCKS HEALTH

Grazing animals have nature to select the plant and grasses based on their taste, color, and which can cover all nutritional requirements and types and nature of grazing management practices will affects both quantity and quality of producing pasture and cattle's behaviors for grazing (Cozzi and Gottardo, 2005; Sanon et al., 2007). Changing forage quantity and quality due to seasonal variations the most important constraint for livestock's health and productivity (Ngwa et al., 2000; Osoro et al., 2013; Manousidis et al., 2016). Spring season is considered for the higher quantity and quality of the selection of grasses for the goats in semi-arid scrubland in Argentina (Pisani et al., 2000). Physical structure and chemical composition will be vary due to variation in seasonal climate (Bennett et al., 2007; Kennedy et al., 2007; Yayneshet et al., 2009; Safari et al., 2011). Permission and integration of cattle's with goats and sheep are good strategies for grazing management which has potential to maintains forage quality and opening the great opportunities for survival of both large and small ruminant species together in heterogeneous and varying environment in the semi-arid environments of the African continent.

There is a great nexus between cattle's frequent grazing on poor quality pastures (grasses) and their adaptation to utilize the fibrous feeding materials based on cattle's morphology and physiological parameters. For example, these species can feed and digest fibrous plants due to presence of large gut capacity in relation to their metabolic requirements (Rook et al., 2004) whereas goats could not able to digest easily and not preferred these fibrous plants utmost. That why small goat and sheep (small ruminants) require relatively more energy than cattle's (large ruminants) and prefers to selects only high-quality feeding grasses. Presence of high nutritive substances along with chemical compounds is the peculiar characteristics of *Acacia* species (Rubanza et al., 2007). However, due to presence of amino acid like proline in saliva of small ruminants and browsing animals (goats and sheep), they are easily digest and tolerate to metabolize such toxic chemicals into less and non-toxic derivatives (Hoven, 2010).

However, animals prefer only good quality forage plants and their eating choice relates to a function of the availability of that forage plant species (Fehmi et al., 2002; Sebata and Ndlovu, 2012). Although, animal's may save their energy and times for searching high quality species through making trade off quantity and quality switch to less preferred species. However, feed plan and its selection by animals are not only determined by preferred quality of forage and grasses but also affected by its availability, accessibility, and cost incurred in gathering the feeds (Distel et al., 1995; Baumont et al., 2000). Thus, this is a good strategy behind success of animals grazing in natural grasses and pastureland which can effectively balancing the proportion of offered and incurred energy. Also, energy required during locomotion distances covered by grazing animals on natural pasture land will substantially related to energy expenditure (Lachica and Aguilera, 2005). Similarly, poor quality of forage plants may change the feeding behaviors of grazing animals (Ngwa et al., 2000). Therefore, distance covered, varying quality of forage and physical characteristics along with land topography will affects the animal's decisions on selection of patch on grazing on pasture/grass land (Piasentier et al., 2007). Beside it, the process of urination and defecation will also affect the patch selection. In addition, patch selection can also be enhanced through variability caused by urination and defecation (Jaramillo and Detling, 1992).

4.6 CONTRIBUTION OF LIVESTOCK'S IN GREENHOUSE GAS EMISSION AND CLIMATE CHANGE

Today, emission of GHGs due to some anthropogenic and natural activities resulted most dangerous picture of our earth by changing our climate and extreme weather conditions due to uncertain rainfall and higher temperature. It is very well known that livestock deserves a utmost position in rural communities and a good property of the poor farmers people but the event like extreme weather and changing climate are affecting the health and productivity of these animals (Easterling et al., 2007; FAO, 2007; Thornton et al., 2007; Gill and Smith, 2008; Calvaba, 2009). It is surprising to know that agriculture itself promoting climate change due to unstoppable application of inorganic, chemical fertilizers, faulty land conversions and practices and flooded rice cultivation etc and also affected by the resulted extreme weather and climate change (Aydinalp and Cresser, 2008).

However, livestock's play an viable role in providing tangible and intangible benefits as food, employment, incomes, and risk insurance to poor farmers (Perry and Sones, 2007; Herrero et al., 2009). Although, there are several farming practices and livestock's activity such as Uncontrolled grazing and desertification, enteric fermentation, biomass burning and carbon releasing from feed production, faulty land use management and emission from post harvest technology that resulted climate change and global warming (Figures 4.3–4.5) (Steinfeld et al., 2006; Gerber et al., 2013). Similarly, production from the diary industries are contributed in greenhouse gases emissions (Weiske et al., 2006). Similarly, 80% of methane emissions are contributed by enteric fermentation and manure activities in agricultural land (FAO, 2006).

Well, because of unavailability of accurate data of livestock's populations there is a dearth of information on overall GHGs emissions through the activities of livestock's. Livestock activities and its production release 51% of anthropogenic greenhouse gas emissions and reported that CO_2 while releasing from animals respiration, is not considered under GHGs (Goodland and Anhang, 2009) whereas the same author has contradicted the FAO (2006) report on anthropogenic greenhouse gas and the emission's figure is 18% from livestock's industry. Similarly, livestock's activities and its industry released 44, 29 and 27% of CH_4, NO_2, and CO_2, respectively whereas 5% (9.2 Gt CO_2-eq of CO_2 per year) of anthropogenic emissions of CO_2 is contributed by livestock supply chains (IPCC, 2007b). Although, cattle, and other animals (poultry, pigs, small ruminants and buffaloes) are contributed 65 and 7–10% of the total livestock emissions whereas milk from the cattle, beef production, milk from the small ruminants, pork production, buffaloes meat and milk and chicken meat and eggs contributed 20, 41, 6, 9, 8 and 8%, respectively of total livestock's emissions. The remaining is emitted by inedible and other poultry species (MacLeod et al., 2013a, 2013b) (Table 4.3).

There is another estimate on role of livestock's industry in GHGs emissions, for example, the activity and production of livestock's contributed 44, 27 and 29% value of CH_4, CO_2 and NO_2, respectively whereas 65% emissions are contributed by cattle production. Moreover, the activities like feed processing and production and ruminants enteric fermentation are contributed 45 and 39% of total emissions, respectively whereas the rest of the percentage are shared by the activities of storage and processing of manures and animal processing and its transportation (Sarkwa et al., 2016). Globally, 14.5% of GHGs emissions are contributed by livestock sector that enough to promoting the further climate change phenomenon (Rojas-Downing et al., 2017). Thus, there should be given bit more considerations in management of livestock's and related farm activities for controlling and minimizing continuous emissions of GHGs resulting climate change and global warming phenomenon.

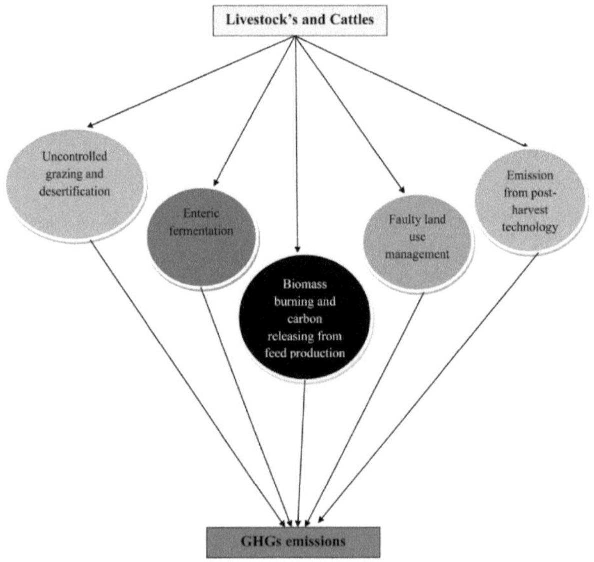

FIGURE 4.3 Livestock's contributions in GHGs emission.

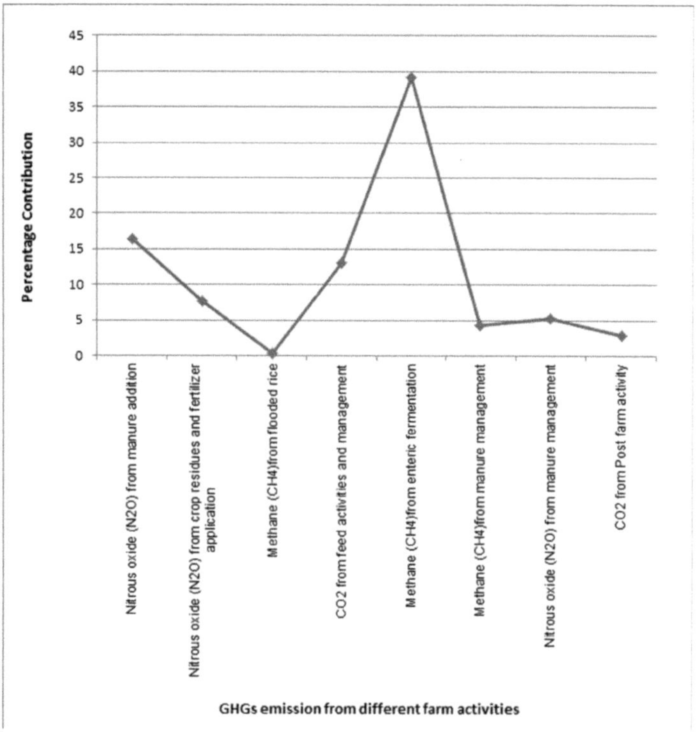

FIGURE 4.4 The major categories of anthropogenic GHG emission for livestock and livestock farm-related activities (Adapted from Steinfeld et al., 2006).

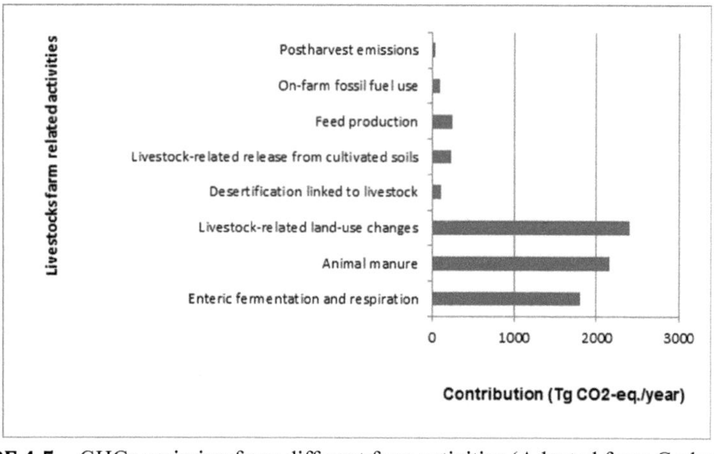

FIGURE 4.5 GHGs emission from different farm activities (Adapted from Gerber et al., 2013).

TABLE 4.3 GHGs Emissions (%) Through Livestock's Production and its Activity

Livestock's Production and its Activity	GHGs Emissions (%)
Cattle milk production	20
Beef production	41
Small ruminants milk	6
Pork production	9
Buffaloes meat and milk	8
Chicken meat and eggs	8
Inedible and other poultry species	8
Total	100

(Adapted from MacLeod et al., 2013a, 2013b).

4.7 CARBON SEQUESTRATION THROUGH PASTURE-BASED AGROFORESTRY

The long term storage and fixation of carbon dioxide from the atmosphere into vegetation and soil is termed as carbon sequestration that has been gaining wide recognition in term of mitigation global warming and climate change phenomena. However, the potential of pasture-based farming systems are widely known in term of storing and sequestering the carbon dioxide as sink into vegetation and soils which helps in enhancing biomass productivity along with improving fertility status of soils. Moreover, silvopastoral system (a component of agroforestry system) has potential to store more organic carbon in soils due to greater net sequestration of carbon. Integration of some perennial woody N_2 fixing MPTs with pastures/grasses would be great option and perfect strategy for better storage and sequestration of atmospheric CO_2 which helps in mitigating climate change impacts on biodiversity and maximize the overall biomass production and soil health (Jhariya and Raj, 2014; Jhariya et al., 2018). Many studies have be conducted in this direction, i.e., potential role of silvopastoral system in offsetting the climate change impacts by efficiently participating in storage and sequestration of carbon into soil and enhancement in overall biomass as compared to sole cropping systems of pastures/grasses. For example, both neem (*Azadirachta indica*) + *Cenchrus ciliaris* based silvopastoral system has greater potential to added higher biomass (34.9%) as compared to neem only (sole) having 33.8%. Similarly, 36.3 to

60.0% increment in organic carbon stocks in soils have been seen under combine systems rather sole system (Mangalassery et al., 2014).

However, the carbon storage and sequestration value of different silvipasture systems in the world is depicted in Table 4.4. Thus, the potential of carbon absorption and sequestration into soils pool are depends on type and nature of silvopastoral systems and its suitability in locations. Therefore, silvopastoral systems are a type of agroforestry system and its improved management practices has potential to sequester 0.012 TgC ha^{-1} and while converting degraded croplands and grassland of 630 million ha into well managed agroforestry system can sequester 0.59 Tg C ha^{-1} by 2040 (Kirby and Potvin, 2007; Ghosh and Mahanta, 2014). Thus, the hypothesis of incorporating woody perennial trees in agricultural field and pasture/grass land would be a good strategies to offset the deleterious impact of extreme climate through the greater potential of carbon sequestration in both above and belowground (Haile et al., 2008; Singh and Gill, 2014).

TABLE 4.4 Carbon Values of Silvipasture Systems in the World

Region	Systems	Carbon Value	References
Nicaragua, Central America	Silvopastoral systems having sole grass of *Hyparrheniarufa*	The value of soil carbon stock was 150 Mg C ha^{-1}at 0.6 m depth of soil	Ruiz et al., 2004
	Brachiariabrizantha (palisade grass), *Cynodon dactylon* (dhoob grass) and *Panicum maximum* (Guinea grass)based silvopastoral systems	Carbon stock value was 158 Mg C ha^{-1}	
Costa Rica	*Brachiariabrizantha* (palisade grass)and *Hyparrheniarufa* based silvopastoral system	The value of carbon was 3.5 (in treeless *Hyparrheniarufa*) and 12.5 Mg C ha^{-1} in *Brachiariabrizantha* (palisade grass) with tree of *Diphysarobinioides*	Andrade et al., 2008
Brazilian Cerrado	Integration of fodder grass under Eucalyptus trees along with animals/livestock's and sole Eucalyptus plantation	Soil organic carbon values varied from pasture (461 Mg ha^{-1}) to sole eucalyptus plantation (393 Mg ha^{-1})	Tonucci et al., 2011

TABLE 4.4 *(Continued)*

Region	Systems	Carbon Value	References
Central-western Spain	Dehesa cork oak based silvopastoral system	The carbon value was 50.2, 37, and 26.5 Mg ha^{-1} as per distance of 2, 5 and 15 meter from cork oak tree at one-meter depth, respectively.	Howlett et al., 2011
Ecuadorean Andes	Pasture crop of *Setaria sphacelata* and N_2 fixing Inga tree species (commonly known as shimbillo)	Open pasture crop of *Setaria sphacelata* having increased carbon value of 7.9% but 11.4% was found under the canopies of N_2 fixing Inga tree species. Similarly, Inga comprised 20 Mg C/ha at 15 cm upper compared to open pasture system.	Rhoades et al., 1998
	System comprised natural grassland and silvipastoral system comprising trees (*Albizia amara, Leucaena leucocephala and Dichrostachyscinerea*) and some grasses species (*Chrysopoganfulvus* and *Stylosantheshamata* along with *Salvia scabra* as legume)	Carbon stored value in silvopastoral systems was two times more (6.72 ton carbon/ha/yr) than natural grassland system comprised 3.14 ton carbon/ha/yr from natural grassland.	NRCAF, 2007
Semi arid region of Uttar Pradesh, India	Silvopastoral system in a natural grassland species of *Emblica, Dalbergia, Albizia,* and *Eucalyptus*	Slivopastoral system comprised 1.89–3.45 $tCha^{-1}$as compared to pure pasture having 3.94 $tCha^{-1}$	Rai et al., 2009
	Silvopastoral system in degraded land	Increment in soil carbon stock value from 24.3Pg to 34.9 Pg	Narain, 2008
Regional research station, Kukma (Gujarat)	Integration of *Acacia tortilis* tree with grass of *Cenchrus setegerus* under silvopastoral system	Belowground carbon stock value was 1.6 t ha^{-1} which is about 23.4% of the total carbon stock value	Kumar, 2010

4.8 SOCIOECONOMICS OF PASTURE-BASED FARMING SYSTEMS

Indeed, pasture-based farming systems (silvopastoral system as agroforestry) has also provided greater economic returns to farmers besides providing diversifying products and carbon sequestration potential (CSP). Many authors have reported the economic benefits through establishment of pasture-based farming systems. For example, several species such as *Grewia optiva, Bauhinia vahlii, Bauhinia retusa, Quercus leucotrichophora, Flacourtia ramontchii, Ougenia oogeinensis, Celtis australis, Lantana camara, Woodfordia fruticosa, Mangifera indica, Butea monosperma* and *Syzigium cuminii,* etc., are reported and consumed as fuelwood whereas *C. australis, B. vahlii, Q. leucotrichophora, W. fruticosa, Murrayakoenigii, F. ramontchii, Chrysopogon fulvus,* and *Apluda mutica* are used as a fodder resource for livestock which helps in strengthening the socioeconomic status of poor farmers in Arnigad watershed of Mussoorie hills, Uttarakhand (Chandra et al., 2008).

4.9 FUTURE STRATEGIES FOR IMPROVED PASTURE AND LIVESTOCK

Integration and management of good quality fodder/grasses crops in leguminous and non-leguminous trees along with adoption of small and large ruminants in the pasture-based farming system should be managed in such a way that whole system can provides ecosystem services in term of the provision of both tangible (timber, NTFPs, fuelwood, quality fodder for cattle's) and intangible benefits (microclimate improvement, soil & water conservation, nutrients enrichment, maintaining environment quality, socioeconomic enhancement of farmers and whole ecosystem and landscape development with sustainable production and natural resource conservation, etc.).

Therefore, extensive research should be done in this direction to make socially acceptable, economically viable and ecologically acceptable. Raise awareness among people and farmers for model benefits and its importance of products through the chain of value addition process. Location specific models should be emphasized for better dry matter production, livestock's health and good economic return in varying Agroclimatic zones. New technology should be strengthen and included in the scientific R&D for better response of any silvopastoral models in any given regions this would not only minimize the problem of food insecurity but also helps to maintain overall yield and livestock's health along with strengthening farmers life's and environmental security. Thus, building an effective policy for scientific R&D in such a way which deals with the complex problem of crop-animals productivity and soil-water erosion under threatening climate change issues in the present and upcoming era is essential.

4.10 CONCLUSIONS

Today's pasture-based farming systems are gaining recognitions due to its wider applicability in different agroclimatic zones and its great ecosystem services in term of providing direct (yield and biomass production) and indirect functions such as soil water conservation, efficient close nutrient cycling, climate change mitigation through carbon sequestration, maintenance of soil fertility and enhancement of environmental quality etc. Integrating livestock's in pasture and cropland and its proper scientific management would enhance the quality and quantity of feed materials and health and productivity of grazing animals. Moreover, healthy pastures/ grasses are prerequisites for healthy and productive livestock's which can strengthen the framers' income by providing various consumable and non-consumable products, etc. Thus, building an effective policy for better scientific research and involvement of government, NGOs, and private agency can play an important role in both establishment and management of location specific pasture-based farming models in different Agroclimatic region.

KEYWORDS

- **cropland**
- **environment**
- **farming system**
- **livestock's**
- **multipurpose tree species (MPTs)**
- **pasture**

REFERENCES

Andrade, H. J., Brook, R., & Ibrahim, M., (2008). Growth, production and carbon sequestration of silvopastoral systems with native timber species in the dry lowlands of Costa Rica. *Plant Soil, 308*(1/2), 11–22.

Aydinalp, C., & Cresser, M. S., (2008). The effect of global climate change on agriculture. *American-Eurasian Journal of Agricultural Environmental Science, 3*, 672–676.

Bahamonde, H. A., Peri, P. L., Alvarez, R., & Barneix, A., (2012). Producción y calidad de gramíneas en ungradiente de calidades de sitio y coberturas en bosques de Nothofagusantarctica (G. Forster) Oerst. en Patagonia. *Ecología Aust, 22*, 62–73.

Basu, A. K., & Bandhyopadhyay, P. K., (2004). The effect of season on the incidence of ticks. *Bull. Anim. Health Prod. Afr., 52*(1), 39–42.

Baumont, R., Prache, S., Meuret, M., & Morand-Fehr, P., (2000). How forage characteristics influence behavior and intake in small ruminants: A review. *Livest. Prod. Sci., 64*, 15–28.

Bennett, J., Lent, P. C., & Harris, P. J. C., (2007). Dry season foraging preferences of cattle and sheep in a communal area of South Africa. *Afr. J. Range Forage Sci., 24*, 109–121.

Birthal, P. S., & Taneja, V. K., (2006). *Livestock Sector in India: Opportunities and Challenges for Smallholders, Smallholder Livestock Production in India: Opportunities and Challenges*. Proceedings of an ICAR-ILRI International Workshop. pp. 32.

Broom, D. M., Galindo, F. A., & Murgueitio, E., (2013). Sustainable, efficient livestock production with high biodiversity and good welfare for animals. *Proceedings of the Royal Society B, 280*, 2013–2025.

Calle, Z., Murgueitio, E., & Chará, J., (2012). Integrating forestry, sustainable cattle-ranching and landscape restoration. *Unasylva, 63*, 31–40.

Calvasa, C., Chuluunbaatar, D., & Fara, K., (2009). *Livestock Thematic Papers- Tools for Project Design*. International Fund for Agricultural Development (IFAD) publication. Retrieved from: www.ifad.org/lrkm/factsheet/cc.pdf (Accessed on 4 August 2019).

Chandra, R., Soni, P., & Yadav, V., (2008). Fuel wood, fodder and livestock status in Himalayan watershed in Mussoorie hills. *Indian Forester, 134*(7), 894–906.

Cozzi, G., & Gottardo, F., (2005). Feeding behavior and diet selection of finishing Limousin bulls under intensive rearing system. *Appl. Anim. Behav. Sci., 91*, 181–192.

Cuartas, C. A., Naranjo, J. F., Tarazona, A. M., Murgueitio, E., Chará, J. D., Ku, J., Solorio, F. J., X Flores, M. X., Solorio, B., & Barahona, R. (2014). Contribution of intensive silvopastoral systems to animal performance and to adaptation and mitigation of climate change. *Rev Colomb Cienc Pecu 27*, 76–94.

Cuartas, C. A., Naranjo, J. F, Tarazona, A., Correa, G., & Barahona, R., (2014). *Estimation of Forage Voluntary Intake by Beef Steers Grazing in Intensive Silvopastoral Systems Based on Leucaena Leucocephala*. (Unpublished).

Dinar, A., Mendelsohn, R., Evenson, R., Parikh, J., Sanghi, A., Kumar, K., McKinsey, J., & Lonergan, S., (1998). *Measuring the Impact of Climate Change on Indian Agriculture* (p. 266). World Bank Technical Paper No. 402, World Bank, Washington DC.

Distel, R. A., Laca, E. A., Griggs, T. C., & Demment, M. W., (1995). Patch selection by cattle: Maximization of intake rate in horizontally heterogeneous pastures. *Appl. Anim. Behav. Sci., 45*, 11–21.

Dutt, T., Taneja, V. K., Singh, A., & Singh, A., (1992). Comfort zone for maximal milk production in crossbred cattle. *Ind. J. Dairy Sci., 45*(3), 119–122.

Dziba, L. E., Scogings, P. F., Gordon, I. J., & Raats, J. G., (2003). Effects of season and breed on browse species intake and diet selection by goats in the False Thornveld of Eastern Cape, South Africa. *Small Rumin. Res., 47*, 17–30.

Easterling, W. E., & Aggarwal, P. K., (2007). Food, fiber and forest products. In: Parry, M. L., Canziani, O. F., Palutikof, J. P., Van Der Linden, P. J., & Hanson, C. E., (eds.), *Climate Change 2007- Impacts, Adaptation and Vulnerability* (pp. 273–313). Contribution of working group II to the Fourth Assessment Report of the Intergovernmental Panel on Climate Change. Cambridge University Press.

Fajardo, D., Johnston-González, R., Neira, L., Chará, J., & Murgueitio, R., (2009). Influencia de los sistemassilvopastoriles en la diversidad de aves en la Cuenca del río La Vieja, Colombia. *Revista Recursos Naturalesy Ambiente, 58*, 9–16.

FAO, (2006). In: Steinfeld, H., Gerber, P. J., Wassenaar, T., Castel, V., Rosales, M., & De Haan, C., (eds.), *Livestock's Long Shadow – Environmental Issues and Options.* Rome. pp. 377.

FAO, (2007). *Adaptation to Climate Change in Agriculture, Forestry, and Fisheries: Perspective.* Framework and priorities. FAO, Rome.

FAOSTAT. ((2019)). *Agricultural Production Database.* Food and Agricultural Organization. http://apps.fao.org (Accessed on 4 August 2019).

Fehmi, J. S., Karn, J. F., Ries, R. E., Hendrickson, J. R., & Hanson, J. D., (2002). Cattle grazing behavior with season-long free-choice access to four forage types. *Appl. Anim. Behav. Sci., 78*, 29–42.

Fertig, M., Hansen, N., & Tejera, L., (2009). Productividad y calidadforrajera en raleos en bosques de ñire Nothofagusantarctica. In: *Proceedings of the 1st National Congress of Silvopastoral Systems, Misiones, Argentina* (pp. 358–363). INTA Editions.

Gerber, P. J., Steinfeld, H., Henderson, B., Mottet, A., Opio, C., Dijkman, J., Falcucci, A., & Tempio, G., (2013). *Tackling Climate Change Through Livestock – a Global Assessment of Emissions and Mitigation Opportunities* (pp. 115). Food and Agriculture Organization of the United Nations (FAO), Rome.

Ghosh, P. K., & Mahanta, S. K., (2014). Carbon sequestration in grassland systems. *Range Mgmt. & Agroforestry, 35*(2), 173–181.

Gill, M., & Smith, P., (2008). Mitigating climate change: The role of livestock in agriculture. *Livestock and Global Change Conference Proceeding.* Tunisia. ISBN 978-0-906562-62-8; pp. 29–30.

Giraldo, C., Escobar, F., Chará, J. D., & Calle, Z., (2010). The adoption of silvopastoral systems promotes the recovery of ecological processes regulated by dung beetles in the Colombian Andes. *Insect Conserv. Divers., 4*, 115–122.

GOI, (2002). *Report of the Working Group on Animal Husbandry and Dairying for the Tenth Five Year Plan (2002–2007)* (p. 214). Working Group Sr. No. 42/2001, Government of India, Planning Commission.

Goodland, R., & Anhang, J., (2009). *Livestock and Climate Change* (pp. 10–19). What if the key actors in climate change were pigs, chickens and cows? Worldwatch November/December 2009, Worldwatch Institute, Washington, D C, USA.

Hahn, G. L., (1999). Dynamic responses of cattle to thermal heat loads. *J. Animal Science, 77*(2), 10–20.

Haile, S. G., Nair, P. K. R., & Nair, V. D., (2008). Carbon storage of different soil-se fractions in Florida silvipastoral systems. *Journal of Environmental Quality, 37*, 1789–1797.

Hansen, N., Fertig, M., Escalona, M., Tejera, L., & Opazo, W., (2008). Ramoneo en regeneración de ñire y disponibilidadforrajera. In: *Proceedings Segunda Reuniónsobre Nothofagus en la Patagonia – EcoNothofagus 2008, Esquel* (pp. 137–142). Chubut, Argentina.

Hansen, N., Rios, C., Dromaz, M., De María, G., Rôo, G., von Müller, A., Lloyd, C., Raso, M., Buduba, C., & Fetherson, S., (2013). Ejecución de prácticas de manejosilvopastoril en bosque de ñire (*Nothofagusantarctica*). In: Peri, P. L., (ed.), *2nd Congress on Patagonian Agroforestry (International), El Calafate* (p. 92). Argentina. INTA Editions.

Herrero, M., Thornton, P. K., Gerber, P., & Reid, R. S., (2009). Livestock, livelihoods and the environment: Understanding the trade-offs. *Current Opinion in Environmental Sustainability, 1*, 111–120.

Hoven, W. V., (2010). The natural diet of wild herbivores. In: Bothma J. P., & Toit, J. D., (eds.), *Game Ranch Management* (pp. 347–383). Van Schaik, Pretoria.

Howlett, D. S., Moreno, G., Mosquera, L. M. R., Nair, P. K. R., & Nair, V. D., (2011). Soil carbon storage as influenced by tree cover in the Dehesa cork oak silvopasture of central-western Spain. *J. Environ. Monit., 13*(7), 1897–1904.

Hristov, A. N., Oh, J., Lee, C., Meinen, R., Montes, F., Ott, T., et al. (2013). Mitigation of greenhouse gas emissions in livestock production – a review of technical options for non-CO_2 emissions. In: Gerber, P., Henderson, B., & Makkar, H., (eds.), *FAO Animal Production and Health* (p. 177). FAO, Rome.

IPCC, (2007b). Climate change 2007: mitigation. In: Metz, B., Davidson, O. R., Bosch, P. R., Dave, R., & Meyer, L. A., (eds.), *Contribution of Working Group III to the Fourth Assessment Report of the Intergovernmental Panel on Climate Change*. Cambridge University Press, Cambridge, United Kingdom and New York, NY, USA.

Jaramillo, V. J., & Detling, J. K., (1992). Small-scale heterogeneity in semi-arid North American grassland II. Cattle grazing of simulated urine patches. *J. Appl. Ecol., 29*, 9–13.

Jhariya, M. K., & Raj, A., (2014). Human welfare from biodiversity. *Agrobios Newsletter, XIII*(9), 89–91.

Jhariya, M. K., Banerjee, A., Yadav, D. K., & Raj, A., (2018). Leguminous trees an innovative tool for soil sustainability. In: Meena, R. S., Das, A., Yadav, G. S., & Lal, R., (eds.), *Legumes for Soil Health and Sustainable Management* (pp. 315–345). Springer, ISBN 978–981–13–0253–4 (eBook), ISBN: 978–981–13–0252–7 (Hardcover). https://doi.org/10.1007/978–981–13–0253–4_10 (Accessed on 4 August 2019).

Kale, M. M., & Basu, S. B., (1993). Effect of climate and breed on the milk production of crossbred cattle. *Ind. J. Dairy Sci., 46*(3), 114–118.

Kaur, H., & Arora, S. P., (1982). Influence of level of nutrition and season on the estrus cycle rhythm and on fertility in buffaloes. *Trop. Agric., 59*(4), 274–278.

Kennedy, M., Donovan, O., Murphy, J. P., Delaby, L., & O'Mara, F. P., (2007). Effect of spring grazing date and stocking rate on sward characteristics and dairy cow production during midlactation. *J. Dairy Sci., 90*, 2035–2046.

Kirby, K. R., & Potvin, C., (2007). Variation in carbon storage among tree species: Implications for the management of small scale carbon sink project. *Forest Ecology and Management, 246*, 208–221.

Kumar, A. K., (2010). Carbon sequestration: Underexplored environmental benefits of Tarai agroforestry. *Indian Journal of Soil Conservation, 38*, 125–131.

Kumar, B. M., Suman, J. G., Jamaludheen, V., & Suresh, T. K., (1998). Comparison of biomass production, tree allometry and nutrient use efficiency of multipurpose trees grown in woodlot and silvopastoral experiments in Kerala, India. *For. Ecol. Manage., 122*, 145–163.

Kumar, S., Prasad, K. D., & Deb, A. R., (2004). Seasonal prevalence of different ectoparasites infecting cattle and buffaloes. *B.A.U.J. Res., 16*(1), 159–163.

Lachica, M., & Aguilera, J. F., (2005). Energy expenditure of walk in grassland for small ruminants. *Small Rumin. Res., 59*, 105–121.

Lal, S. N., Verma, D. N., & Husain, K. Q., (1987). Effect of air temperature and humidity on the feed consumption, cardio respiratory response and milk production in Haryana cows. *Indian Vet. J., 64*(2), 115–121.

Le Quere, C., Raupach, M. R., Canadell, J. G., Marland, G., Bopp, L., Ciais, P., et al. (2009). Trends in the sources and sinks of carbon dioxide. *Nat. Geosci., 2*(12), 831–836.

MacLeod, M., Gerber, P., Mottet, A., Tempio, G., Falcucci, A., Opio, C., Vellinga, T., Henderson, B., & Steinfeld, H., (2013a). *Greenhouse Gas Emissions from Ruminant Supply Chains – A Global Life Cycle Assessment.* Food and Agriculture Organization of the United Nations (FAO), Rome. pp. 191.

MacLeod, M., Gerber, P., Mottet, A., Tempio, G., Falcucci, A., Opio, C., Vellinga, T., Henderson, B., & Steinfeld, H., (2013b). *Greenhouse Gas Emissions from Pig and Chicken Supply Chains – A Global Life Cycle Assessment.* Food and Agriculture Organization of the United Nations (FAO), Rome. E-ISBN 978-92-5-107944-7; pp. 171.

Mandal, D. K., Rao, A. V. M. S., Singh, K., & Singh, S. P., (2002b). Comfortable macroclimatic conditions for optimum milk production in Sahiwal cows. *J. Appl. Zool. Res., 13*(2/3), 228–230.

Mangalassery, S., Dayal, D., Meena, S. L., & Ram, B., (2014). Carbon sequestration in agroforestry and pasture systems in arid northwestern India. *Current Science, 107*(8), 1290–1293.

Manousidis, T., Kyriazopoulos, A. P., Parissi, Z. M., Abraham, E. M., Korakis, G., & Abas, Z., (2016). Grazing behavior, forage selection and diet composition of goats in a Mediterranean woody rangeland. *Small Ruminant Research, 145*, 142–153.

McDowell, R. E., (1972). *Improvement of Livestock Production in Warm Climates* (p. 711). Freeman, San Francisco, CA.

Montagnini, F., Ibrahim, I., & Murgueitio, E., (2013). Silvopastoral systems and climate change mitigation in Latin America. *Bois etForêts des Tropiques, 67*, 316, 3–16.

Murgueitio, E., Chará, J., SolarteA., Uribe, F., Zapata, C., & Rivera, J., (2013). Agroforestería Pecuariay Sistemas Silvopastoriles Intensivos (SSPi) para la adaptaciónganadera al cambioclimático con sostenibilidad. *Revista Colombiana de Ciencias Pecuarias, 26*, 313–316.

Nair, P. K. R., Kumar, B. M., & Nair, V. D., (2009). Agroforestry as a strategy for carbon sequestration. *Journal of Plant Nutrition and Soil Science, 172*, 10–23. doi: 10.1002/jpln.200800030.

Narain, P., (2008). Dryland management in arid ecosystem. *Journal of the Indian Society of Soil Science, 56*, 337–347.

Ngwa, A. T., Pone, D. K., & Mafeni, J. M., (2000). Feed selection and dietary preferences of forage by small ruminants grazing natural pastures in the Sahelian zone of Cameroon. *Anim. Feed. Sci. Technol., 88*, 253–266.

NRC, (2001). *Nutrient Requirement of Dairy Cattle*. National Research Council, National Academy Press, Washington, DC.

NRCAF, (2007). *Perspective Plan Vision 2025 National Research Centre for Agroforestry*. Jhansi, Uttar Pradesh.

Peri, P. L., Hansen, N. E., Bahamonde, H. A., Lencinas, M. V., Von Müller, A. R., Ormaechea, S., et al. (2016). Silvopastoral systems under native forest in Patagonia Argentina. In: Peri, P., Dube F., & Varella, A., (eds.), *Silvopastoral Systems in Southern South America: Advances in Agroforestry* (Vol. 11, pp. 117–168). Springer, Cham. https://doi.org/10.1007/978-3-319-24109-8_6.

Perry, B., & Sones, K., (2007). Poverty reduction through animal health. *Science, 315*, 333–334.

Piasentier, E., Saccà, E., & Bovolenta, S., (2007). Dietary selection and ingestive behavior of fallow deer and sheep grazing on adjacent monocultures of white clover and tall fescue. *Small Rumin. Res., 71*, 222–233.

Pisani, J. M., Distel, R. A., & Bontti, E. E., (2000). Diet selection by goats on a semi-arid shrubland in central Argentina. *Ecol. Austral., 10*, 103–108.

Rai, P., Ajit, C. P. O., Singh, R., & Singh, U. P., (2009). Biomass production in multipurpose tree species in natural grasslands under semi arid conditions. *Journal of Tropical Forestry, 25*, 11–16.

Raj, A. K., Kunhamu, T. K., Jamaludheen, V., & Kiroshima, S., (2016). Forage yield and nutritive value of intensive silvopasture systems under cut and carry systems in humid tropics of Kerala, India. *Indian J. of Agroforestry, 18*(1), 47–52.

Raj, A., Jhariya, M. K., & Bargali, S. S., (2018). Climate smart agriculture and carbon sequestration. In: Pandey, C. B., Gaur, M. K., & Goyal, R. K., (eds.), *Climate Change and Agroforestry: Adaptation Mitigation and Livelihood Security* (pp. 1–19). ISBN: 9789–386546067. New India Publishing Agency (NIPA), New Delhi, India.

Ramarao, D., (1988). Seasonal indices and meteorological correlates in the incidence of foot-and-mouth disease in Andhra Pradesh and Maharashtra. *Indian J. Anim. Sci., 58*(4), 432–434.

Rhoades, C. C., Eckert, G. E., & Coleman, D. C., (1998). Effect of pasture trees on soil nitrogen and organic matter: Implications for tropical montane forest restoration. *Restor. Ecol., 6*(3), 262–270.

Rivera, L., Botero, M., Escobar, S., & Ambrecht, I., (2008). Diversidad de hormigas en sistemasganaderos. In: Murgueitio, E., Cuartas, C., & Naranjo, J. F., (eds.), *Ganadería del Futuro: Investigaciónpara el Desarrollo* (p. 490). CIPAV. Cali, Colombia.

Rojas-Downing, M. M., Nejadhashemi, A. P., Harrigan, T., & Woznicki, S. A., (2017). Climate change and livestock: Impacts, adaptation, and mitigation. *Climate Risk Management, 16*, 145–163.

Rook, A. J., Dumont, B., Isselstein, J., Osoro, K., WallisDeVries, M. F., Parente, G., & Mills, J., (2004). Matching type of livestock to desired biodiversity outcomes in pastures – a review. *Biol. Conserv., 119*, 137–150.

Roquet, C., Dumont, B., & Prache, S., (1998). Selection and use of feeding sites and feeding stations by herbivores: A review. *Ann. Zootech., 47*, 225–244.

Rubanza, C. D. K., Shem, M. N., Bakengesa, S. S., Ichinohe, T., & Fujihara, T., (2007). Effects of *Acacia nilotica*, *A. polyacantha* and *Leucaena leucocephala* leaf meal supplementation on performance of Small East African goats fed native pasture hay basal forages. *Small Rumin. Res.*, *70*, 165–173.

Ruiz, A., Ibrahim, M., Locatelli, B., Andrade, H. J., & Beer, J., (2004). Fijación y almacenamiento de carbono en sistemassilvopastoriles y competitividadeconómica de fincasganaderas en Matiguás, Nicaragua. *Agrofor Am*, *41/42*, 16–21.

Safari, J., Mushi, D. E., Kifaro, G. C., Mtenga, L. A., & Eika, L. O., (2011). Seasonal variation in chemical composition of native forages, grazing behavior and some blood metabolites of Small East African goats in a semi-arid area of Tanzania. *Anim. Feed. Sci. Technol.*, *164*, 62–70.

Sarkwa, F. O., Timpong-Jones, E. C., Assuming-Bediako, N., Aikins, S., & Adogla-Bessa, T., (2016). The contribution of livestock production to climate change: A review. *Livestock Research for Rural Development, 28,* http://www.lrrd.org/lrrd28/3/sark28037.htm (Accessed on 4 August 2019).

Sebata, L., & Ndlovu, R., (2012). Effect of shoot morphology on browse selection by free ranging goats in a semi-arid savanna. *Livest. Sci., 144*, 96–102.

Sharma, S. K., Singh, G. R., & Pathak, R. C., (1991). Seasonal contours of foot and mouth disease in India. *Indian J. Anim. Sci., 61*(12), 1259–1261.

Singh, A. P., Singla, L. D., & Singh, A., (2000). A study on the effects of macroclimatic factors on the seasonal population dynamics of *Boophilus microplus* infesting the crossbred cattle of Ludhiana district. *Int. J. Anim. Sci., 15*(1), 29–31.

Singh, B., & Gill, R. I. S., (2014). Carbon sequestration and nutrient removal by some tree species in an agrisilviculture system in Punjab, India. *Range Management and Agroforestry*, *35*, 107–114.

Singh, K. B., Nauriyal, D. C., Oberoi, M. S., & Baxi, K. K., (1996). Studies on occurrence of clinical mastitis in relation to climatic factors. *Ind. J. Dairy Sci.*, *49*(8), 534–536.

Solomon, S., Qin, D., Manning, M., Alley, R. B., Berntsen, T., Bindoff, N. L., et al. (2007). Technical summary. In: Solomon, S., et al. (eds.), *Climate Change 2007: The Physical Science Basis: Contribution of Working Group I to the Fourth Assessment Report of the Intergovernmental Panel on Climate Change*. Cambridge University Press, Cambridge/New York. (ISBN 978 0521 88009-1 Hardback; 978 0521 70596-7 Paperback); pp. 996.

Solorio-Sánchez, F. J., Bacab-Pérez, H. M., & Ramírez-Avilés, L., (2011). Los Sistemas Silvopastoriles Intensivos: Avances de Investigación en el Valle de Tepalcatepec, Michoacán. Memorias III Congresosobre Sistemas Silvopastoriles Intensivos, para la ganaderíasosteniblede lsiglo XXI. Morelia, México: Fundación Produce Michoacán, COFRUPO, SAGARPA, Universidad Autónoma de Yucatán – UADY.

Steinfeld, H., Gerber, P., Wassenaar, T., Castel, V., Rosales, M., & de Haan, C., (2006). *Livestock's Long Shadow, Environmental Issues and Options* (pp. 377). Livestock, environment, and development initiative. United Nations Food and Agriculture Organization, Rome.

Sutherst, R. W., Yonow, T., Chakraborty, S., O'Donnell, C., & White, N., (1996). A generic approach to defining impacts of climate change on pests, weeds and diseases in Australia. In: Bouma, W. J., Pearman, G. I., & Manning, M. R., (eds.), *Greenhouse: Coping with Climate Change* (pp. 281–307). CSIRO, Melbourne.

Tailor, S. P., & Nagda, R. K., (2005). Conception rate in buffaloes maintained under sub-humid climate of Rajasthan. *Ind. J. Dairy Sci., 58*(1), 69–70.

Thornton, P., Herrero, M., Freeman, A., Mwai, O., Rege, E., Jones, P., & McDermott, J., (2007). Vulnerability, climate change and livestock – Research opportunities and challenges for poverty alleviation. *SAT eJournal*, *4*(1), 1–23.

Tonucci, R. G., Nair, P. K., Nair, V. D., Garcia, R., & Bernardino, F. S., (2011). Soil carbon storage in silvopasture and related land-use systems in the Brazilian cerrado. *J. Environ. Qual., 40*(3), 833–841.

Varsha, K. M., Raj, A. K., Kurien, E. K., Bastin, B., Kunhamu, T. K., & Pradeep, K. P., (2017). High-density silvopasture systems for quality forage production and carbon sequestration in humid tropics of Southern India. *Agroforest. Syst.*, 1–14. https://doi.org/10.1007/s10457–016–0059–0 (Accessed on 4 August 2019).

Wathes, C. M., Jones, C. D. R., & Webester, A. J. F., (1983). Ventilation, air hygiene and animal health. *Vet. Rec.*, *113*(24), 554–559.

Weiske, A., Vabitsch, A., Olesen, J. E., Schelde, K., Michel, J., Friedrich, R., & Kaltschmitt, M., (2006). Mitigation of greenhouse gas emissions in European conventional and organic dairy farms. *Agriculture Ecosystems and Environment*, *112*, 221–232.

Yayneshet, T., Eik, L. O., & Moe, S. R., (2009). Seasonal variations in the chemical composition and dry matter degradability of exclosure forages in the semi-arid region of northern Ethiopia. *Anim. Feed Sci. Technol.*, *148*, 12–33.

CHAPTER 5

Role of Higher Plants for Air Pollution Mitigation in Urban Industrial Areas

SHIBORAM BANERJEE[1] and DEBNATH PALIT[2]

[1]PG Department of Conservation Biology, Durgapur Government College, Durgapur, West Bengal, India, Mobile: +91-9153083631, E-mail: shiboramb@gmail.com

[2]Department of Botany, Durgapur Government College, Durgapur, West Bengal, India

ABSTRACT

Recently air pollution poses a serious threat on mankind all over the world by damaging different ecosystem process. The main polluting substances are sulfur oxides, nitrogen oxides, carbon oxides, ozone, particulate matter, and heavy metals that cause damage to the surrounding environment. Photosynthetic plants are the one and only natural barrier and buffer against air pollution. Herbs, shrubs, and trees all are responsible for cleaning air and greening environment by their internal microanatomy, physiology, and biochemistry. In an area where a large forestland was removed for big and many industrial setups, there a scientific plantation of well air pollution tolerated plant species will be the main strategy to overcome the air pollution problem. A large tree, may be Gymnosperm or Angiosperm, plays a good role to combat against air pollution by not only soak or metabolized the polluted gases but also it can reflect and precipitate the particulate matter. In this context, appropriate research work need to be designed an establishment of structural green belt by well air pollution tolerant plant species.

5.1 AIR POLLUTION: AN INTRODUCTORY REMARK

Our earth is submerged in an ocean of air. It is a mixture of mainly two gaseous components, one is the active component, namely oxygen and another one is the inactive component namely nitrogen. In a 5 liters of air, 4 liters is of nitrogen (i.e., 75%) and only 1 liter is oxygen (i.e., 25%). Oxygen is active in the sense that it is directly responsible for existing life on earth as it is essential at the last step in electron transport system (ETS) of cellular metabolism for production of ATP molecules. On the other hand, nitrogen is inactive because it is not supporting life directly but indirectly it has a lot of significance to construct living biomass by N_2-cycling between living and non-living world. Finally, the air is the sum product of 78.09% nitrogen, 20.95% oxygen, and 0.1–4% water vapor showing their high percentage in the atmosphere.

In contrast the components present in low percentage are argon (0.93%), carbon dioxide (0.0384%) and the remaining present in very low percentage such as neon ($18X\ 10^{-4}$%), helium ($52X10^{-5}$%), methane ($2X10^{-4}$%), krypton ($11X10^{-5}$%), nitrogen monoxide ($2X10^{-5}$%), hydrogen ($5X10^{-5}$%), xenon ($87X10^{-6}$%), sulfur dioxide ($2X10^{-5}$%), ozone ($2X10^{-6}$%), nitrogen dioxide ($2X10^{-6}$%), ammonia ($1X10^{-6}$%), and carbon monoxide ($1X10^{-6}$%). Among the mixture components sulfur dioxide, nitrogen dioxide, nitrogen monoxide, methane, and carbon monoxide are gaseous pollutants and carbon dioxide though it is an air pollutants, it has been considered as the raw items for photosynthesis, producing biomass in flora through carbon assimilation. Air is a basic need and essential for living and life could not exist if air absent on earth. Biota cannot live without air for a few minutes. But in recent time after the industrial revolution, the purity of air becomes alarmingly deteriorated worldwide due to excessive increase of air pollutants into the atmosphere. Though, it is diluted into the air or is absorbed majority in different pockets on earth like the ocean, forest, soil, and in mudflats as a secondary byproduct, yet it has been increasing with pace. Hence, the concentration of pollutants in air should be kept as low as possible.

The quality of air has been deteriorated immensely with the industrialization, urbanization, and destruction of forest ecosystem. These have created a new danger for mankind. Human themselves are one of the major factors behind such deterioration of the environment. With alarming population increase, increases in industries and vehicular load, constructions have become the major polluting sources. Day by day with a pacing influence, the life-sustaining air becomes life-damaging.

5.2 THE MAIN SOURCES AND CONSEQUENCES OF AIR POLLUTION

Human population is the main source of various types of pollution. Various activities of man like industrial activity, use of vehicles, combustion of huge amount of fossil fuels and solid wastes, civil construction, deforestation cause air pollution. Sources of air pollution can be stationary or mobile. Stationary sources can be point sources such as industrial units or area sources such as domestic combustion. Mobile sources are classified as line sources or aerial sources. Highway vehicles, trains, etc. are examples of line mobile sources whereas aircrafts, port vessels are examples of aerial mobile sources. It depends on factors such as design and operation of vehicular engines which is not quite enough to completely burn the fossil fuel due to insufficient oxygen. Among the five oxides of nitrogen (NOx) only two that is nitrogen dioxide (NO_2) and nitric oxide (NO) have been produced significantly from automobiles and industrial sectors due to fossil fuel burning. Other than these, NOx, carbon dioxide (CO_2), lead (Pb), various carcinogenic compounds such as polycyclic aromatic hydrocarbons (PAH), dioxin etc. are the significant pollutants emitted by automobiles. Combustion of fossil fuels by human activities released annually about 20 million metric tons of sulfur dioxides. It was estimated that on an average 60% of air pollution is due to automobile pollution in metropolitan cities. In the atmosphere the sulfur dioxide is produced naturally by the oxidation of hydrogen sulfide which is released from volcanoes and hot springs.

Exhaust emissions produce very fine particles. Particulate matters such as smog, fly ash, cement dust, soot, oil smoke, soil dust of different size and shape have been produced largely from transport and industrial sector and from civil construction site in urban areas contributing major air pollution problem along with gaseous air pollutants. Natural phenomenon like volcanic eruption, prescribed or un-prescribed forest fires, asteroid impacts are observed from many parts of the world, are also contributing hazardous gases and fine particles that deteriorating the atmosphere. Fly ash is one of the end products of power and steel producing companies where the coal is used mainly for fuel purpose, is removed from the chimney and dumped in the wasteland for year after year. During heavy airflow or during the cyclone, the mass dissemination of fly ash has occurred in the surrounding environment and harm to biota. Plume from the large industrial sectors has been spread over the area in different directions depending on the airflow and wind direction. These types of particulate contaminants are known as total suspended particulate matter (TSPM) and the subtype of this TSPM are suspended particulate matter (SPM) and respirable suspended particulate matter (RSPM). Particles having the diameter between 0.5 to 10 μm are under RSPM (PM10) and particles having the aerodynamic size up to 100 μm known as SPM.

5.3 AIR POLLUTION SCENARIO IN URBAN INDUSTRIAL AREA

Urban Industrial Township has been established mainly by removal of large forestland and/or by stopping the production of occupied agricultural land from farmers. Steel plants, thermal power plants, cement industries, alloy steel plants, sponge iron, mineral-based industries; refineries are the major polluting industries in urban areas. In the third world country, the industrial sectors had flourished with massive utilization of coal and other fossil fuels as a raw material for the production of power, steel, alloy, etc. Subsequent development of urban area, therein transportation sectors and increasing demographic pressure based on the industries has further damaged the air quality. These big and other small ferroalloy steel factories regurgitate million tons of carbon, poisonous gases, dust, fly ashes, and volatile heavy metals everyday into the air. Day by day the city's internal traffic load and twenty-four hours working big and small industrial sectors intensifies the air pollution problem. Although huge quantities of fossil fuels are burnt in steel plants, power sectors and other factories for industrial manufacturing purpose, automobiles are another significant source of air pollution in urban area contributing carbon monoxide (CO), nitrogen oxides (NOx), and volatile organic compounds (VOCs). The carcinogenic automobile exhaust Dioxin is recognized in urban atmosphere. Apart from industrial and automobile emission in urban area, there mass burning of solid waste has been observed around the outskirt of the city. These prescribed solid waste burning continuously pushing toxic gases, fine, and ultrafine particles into the air. The higher plants growing along the roadside, around industries, in several vacant places and in home garden have been showered continuously with those pollutants (Bakiyaraj and Ayyappan, 2014).

5.4 MONITORING OF AMBIENT AIR QUALITY

Air quality monitoring is the most important and necessary step to quantify the gaseous and particulate matter load in urban air. In 1984 the CPCB (Central Pollution Control Board) had started National Ambient Air Quality Monitoring program for assessing the ambient air quality in several monitoring stations in India. Furthermore, the air quality data has also been monitored by SPCB (State Pollution Control Board) on a regional basis. The harmful gaseous components that have been measured are SO_2, NOx, O_3, CO_2, CO, etc. and the particulate matters of different shape and sizes. The airborne particles that have the aerodynamic size of about 0.01 μm to 100 μm or larger are known as SPM. A fraction of SPM has been found in the atmosphere of about 10 μm in size and is identified as Inhalable Suspended Particulate Matter (ISPM), which has the capability of being respired significantly into the respiratory system by human beings. This particle is also known as coarse particle or PM_{10}. Another subset of SPM having the aerodynamic particle diameter of less than 2.5 μm referred to as RSPM which could reach alveoli of the human respiratory system and are responsible for various types of cardiopulmonary disorder. This particle is also known as fine particle or $PM_{2.5}$. This different subset of particulate matter and the harmful gaseous components have been assessed by high volume sampler. In India, respirable dust has been included in the NAAQS (National Ambient Air Quality Standards) since 1994.

The gaseous pollutants have been analyzed by collecting the air forcefully into a chamber that contained the absorbing solution of specific pollutants. The pollutants then get mixed with this absorbing solution by bubble formation. The air has been collected at a flow rate of about 1.5 ml per minute into the glass made impingers. Inside the High Volume Sampler, there have a number of impingers each containing specific absorbing liquid of 25 ml volume for absorption of specific polluted gas of the atmosphere. SO_2 has been absorbed in sodium tetrachloromercurate (NaK_2HgCl_4); the mixture of sodium hydroxide (NaOH) and sodium arsenite ($NaAsO_2$) have been used to absorb NO_2 and potassium iodide (KI) solution is used as absorbing solutions for O_3. The sampling has been done on 24 hr basis (Figures 5.1 and 5.2). After sampling the samples, i.e., the absorbing solution which contains different concentrations of solubilized harmful gaseous components, brought to the laboratory in a cold ice container. The gases become more stable at a lower temperature. In the laboratory, the samples have been analyzed quickly by the standard method.

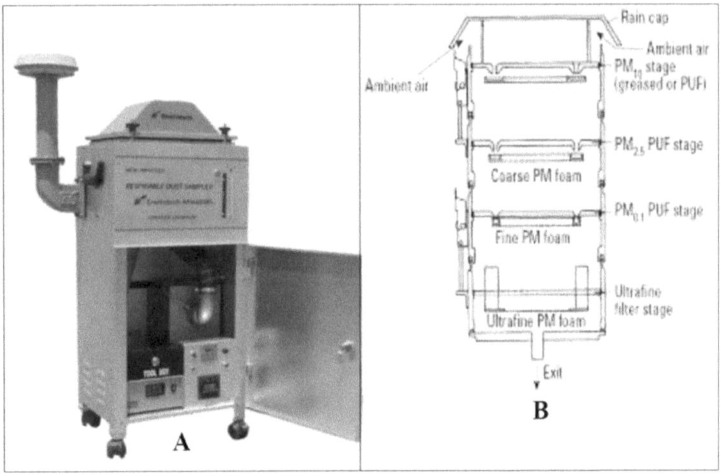

FIGURE 5.1 (A) High volume air sampler, (B) Airflow in the sampler. (Source: Google).

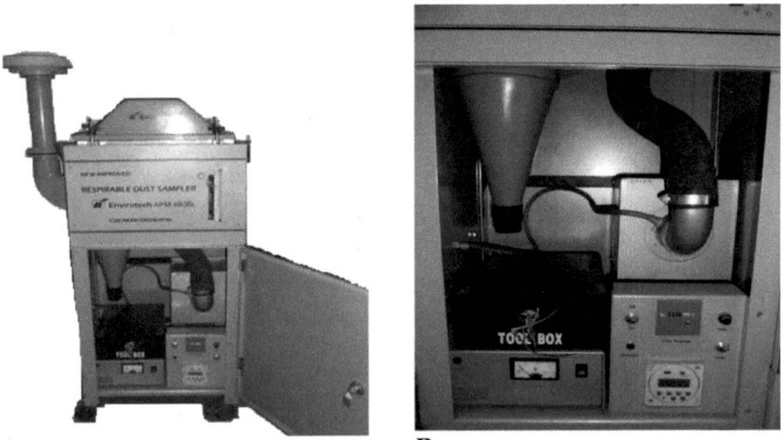

A B

FIGURE 5.2 (A) Respirable Dust Sampler (Envirotech Model APM 460 BL) (B) Internal machinery of the Dust Sampler.

For determination of particulate matters, the ambient air has been pooled forcefully through a selective pore of the sampler and filtered by 25.4 into20.3 cm filter paper with 1132 liter per minute airflow rate. The particles size smaller than the inlet pore of machine could enter into the chamber through the filter paper. The amount of the particulate matters has been determined by the differences of weight of the filter paper, before, and after the sampling. At first, the volume of air sampled has been calculated by the following formula:

$$\text{Volume of air sampled } (\mu g/m^3) = \frac{(Qi+Qf)\times T}{2}$$

where, Q_i = rate of airflow at first (m^3/minute), Q_f = rate of airflow at end of sampling (m^3/minute), and T = period of sampling (minute).

Then the amount of particles or the SPM load has been evaluated by the following formula:

$$\text{Load of SPM } (\mu g/m^3) = \frac{(Wf\text{-}Wi)\times 106}{V}$$

where, V = sampled air volume, W_f = final weight of the filter paper (gram), and W_i = initial weight of filter paper (gram).

The standard air quality of three major air pollutants have been recommended by the CPCB are 60μg/m^3 for sulfur dioxide, 60μg/m^3 for NOx and 140 μg/m^3 for SPM. An index of air pollution was developed by measuring the SO_2, NO_x, and SPM level from samples collected at any spot (Rao and Rao, 1989). The ratio of the pollutants have been made with their respective standards and then added together. An average was prepared of these pollutants and the resultant finally multiplied with 100 to get the air pollution index (API) value.

$$API = 1/3[(SPM)/(S_{SPM}) + (SO_2)/S_{SO2}) + (NO_X)/(S_{NOX})]X\ 100$$

Where, standards of SO_2, NOx, and SPM have been identified as S_{SO2}, S_{NOX}, and S_{SPM}

The API values which are in between 0 to 25 indicated as clean air of the area. The values which have been followed the range 26 to 50, subjected as light air pollution zone. Moderate air pollution zone have been identified if the API values in between 51 to 75. If the calculated API values in 76–100, then the site is denoted as heavy air pollution zone. Finally, if it is >100, then the region has been spotted as severe air pollution station (Tripathi and Gautam, 2007).

5.5 PHOTOSYNTHETIC PLANTS AS NATURAL BUFFER AGAINST AIR POLLUTION

Worldwide air pollution has posed a serious threat to humankind and there is no known field-based scientific method or strategy to reduce the air pollution in urban atmosphere. Green plants are one and only natural shield of the environment that can soak all types of polluted gases and particulate matter. The pollutants become adsorbed, absorbed, accumulated, detoxified, and metabolized maximally through foliar structure and precipitated by the physical configuration of plant body (Beckett et al., 1998; Shannigrahi et al., 2004). Inside it, the pollutants changed into less toxic metabolic byproduct by physiological and biochemical processes. These end products either are stored inside the plant body or it is excreted out. Atmospheric particulate matters are of different shape and sizes measuring of about 1 μm to 100 μm in diameter. These particulate matters have been deposited or precipitated depending upon the influence of gravity, velocity of air, particle's own shape and size and ultimately on the direction of plant leaves. The penetration of pollutants into plant leaves is directly proportional to the higher airflow. Particulate matters are more efficiently aggregated along the leaf petioles than leaf lamina and other portion of the plant parts (Ingold, 1971). The degree of impaction has also been influenced by the architecture of canopy, geometry of leaves, plenty of fine hairs present on leaf surface. Particle diameter ranging of about 1–5 μm is not efficiently entangled on laminar surface (Tewari, 1994).

Stomata are the main gateway of atmospheric gases that's could enter into plant leaves from exterior. The exchange of gases has been occurred via stomatal openings. If the outer surface of plant leaves is wet, some water-soluble gases like NO_2, SO_2, O_3, HF (hydrogen fluoride), etc. become dissolved in it and then slowly absorbed into the leaf interior. This uptake of these gaseous air pollutants has been increased with their solubility in water. The more water content in plant leaves indicates that it has greater potentialities to dissolve and absorb pollutants by leaf surface. On the other hand, carbon monoxide (CO) and nitric oxide (NO) are very insoluble in water and are not absorbed well by all plant species. The more water-soluble SO_2 has been dissolved in the apoplastic water after uptake through the plant leaves and produced sulfite (SO_3^{2-}), bisulfite (HSO_3^-) and H^+ ion, keeping inside more acidic (Legge and Krupa, 2002). The phytotoxicity has been arisen due to those SO_3^{2-} and HSO_3^- ions. Most of the plant species cope up the situation by detoxifying those produced anions (when their concentration are not high excessively) into less toxic sulfate ion (Rao, 1992).

$$SO_2 + H_2O \longrightarrow HSO_3^- + H^+$$

$$HSO_3^- + .OH + H^+ \longrightarrow HSO_3\cdot + H_2O$$

$$HSO_3\cdot + O_2 \longrightarrow SO_3^{2-} + O_2^- + H^+$$

Nitrogen dioxide the other water-soluble gas has been dissolved into the extracellular water of substomatal cavity and yield nitrate (NO_3^{2-}), nitrite (NO_2^{2-}) and proton (H^+). In leaf cell interior the nitrate becomes reduced to ammonia.

$$NO_2 + H_2O \longrightarrow HNO_3$$

$$NO + H_2O \longrightarrow HNO_2$$

$$HNO_3 + HNO_2 \longrightarrow NO_3^{2-} + NO_2^{2-} + H^+$$

Ozone is another most important secondary photochemical air pollutant that is abundant in stratosphere and function as a protective shield for extraterrestrial ultraviolet rays. But its tropospheric concentration has been increasing alarmingly during the past 60 years (Anfossi et al., 1991). Though a beneficial effect of it has been observed on stratosphere but in troposphere harmful effect is detected on agricultural crops, forests cover, and vegetation and on ornamental plants. It reacts with the extracellular water of plant leaves produces many free radicals like superoxide ($\cdot O_2^-$), peroxyl ($OH_2\cdot$) and hydroxyl ($OH\cdot$) radicals (Rao et al., 2000), and the reaction is going on at higher or neutral pH.

$$O_3 + OH\cdot \longrightarrow {}^-O_2\cdot \text{ (Superoxide radical)}$$

$$.O_2^- + H^+ \longrightarrow HO\cdot_2 \text{ (Peroxy radical)}$$

The buffering capacity is lower in deciduous trees because they shed their leaves in winter, in contrast, the evergreen trees have been enhanced the air quality precisely all over the year. Higher plants regulate the local climate through evapotranspiration and direct shedding by lowering the air temperature. The secondary air pollutants, which are the byproduct of different primary pollutants and primary pollutants with indigenous aerial materials are dependent completely on air temperature. The green plants mediated reduced air temperature slows down the photochemical reactions in air and formation of secondary product (Nowak et al., 2000). Thick vegetation cover along the roadside, national highways, and industrial surroundings could reduce the noise pollution by acting as a buffer (Huddart, 1990). Not all the plant species have the capacity to tolerate all types of environmental pollution, only a few of them with their extraordinary physiological, biochemical, micromorphological, and biological characteristics could battle against adverse situations.

5.6 BIOLOGICAL INTERVENTION TOWARDS SCREENING OF POLLUTION TOLERANT SPECIES

There are different biochemical and biological screening techniques for isolation of pollution tolerant plant species from sensitive ones. Plants have the adaptive strategies to overcome the harsh environmental situation by changing their biochemical and physiological characteristics. There is no way to escape the polluted environment by plants as like animals because plants are static and they become adapted with their vast genetic heterogeneity in all type of air, soil, and water pollution. But not all plants play equally well when they come in contact with the pollutants. Only the tolerant ones can survive well with their superior biochemical, physiological, and biological characteristics in polluted sites. In polluted air conditions, the potentialities of higher plants have been evaluated by using some physiological and biochemical parameters of their leaves. By using these parameters, Singh and Rao (1983) was developed an index of tolerance level of higher plants against air pollution, known as air pollution tolerance index (APTI). Ascorbic acid (A), total chlorophyll (T) are two biochemical and pH (P), relative water contents (R) are two physiological parameters of plant leaves. In the APTI formula total chlorophyll have been added with leaf extract pH and multiplied with ascorbic acid. The resultant again added with relative water content (RWC) and the outcome is divided by 10 to get the APTI formula.

$$\text{APTI} = \frac{(A(T+P)+R)}{10}$$

After getting the APTI values for each plant species the values should be compared with the scale of APTI which was recommended by Mashita and Paise (2001). If the APTI values is less than 1, then the plant species marked as highly sensitive; within 1–16 are registered as sensitive; between the range 17–30 has been considered intermittently tolerable and finally the APTI values in between 30–100 is branded as the species is air pollution tolerant. Ascorbic acid is small molecular weight antioxidant which could protect the plant body from reactive oxygen species (ROS) which has been produced unavoidably by photosynthetic and normal metabolic processes (Smirnoff, 1996). Higher ascorbic acid means greater protection of plants from air pollution. Similarly, higher chlorophyll concentration, RWC, and high pH reflect better tolerance capacity of plant species against air pollution (Singh and Verma, 2007).

Another approach by which the selection procedure of tolerant plant species is going on is the determination of Anticipated Performance Index (API). The API has been analyzed by the addition of socioeconomic and biological characters of plant species with their calculated APTI values. Plant types, plant height, the architecture of canopy, and foliar structure are considered as macro-morphological characteristics of a plant. When APTI values of a plant have been added with their morphological characteristics along with its socio-economic values, then we get API percentage, and according to percentage marks, each plant has been subjected with numerical grade points. This grade point is different in different plant species. The grade point of this particular plant species has been compared with the maximum possible grade point with other plants by the simple unitary formula.

$$\frac{\text{API}(\%)=\text{Plant get their own grade}}{\text{recommended maximum grade of any plant specie}}\times 100$$

5.7 PHYSIOLOGICAL AND BIOCHEMICAL CHANGES IN PLANT DUE TO AIR POLLUTION

All green plants have been played a crucial role to serve many form of life on earth by producing oxygen atom. O_2 is needed at the last step of cellular metabolism for ATP production. Though oxygen production is the unique characteristics for all green plants but not the only criteria to judge a plant species with others for superiority against many adverse atmospheric conditions. There is some biochemical and physiological characteristics which could make some plants with better or tolerant varieties and some are inferior or sensitive varieties against air pollution stress. In many previous studies, it was revealed that the biochemical and physiological parameters deteriorated very rapidly in sensitive plant species than in tolerant ones. So the sensitive plants are now using as a living tool for bio-monitoring of pollution affected areas. The biochemical variation has been detected very early before the visible anatomical injury has been arrived (Tripathi et al., 2009; Rai, 2016 a, b). Tolerant plant varieties considering their inherent physiological and biochemical status could reduce the overall pollution load. Laminar structure like size, texture, and hardiness of a plant leaf; petiole length (Ingold, 1971); arrangement and position of a leaf, all of these biological characteristics of plants are responsible to reduce the air pollution. The main parameters that have been used to identify the status of plants are ascorbic acid, total chlorophyll (chl.-a and chl.-b), carotenoid, pH, RWC, total soluble sugar, amino acid, protein, proline, phenol, cysteine content, etc.

1. **Ascorbic Acid:** There are several unfavorable biotic and abiotic factors which could create reverse impact on the floral kingdom by producing different types of ROS. These wide ranges of stress are pathogen attack, aerobic metabolism, photosynthetic process, oxygen deprivation, herbicide infestation, UV attack and number of atmospheric pollutants. The stress-dependent free radicals have been involved to damage many cellular components like carbohydrate, protein (destruction of quaternary, tertiary, and secondary structure), fats (unsaturated fatty acid peroxidation in cell membrane), nucleic acids and the cellular organelles (chloroplast) (Blokhina et al., 2003). This ROS production is prevented by a number of antioxidant molecules and ascorbic acid is one of them endogenous low molecular mass antioxidant which has been played a significant role in pollution mitigation and control other stressful conditions. It is present in many types of plant cells, chloroplast (90% in reduced form) and in apoplast (Smirnoff, 2000).

The plant cell wall contains ascorbate and ascorbate oxidase (AO) which has the capabilities of cellular growth and development. Plant's cell division has been operated through pooling G_1phase to S phase of cell (Smirnoff, 1996) and the elongation of cell morphology has also been regulated by ascorbic acid (De Tullio, 1999). It is also known as vitamin

C having an influential role to prevent and protect the whole biotic life from various complicated diseases. The footprint of intracellular ascorbic acid biochemistry was not detected properly, although many cell types in plants shows higher level of it. Guinea pig, primates, and human beings are dependent on external ascorbate intake because they all are unable to synthesize the L-Gulono-1, 4- Lactone Oxidase the final effective enzyme that is responsible to synthesize ascorbic acid (Smirnoff, 1996). It has significant characteristics to play many physiological, biochemical, and defensive purpose with increased reducing power which is positively correlated with its internal concentration (Lewin, 1976; Seyyednejad et al., 2011).

The hexose sugar has been converted to ascorbic acid, the reaction of which is enhanced by higher leaf pH (Bora and Joshi, 2014) and in contrast is down regulated by lower pH level of leaf (Conklin, 2001; Pasqualini et al., 2001; Escobedo et al., 2008, Jyothi, and Jaya, 2010). The leaf endogenous pH is regulated by species own inheritance and also by status of the surrounding environment. It was observed that, pH becomes lower or alarmingly low in plants located in the severely polluted area. But the same plant has been showed higher leaf pH when we found it in relatively less polluted area. Though a direct relationship existed between the pH and ascorbic acid, yet lower pH not always prevent the formation of ascorbic acid. So pH is not the only regulatory environment in leaf for control the concentration of ascorbic acid. A biochemical and physiological plasticity have always been presented in plant life for all parameters and this is known as adaptive response. This biochemical plasticity is detected more in tolerant plant species which has the capabilities to cope up with the adverse environmental conditions. But the sensitive plant species could not balance properly their biochemistry and physiology in different environmental situations and so they have been used as a bio tool in areas with various grade of air pollution (Singh and Rao, 1983; Nrusimha et al., 2005).

2. **Total Chlorophyll:** One of the most important photoreceptor is chlorophyll which is found in chloroplast of all living green plants on earth. This green pigment is polycyclic, resembling the planar structure of protoporphyrin of hemoglobin and the central portion is occupied by Mg^{2+} instead of Fe^{2+}. Chlorophyll present in the thylakoid compartment of cell and there the chlorophyll a has a higher percentage than chlorophyll b, both are considered as total chlorophyll. Chl.-a has been synthesized more than chl b in higher plant species and the tolerance capacity of the plants are

dependent on higher level of chl a than chl b. The higher values of this ratio could reflect higher tolerance level against polluted environment (Konodo et al., 1980). The sum of chlorophyll has been decreased significantly in plants growing in the polluted area (Raina and Sharma, 2003). The role of chlorophyll molecule is to perform the light phase of photosynthetic reactions in its reaction centers. The by-products of the light phase reaction (ATP and NADPH) reduce CO_2 and H_2O, the raw materials of photosynthesis and producing many metabolic intermediates like polysaccharides (starch, cellulose), DNA, RNA, protein, and lipid. This is known as CO_2 assimilation which forms the biomass in plants. The air pollution has been severely damaged the chlorophyll molecule (Prustyet al., 2005; Joshi and Bora, 2011). In spite of air pollution load the higher level of chlorophyll is indicative of higher tolerance level (Jyothi and Jaya, 2010). In apoplastic water of leaves the SO_2 becomes dissolved after its absorption through stomata and form sulfate (SO_3^{-2}), bisulfite (HSO_3^-) and H^+ ion and finally lower the pH of the medium (Legge and Krupa, 2002). Air pollution leads to decrease the leaf internal pH (2 and 3.5) which has been displaced the Mg^{2+} ion from the central position of chlorophyll molecule by H^+ and form phaeophytin, a degraded product of chlorophyll. For displacement of Mg^{2+} ion the acidic environment is provided by splitting of H_2SO_3 into SO_3^{-2} and H^+. The final product phaeophytin no longer has the capability to receive the solar photon particle for photosynthesis (Rahmawati et al., 2014). Variation in chlorophyll molecule has been observed from area to area considering abiotic and biotic conditions and level of pollution, from species to species, leaf age (Rai and Panda, 2015).

3. Carotenoid: It is the secondary light absorbing accessory pigment which is fat soluble in nature and is present in thylakoid compartment of plant chloroplast. It is present in green algae, photosynthetic bacteria and plants where it has been performed functions in photosynthetic processes. Carotenoid act as an antioxidants and protect chlorophyll from destruction by photo-oxidation (Sifermann-Harms, 1987). Free radical superoxide has been produced during SO_2 metabolism in plant tissue at pH 3.5 or above by HSO^{3-} and carotenoid becomes oxidized. This oxidized carotenoid has no longer power to protect chlorophyll molecule and plant chlorophyll become oxidized in absence of carotenoid. Finally, the photosynthetic ability of plants has been reduced with great loss of plant biomass (Yu, 1988). The response of carotenoid towards SO_2 is greater than chlorophyll and deteriorated very quickly (Shimazaki et al., 1980). It is oxidized through light-catalyzed reaction and form epoxide which in the dark phase become reduced to enzyme-catalyzed reaction (Calvin, 1955). In summary, a decreased carotenoid concentration has been observed in area affected with air pollution.

4. pH: It is one of the most important physiological parameter which has been regulated the plants internal vulnerability towards air pollution. Variation of pH level has been detected from species to species but environmental pollution also controls the leaf pH. Every biochemical reactions and processes are under control of pH. Slightest variation of it which is mainly pollution induced could reflect abnormality in many biochemical processes and in product formation. A significant correlation has been observed in decreased leaf internal pH and the surrounding air pollution (Das and Prasad, 2010). SOx, NOx, and other gaseous pollutants diffused through stomatal pore and form many acid radicals by dissolving into the cellular water of leaf matrix (Turk and Wirth, 1975; Das and Prasad, 2010). Reduced leaf extract pH has been detected more in sensitive plant species rather than tolerant ones. Sulfur dioxide has been entered into the leaf matrix through stomatal pore, dissolved in intercellular water of mesophyll cells and dissociated to form H+, sulfite (SO_3^{2-}) and bisulfite (HSO_3^-). This event caused the surrounding more acidic and the acidic pH is an indicator of SO_2poisoning (Puckeet et al., 1973). This $SO_3^-$and HSO_3^- ions have been formed the floral toxicity (DeKok, 1990).

5. Relative Water Content (RWC): Water has been maintained the cells and the intercellular space moist, keeping active of all biochemical processes and transportation of structural and functional organic molecules. Relative water means it is the water of plant body after absorption through roots and transpiration through leaves (Jones, 1994). Higher RWC has been conferred protection to plant body from all types of air pollution stress (Singh and Verma, 2007) by dissolving and diluting obnoxious gaseous pollutants (Palit et al., 2013). The total leaf area though it is inherent characteristics but somewhat dependent leaf internal RWC (Schuppler et al., 1998) and air pollution level. The leaf relative water of plants is responsible for transpiration, respiration, and plant growth (Dhankhar et al., 2015). The plants could not function well if the transpiration through leaves becomes interrupted by blocking the stomatal pore by environmental pollutants. So the movement of water and nutrient molecule from root to leaves has been hampered (Lohe et al., 2015; Sisodia and Dutta, 2016). Air pollution has been increased cellular permeability (Keller, 1986) by peroxidation of polyunsaturated fatty acids (PUFAs) of cell membrane resulting in water and nutrient loss (Masuch et al., 1988). The tolerant plant species which are capable to battle against air pollution have higher percentage of relative water (Jyothi and Jaya, 2010; Kuddus et al., 2011; Krishnaveni et al., 2013).

6. Total Soluble Sugar: One of the most important energetic organic molecules is sugar which is water-soluble and produced during photosynthesis. This substance provided energy to all biotic life on earth through its own metabolic breakdown, and its standard concentration is indicative of pollution tolerance (Prado et al., 2000; Finkelstein and Gibson, 2001). In the plant leaves, the concentration of sugar has been depleted in the effect of environmental pollutants like SO_2, NO_2, and O_3 (Rai and Panda, 2015). These pollutants are readily involved to inhibit the photosynthetic processes in one side and on the other side respiration rate associated with ATP production has also been increased (Davison and Barnes, 1986; Tzvetkova and Kolarov, 1996; Panda and Rai, 2015) for additional activities of pollutants removal. As a result, the total soluble sugar content has been depleted in plant leaves in pollution affected areas (Seyyednejad and Koochak, 2011). In an experiment, gaseous mixture of SO_2, NO_2 and O_3 had fumigated on the young and mature leaves and it was found that the level of soluble sugar had been decreased significantly by increased metabolic consumption of sugar molecule (Lorenc-Plucinska, 1982). Atmospheric out of range dust particles has been precipitated on the upper surface of plant leaves and formed different layer depending upon the severity of the pollution level. Photosynthesis and side by side other metabolic product formation become hampered due to inability of functional light frequency penetrated well inside the plant leaves. The aldehyde of aldohexose and ketone of ketohexose of carbohydrate molecule has been affected by sulfite (SO_3^{2-}), a reaction intermediate of SO_2 (Tripathi and Gautam, 2007; Rai, 2016). This scenario has been observed mainly in pollution sensitive plant species, but the tolerant have the potentialities to constant their sugar level or have been kept the level higher for battle against air pollution or other stressful conditions. The soluble sugar content of *Callistemon citrinus* and *Albizia lebbeck* has been increased significantly at air pollution affected region (Seyyednejad et al., 2009).

7. Amino acid: It is nitrogen-containing biological macromolecule arranged one by one through peptide bond as like beads on a string to form the primary structure of protein. There are thousands of proteins molecules which have been produced by different combinations of amino acids. Twenty different amino acids are present in nature and each amino acid contained a distinct side chain or R group which varies in size, structure, ionic charge and has the property with different solubility potentialities of amino acid in water (Nelson and Cox, 2005). These amino acids are the biochemical indicator molecule in plants produced during environmental stressful conditions like air pollution and other adverse situations. Gradual increase of amino acid has been observed in plant's leaves as the level of air pollution increased. Proteins become denatured to form more free amino acid after degradation of quaternary, tertiary, secondary structure with breakdown of associated covalent, non-covalent, and ionic bonds. Accumulation of free amino acid is an indicative of impact of stress factor which may be biotic or abiotic, such as air pollution stress.

8. Protein: These are the most essential organic macromolecule which is present in all cell types and construct a basic form of life. The primary structure of protein is made by different combination of amino acid and which is regulated through specific gene expression. The functional product of protein is the quaternary structure which is made up of different polypeptide chain with many covalent and non-covalent linkages between them. Enzymes are the functional protein which has been regulated many biochemical reactions in plants. The level of total protein content have been reduced due to higher level of SO_2 and NO_2 pollutants (Agarwal and Deepak, 2003; Rai and Panda, 2015a). The level of it becomes decreased due to denaturation in many adverse situations (Prasad and Inamdar, 1990). Foliar protein has been reduced either due to inhibition of protein synthesis or due to denaturation of existing protein content (Constantinidou and Kozlowski, 1979; Singh and Jyothi, 1999; Iqbal et al., 2000). It is also reduced under the activation of protease enzyme which catalyzes the polypeptide chain of protein molecule to cope up the adverse environmental conditions (Doheman et al., 1990). Automobile emission also made an adverse impact on protein synthesis in proper way (Kumar and Dubey, 1988). Industrial fly ash significantly reduced the protein content in few plant species (Trivedi and Singh, 1995). There are some tolerant varieties of plant species which could resist the stress conditions to some extent and capable to kept the protein level standard.

9. Proline: It is a nonpolar, aliphatic amino acid produced from α-ketoglutarate and is used as a bio-indicator molecule in plants in effect of air pollution. In stressful environmental conditions the higher level of proline has been observed in plant foliar extract (Seyyednejad and Koochak, 2011). Carbon, organic nitrogen, energy, and other metabolic by-product have been produced from Proline (Stewart and Boggess, 1978; Sairam and Tyagi, 2004). It is metabolized in cellular respiratory pathway (Britikov et al., 1965) and act as an osmolyte (Singh et al., 1972; Wyn Jones and Storeys, 1978). Proline is mainly responsible to reduce the level of free radicals which is produced due to pollution induction by some cellular metabolic processes and protect the plants from oxidative damage (Liang et al., 2013). The peroxidation of PUFAs of cell membrane has been operated by ROS and also damages the carbohydrate, protein, lipid, and nucleic acid of cell (Blokhina et al., 2003). A relationship between proline accumulation and lipid peroxidation has been documented in various adverse atmospheric conditions (Wang et al., 2009). Proline accumulation takes place in variety of plants under the following stressful conditions, e.g., in SO_2 fumigation (Tankha and Gupta, 1992), air pollution (Godzic and Linsken, 1974), in heavy metal stress (Aslam, 2017), salt stress (Basu et al., 2002), moisture stress (Pokhriyal and Raturi, 1984) and also in plant defense mechanism (Khattab, 2007). These phenomenon has been reflected the protective mechanism of proline in higher plant species under the challenging environmental conditions.

10. Phenol: The organic alcohol phenol act as an antioxidant to protect the plants from all types of oxidative damage (Cook and Samman, 1996). The floral growth, development, and metabolic processes have been regulated by total phenol concentration (Lewis and Yamamoto, 1990). It has been observed that, the level of phenolic substances have been increased under the impacts of different types of biotic and abiotic stresses like environmental pollution (Zobel, 1996). In *Salix alba* leaf tissue a dark content were deposited in the mesophyll cells and that were probably the phenolic materials (Gostin and Lacramioara, 2007).

11. Cystein: It is a polar, uncharged amino acid and is hydrophilic, i.e., more soluble in water. Cysteine has been well documented in tolerant plants under air pollution stress, i.e., its higher level could indicate a positive response. A positive trend of increase had observed in plants that were exposed to especially SO_2 pollutants (DeKok, 1989; Tausz et al., 1998). In plant cells, interior the cysteine could act as a buffering substance of SO_2among several primary and secondary pollutants of air (Smith and Raven, 1979). Cysteine has been produced by sulfide (S^{2-}) reductive pathway and the source of S^{2-} is the environmental sulfite (SO_3^{2-}). After production, it is accumulated inside the vacuole and also translocated at intracellular space (DeKok et al., 2000).

5.8 MICRO-ANATOMICAL PREVIEW OF PLANT LEAVES UNDER AIR POLLUTION STRESS

Plants are the immovable life on earth and could not escape from bad polluted air to the fresh environment for their own rejuvenation. As a result they become continuously exposed by different toxic gases, particulate matter of different size and shape and other polluting substances in air. The air pollutants encounter first with the leaf epicuticular surface distorted its structural configuration and then enters into the leaf interior. SO_2, NO_2, O_3 and other gaseous pollutants enter into the foliar tissue through stomatal opening and dissolve into the water films of intercellular space of spongy mesophyll parenchyma (Gostin and Ivanescu, 2007).

Some plants could overcome this situation by their own micro-anatomical characteristics. The hardiest polysaccharide cutini containing nonporous or partially porous structure is the stem and the canopy portion has billions of green leaves with different arrangement that are mainly porous through which gaseous exchange and transpiration has been possible. The epicuticular surface of leaf are the principal gateway between internal and external environment of different gases and water molecules and this outer surface continuously exposed to various obnoxious air pollutants with different concentrations. The principal barrier against all the stressful factors like heat and cold stress, O_2 deprivation, insect attack, pathogen invasion, air pollutants, is the wax containing cuticle layer of leaf.

The chemistry and composition of the waxy layer was revealed by many authors (Kolattukudy et al., 1981; Martin and Juniper, 1970) and the impacts of air pollutants on the epicuticular waxy layer had been noticed by another group of author (Cape and Fowler, 1981; Godzik and Sassen, 1978; Huttunen and Laine, 1981). Two major hydrophobic components are present in the leaf cuticle layer, a soluble lipid and an insoluble cutin polymer. The cuticular soluble lipids are termed as waxes (Kolattukudy, 1980). These waxes are always present in crystalline or amorphous form in intra-cuticular portion and only crystalline form at epicuticular portion. Many external abiotic factors like air temperature, humidity, solar radiation, acid precipitation and air pollution (Baker and Hunt, 1986) induces a significant impact on the wax layering of the leaf cuticle. The SO_2 in the ambient air dissolve the epicuticular crystalline wax into the amorphous state (Lendzian, 1984) and also induce to open the stomata (Noland and Kozlowski, 1979). The low molecular component of esters and paraffin fractions were decreased with increasing air pollution (Schuck and Schutt,

1972). The cell permeability and membrane integrity has been distorted in effect of air pollution and this was observed early in sensitive plant species (Forooq and Beg, 1980; Keller, 1986). The water has been lost very fast from the stomatal guard cells and finally blocked the stomatal opening by squeezing themselves. On the other hand particulate matter of different size and shape could block the stomatal opening and obstructing the transpiration of water. Furthermore, the air pollutants which are present on the leaf epicuticular surface also have the capabilities to change the wavelength of incident light, and as a result, the specific wavelength-dependent photosynthesis has been affected. Oxygen production, glucose synthesis, and water transpiration are the basic needs of plant survival, but these all are affected severely by air pollution.

The layering pattern of wax in epicuticular outer surface of plant leaf provides important information regarding taxonomic classification. This type of primary waxy arrangement was used to classify the plant species of same taxa. Primary and secondary hydrocarbon deposition above the epithelial cells of leaf has been occurred naturally and termed as immurations. The size, shape, and architecture of the immurations are different in different plant species with their own taxonomic importance. Above the primary structural configuration of waxy covering there are another tertiary and quaternary depositions was recognized which has environmental or ecological importance. This thickening is possible only in stressful reverse environmental conditions and is not observed in all plant species. The tolerant plant species have the capability to produce this type of multiple layering on their leaf surface to battle against the adverse environmental conditions. In non-polluted fresh air environment an undulated epicuticular surface of plant leaf has been recognized with primary and secondary thickening. Furthermore, in the polluted area this uneven surface becomes partially or completely flat with additional tertiary and quaternary layering (Verma and Singh, 2006).

In polluted populations, the dust particles coming out of automobile exhaust were embedded onto the epicuticular wax and thus formed a crust on the foliar surface. The particulate matter emitted from various industrial activities and automobile exhaust with different size and chemical compositions has been attached on the foliar cuticular layer. This dust particles and naturally occurring waxy granular depositions are likely to be the same structural conformations. Their structural differentiation not made very easily even under Scanning Electron Microscopy (SEM). In non-polluted fresh air environment a well differentiated granular epicuticular depositions were recorded but in severely or highly polluted area this waxy granules were distorted or withered. The total stomatal arrangements are severely damaged with higher air pollution condition. The stomatal openings are clogged with different types of air polluting microparticle (1–2.5 μm). The length, breadth, and size of the stomata have been changed in stressful polluted environment. Air pollution is lead to increase the stomatal size with lower number per unit area and vice versa. The transverse section of plant leaves reflected many black phenolic deposits in palisade and spongy parenchyma and the amount of this deposition is increasing in the area with higher atmospheric pollutants (Zobel, 1996).

5.9 ROLE OF FREE RADICALS AND ANTIOXIDANTS IN RESPONSE TO AIR POLLUTION

In cellular respiration, the atmospheric O_2 is required at the last step of metabolic processes as a terminal electron acceptor which provides more energy than the fermentation process (Bolwell and Woftastek, 1997). In plants, the free radicals or ROS has been produced unavoidably by normal cellular aerobic metabolism as a secondary byproduct (Puntarulo and Boveris, 1988). The biochemical characteristics of ROS are their hyper-reactivity with other viable macromolecules of cells like protein, lipid, nucleic acid, etc. It can also damage the cell membrane, cellular organelles by peroxidation of PUFAs; nonfunctioning some enzymes which ultimately lead to cell death (Foyer and Noctor, 2005). This ROS must be eliminated out from the plant's body by their own enzymatic or non-enzymatic regulatory mechanisms.

The biotic factor-like pathogen invasion and abiotic stresses like drought, salinity, higher temperature, heavy metal load, air pollution have been induced an inhibitory role or delayed the process of removal of ROS from plant body. In plants, there are several types of ROS that have been produced continuously. As a result, they have been accumulated inside the plant body in stressful environmental conditions and can activate the process of oxidation of other important bimolecular resulting retardation of plant growth and development (Ogawa and Iwabuchi, 2001; Hernandez-Jimenez et al., 2002). In spite of their oxidative property, the molecules of ROS would also have to act as a secondary messenger inside the cell (Desikan et al., 2001). It is in the form of a reactive molecule or in ionic form and derived from O_2. In plants 1–5% consumed $O_2$is required for ROS production (Asada and Takahashi, 1987; Moller, 2001).

In plants there are different types of radical ROS, namely superoxide anion ($O_2^{\cdot -}$) which is most common, hydroxyl radical ($OH^{\cdot}$), peroxy radical ($ROO^{\cdot}$), perhydroxy radical (HO_2^-) and non-radical ROS like hydrogen peroxide (H_2O_2), singlet oxygen ($^1O_2$), organic hydroperoxide (ROOH), etc. These radical and non-radical ROS have been produced from different cellular locations such as in cell wall, plasma membrane, mitochondria, chloroplast, endoplasmic reticulum, peroxisome, and in apoplast (Sharma et al., 2012). At day time, the main ROS producing sites are chloroplast and peroxisome (Foyer et al., 1997) and at night it has been produced from mitochondria. These highly reactive free radicals could bind themselves at the site of their origin because they possess extreme power of reactivity in a very short duration of time with other molecules (Slater, 1984).

It has been estimated that, in biological system, the half-life of free radical such as OH and CCl_3 is of few microseconds (Borg and Schaich, 1984). The radius of its diffusion from its origin point is less than on average 100 nm (Slater, 1976). In plant body, the main ROS producing organelles are mitochondria, chloroplast, and peroxisome and they are responsible to control the ROS formation by producing some isoform of antioxidants. There are some nonenzymatic antioxidants which have low molecular mass such as glutathione and ascorbic acid (water-soluble); tocopherol and carotenoid (lipid-soluble) (Mittler et al., 2004, Gill et al., 2011); on the other hand, some high molecular ROS scavenging enzymes are Superoxide dismutase (SOD), Ascorbate peroxidase (APX), Catalase (CAT), Peroxidase, Glutathione peroxidase (GPX), etc. It was observed that the some plants growing in stressful environmental conditions have been maintained their own internal antioxidants level high and cope up the adverse situation (Zaefyzadeh et al., 2009; Chen et al., 2010).

5.10 ROS SCAVENGING NON-ENZYMATIC ANTIOXIDANTS

Ascorbate, glutathione are water-soluble, and tocopherol, carotenoid, alkaloids, flavonoids are water-insoluble non-enzymatic antioxidants found in a different portion of plant parts. In them, the most potent antioxidants are ascorbic acid which is ubiquitous in all plant tissue and efficiently removes the ROS produced subcellular compartment during stress conditions. It is present in reduced form in chloroplast of matured leaves under normal physiological conditions (Smirnoff, 2000). In cytosol 20 mM and in chloroplast stroma 20–300 mM of ascorbic acid has been detected (Foyer and Lelandais, 1996). In many enzymatic and non-enzymatic reactions ascorbic acid has the capability to donate electron in aqueous phase. Singlet oxygen, superoxide, and hydroxyl radical have been scavenged directly by it and reduce H_2O_2 to water (Noctor and Foyer, 1998).

Glutathione is a tripeptide compound found abundantly in plant tissue and cellular compartment like cytoplasm, mitochondria, vacuole, endoplasmic reticulum (Jimenez *et al.*, 1998). It is main storage form of sulfur due to presence of central nucleophilic cysteine residue which is responsible for reductive capability of glutathione. It could scavenge superoxide radical, singlet oxygen, hydroxyl radical, H_2O_2, and has the capability to produce ascorbic acid by ascorbate-glutathione cycle (Noctor and Foyer, 1998).

Tocopherol is synthesized by algae and plants, present in the cell membrane of all plant parts (Janiszowska and Pennock, 1976). There has α, β, γ, and δ these four types of isomer of tocopherol in which α tocopherol is most powerful antioxidant of cell membrane and is also known as Vitamin E. It could deactivate the singlet oxygen, hydroxyl radical and other ROS by charge transfer process (Fryer, 1992).

5.11 ROS SCAVENGING ENZYMATIC ANTIOXIDANTS

SOD is large molecular enzymatic metalloprotein which is ubiquitous in all subcellular compartments of all aerobic organisms. There are three different SODs which are classified depending on metal cofactor present in its central position, such as, iron-SOD, manganese-SOD, and copper/zinc-SOD. These three isozymes have been distributed in different cellular organelles. In cytoplasm, Cu/Zn-SOD are present, Mn-SOD, and Fe-SOD are found in mitochondria and chloroplast of eukaryotes respectively. SOD catalyzes the superoxide (O_2^-) ion to H_2O_2 by dismutation and the rate of this reaction is 10,000 fold faster than the spontaneous dismutation. The H_2O_2has been detoxified subsequently by CAT, APX, and GPX.

CAT and peroxidise are most important intracellular enzymes which could detoxify the H_2O_2 by forming an intermediate CAT-H_2O_2complex, produces dioxygen and water. This complex reacts with another H_2O_2

molecule and produces hydroxyl radical (OH·), another strong antioxidant which could initiate a radical chain reaction with PUFA of membrane lipid. GPX and APX both are also responsible to detoxify the H_2O_2 to H_2O with higher affinity than CAT and peroxidase. Plants have the ability to scavenge different ROSs with several enzymatic and non-enzymatic biotool to withstand against adverse atmospheric conditions. Air pollution affects the floral community round the clock in urban industrial area and cellular antioxidant machineries are most significant for protection against ROS.

5.12 GREEN BELT DEVELOPMENT THROUGH POLLUTION TOLERANT PLANT SPECIES

It is great important to establish a natural barrier of green belt in and around the industrial region for pollution abatement. It has been defined as it is the structural geometry by a belt or strip of such concerned plant species that when grown around the point stationary sources of air pollution, would significantly lower the level of pollution by assimilating or intercepting the pollutants in a sustainable way (Ruth and Willium, 1994). The evergreen and deciduous both types of plants have been planted for absorbing and filtering the air pollutants in mitigating the purpose (Sharma and Roy, 1997). The other type of vegetation like herb, shrub have also been planted but trees are mostly accepted for green belt development. The factors like height, width, and distance of the green belt from the point sources are responsible for development of optimum green belt. Kapoor and Gupta (1984) had originated a model regarding attenuation of industrial air pollution by calculating attenuation factor (A_f). This factor expressed that, it is a ratio of Qwa and Q_B, where Qwa, and Q_B indicate the distance traveled by the air pollutants without green belt and with green belt.

$$A_{f} = \frac{Qwa}{QB}$$

where, Qwa = mass flux of pollutants reaching without Green belt; and Q_B = mass flux of pollutants reaching with Green belt.

Plants have been responded differently in front of air pollution, some showed higher sensitivity by their early biochemical variation and by later micro or macro morphological anomalism, and some acted as tolerant varieties with their superior characteristics. The most tolerant varieties of plant species have been screened out by considering APTI and some related socioeconomic and biological characters by which API of specific plants has been prepared for development of green belt.

The distribution of gaseous pollutants and particulate matter does not occur radially from the stationary point source, in spite of that it has been dispersed non-uniformly depending on the direction, speed, and stability of air and other atmospheric factors. When the industrial gaseous byproduct has been came out from the long chimneys, known as plume, spread over in different directions without mixing with the ambient air. This plume did not come in contact with the surrounding green belt and disperse hundred to thousand meters away from it. After traveling of such distance the gaseous pollutants get mixed slowly with natural fresh air and diluted its toxic effect. So the green belt must be established in such a distance around the industrial sector that the polluted plume come close to the ground and confront with green belt. Furthermore, the green belt must not be formed in one side of industrial sectors, in spite; it should be structured in circular manner because the polluted plume has been dispersed in different direction in different season depending on airflow. In the green belt not all the plant species and other vegetation are recommended for plantation, but only the tolerant ones should be planted after checking their suitability to agro-climatic conditions of the region. The height and width of the green belt would be such that it absorb, adsorb, and precipitate or infiltrate maximum air pollutants depending on the regional severity of air quality. Inside it polycultural practice would be highly appreciated because it enhances the regional biodiversity along with pollution mitigation.

Green belt has also been mitigated the traffic-induced air pollution and noise pollution arising from automobiles. In urban area increasing traffic load has increased day by day and this sector aggravated the pollution intensities along with industrial emission. Obnoxious gaseous pollutants and higher noise level along roads and the adjacent area create health hazards in human beings specially children as it is not diffused properly into the surrounding environment. The auto emission has been impended by high rise of civil construction which could not soak the pollutants. In contrast, if a strip of green plantation has been developed along roads and highways, then it could improve local environmental status by absorbing, precipitating or reflecting the pollutants (Figure 5.3).

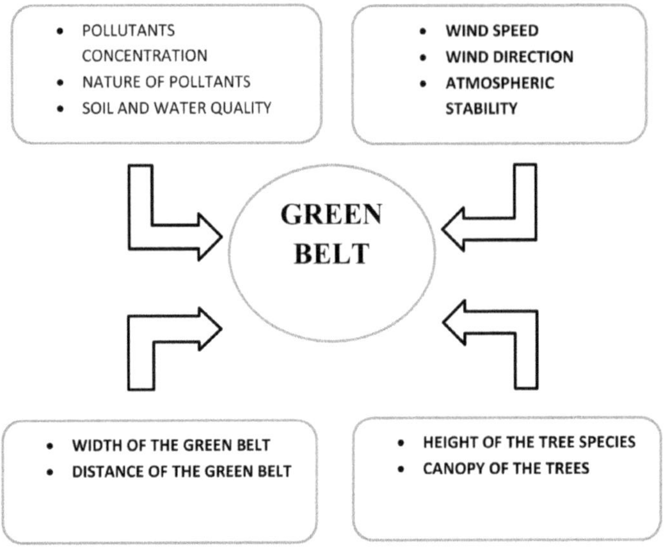

FIGURE 5.3 Model showing the factors affecting green belt.

5.13 POLLUTION TOLERANT SPECIES AND ENVIRONMENTAL SUSTAINABILITY

Pollution tolerant plant species are those plants which by their superior biochemical, physiological, and morphological characteristics could not only strongly battle against air pollution but also showed no severe external injury caused by air pollution. In contrast, some plants have been shown unfavorable biochemical and physiological variation along with external anomaly in air pollution load and are known as sensitive plants which represent them as biological indicators. Tolerant plants have the maximum capability to reduce the injury caused by air pollution.

Today it is important to know about the type of atmospheric pollutants, their concentration in air on seasonal and annual basis, reactivity of pollutants with biota especially green plants for present and future pollution mitigation. Species-specific tolerance to particular air pollutants would also be assessed for healthy and clean environment. The species which could manage some types of air pollutants should have been cultivated more around the periphery of industrial region. Different plants have been exhibited different degree of severity depending on the air pollution load. The morphological features that have to be followed in pollution exposed leaves are length, breadth, and size of stomata; length, frequency, and type of trichomes (Mansfield and Majernik, 1970). Leaf length, breadth, and area; petiole length; width of the midrib; number of epidermal cell per unit area; stomatal index; length of epidermis, palisade, and spongy parenchyma these all macro and micro-morphological characteristics have been regarded as indicators of air pollution. After region-wise classification of gaseous and particulate matter pollution, species-specific tolerant plant species can be selected for plantation. The first objective of this mass plantation or green belt development is to clean the regional atmosphere and the second objective is to enhance the esthetic value of the landscape. *Ficus benghalensis, Ficus mysorensis, Ficus religiosa* and *Albizia lebbeck* have been planted in the periphery of village and urban areas as these plants significantly reduce the air pollutants released by the industry and automobile (Sharma and Roy, 1997). On the basis of experiment and available data from literature review, the following plants have classified according to their degree of tolerance to air pollution.

5.14 CONCLUSION

Green belts are the best possible way to remove out or to attenuate all type of particulate matters and gaseous air pollutants from the surrounding environment of urban industrial area. For green belts not all the plants are significant, only those which have higher APTI and API values, higher internal antioxidant level and favorable biochemical and physiological characters have been regarded as tolerant and would be best for plantation. These screened out tolerant plants with superior internal biochemical and morphological characteristics are act as natural buffers against atmospheric pollution on earth. Furthermore, in industrial and automobile sectors new sustainable processes and technological inventions should have to be implemented for clean and green environment. More priority would be given by utilization of alternative source of energy in both the industrial and transportation sectors. The use of small or large aliphatic hydrocarbon like propane, butane, Compressed Natural Gas (CNG) and Liquefied Petroleum Gas (LPG), these types of fuels have been emitted less pollutants into the atmosphere and should be most accepted in automobiles in spite of petrol, diesel, gasoline, etc. In industrial sectors non-conventional energy sources are most welcome in near future.

So in one side, the aerial pollution load will be minimizes significantly by mass plantation of pollution tolerant plant species and on the other side by more uses of environment-friendly alternative fuels in automobile sector and non-conventional energy sources in industrial sector.

KEYWORDS

- **environment**
- **green belt**
- **pollution mitigation**
- **pollution tolerant plant species**

REFERENCES

Agarwal, M., & Deepak, S. S., (2003). Physiological and biochemical responses of two cultivars of wheat to elevated levels of CO_2 and SO_2, singly and in combination. *Environmental Pollution, 121*, 189–197.

Aslam, M., Saeed, M. S., Sattar, S., Sajad, S., Sajad, M. S., Adnan, M., Iqbal, M., & Muhammad, T. S., (2017). Specific role of Proline in heavy metal toxicity in plants. *Int. J. Pure App. Biosci.*, *5*(6), 27–34.

Baker, E. A., & Hunt, G. M., (1986). Erosion of waxes from leaf surfaces by stimulated acid rain. *New Phytol.*, *102*, 161–173.

Bakiyaraj, R., & Ayyappan, D., (2014). Air pollution tolerance index of some terrestrial plants around an industrial area. *International Journal of Modern Research and Review*, *2*, 1–7.

Basu, S., Gangopadhyay, G., & Mukherjee, B., (2002). Salt tolerance in rice *in vitro*: Implication of accumulation of Na^+, K^+ and proline. *Plant Cell Tiss. Org. Cult., 69*, 55–64.

Beckett, K. P., Freer-Smith, P. H., & Taylor, G., (1998). Urban woodlands: Their role in reducing the effects of particulate pollution. *Environ. Pollut., 99*, 347–360.

Blokhina, O., Virolinen, E., & Fagerstedt, K. V., (2003). Antioxidants, oxidative damage and oxygen deprivation stress: A Review. *Annals of Botany, 91*, 179–194.

Bolwell, G. P., & Woftastek, P., (1997). Mechanism for the generation of reactive oxygen species in plant defense-broad perspective. *Physiol. Mol. Plant Pathol.*, *51*, 347–349.

Bora, M., & Joshi, M. A., (2014). Study on variation in biochemical aspects of different tree species with tolerance and performance index. *Ecoscan.*, *9*(1), 59–63.

Borg, D. C., & Schaich, K. M., (1984). Cytotoxicity from coupled redox cycling of anti-oxidizing xenobiotics and metals. *Israel. J. Chem., 24*, 38–53.

Britikov, E. A., Vladimirtseva, S. V., & Musatova, N. A., (1965). Transformation of proline in germinating pollen. *Fiziologiya Rastenii, 12*, 953–967.

Calvin, M., (1955). Function of carotenoids in photosynthesis. *Nature, 176*, 1211.

Cape, J. N., & Fowler, D., (1981). Changes in epicuticular wax of *Pinus sylvestris* exposed to polluted air. *Silva Fennica, 15*, 457–458.

Chen, Q., Zhang, M., & Shen, S., (2010). Effect of salt on malondialdehyde and antioxidant enzymes in seedling roots of Jerusalem artichoke (*Helianthus tuberosus L.*). *Acta Physiologiae Plantarum, 33*, 273–278.

Conklin, P. L., (2001). Recent advances in the role and biosynthesis of ascorbic acid in plants. *Plant, Cell and Environment, 24*, 383–394.

Constantinidou, H. A., & Kozlowski, T. T., (1979). Effect of sulfur-dioxide and ozone on *Ulmus americana* seedling II; carbohydrate, protein and lipids. *Can. J. Bot., 57*, 176–184.

Cook, N. C., & Samman, S., (1996). Flavonoids- chemistry, metabolism, cardio protective effects, and dietary sources. *Nutritional Biochemistry, 7*, 66–76.

Das, S., & Prasad, P., (2010). Seasonal variation in air pollution tolerance indices and selection of plant species for industrial areas of Rourkela. *Indian Journal of Environmental Protection, 30*(12), 978–988.

Davison, A. W., & Barnes, J. D., (1986). Effects of winter stress on pollutant responses. In: *How are the Effects of Air Pollutants on Agricultural Crops Influenced by the Interaction with Other Limiting Factors?* (pp. 16–32). CEC, Brussels.

De Tullio, M. C., Paciolla, C., Dalla, V. F., Rascio, N., D'Emerico, S., De Gara, L., Liso, R., & Arrigoni, O., (1999). Changes in onion root development induced by the inhibition of peptidyl-prolyl hydroxylase and influence of the ascorbate system on cell division and elongation. *Planta, 209*, 424–434.

DeKok, L. J., (1989). Responses of sulfur metabolism in plants to atmospheric hydrogen sulfide. *Phytonomist, 29*, 189–201.

DeKok, L. J., (1990). In: Rennenberg, H., Brunold, C. H., DeKok, L. J., & Stolen, I., (eds.), *Sulfur Nutrition and Sulfur Assimilation in Higher Plants* (pp. 111–130). SPB Academic Publishing, The Hague.

DeKok, L. J., Westernman, S., Elisabeth, C., Stuiver, E., & Stulen, I., (2000). Atmospheric H_2S as plant sulfur source: Interaction with pedospheric sulfur nutrition – A case study with *Brassica oleracea* L.,' In: Brunold, C., et al. (eds.), *Sulfur Nutrition and Sulfur Assimilation in Higher Plants* (pp. 41–56). Paul Haunt Bern Switzerland.

Desikan, R., Mackerness, S. S. A. H., Hancock, J. T., & Neill, S. J., (2001). Regulation of the *Arabidopsis* transcriptome by oxidative stress. *Plant Physiology, 127*(1), 159–172.

Dhankhar, R., Mor, V., & Narwal, S., (2015). Anticipated performance index of selected plant species in university campus area, Rohtak, Haryana, India. *Int. J. Adv. Multidiscip. Res., 2*(2), 32–41.

Dohmen, G. P., Loppers, A., & Langebartels, C., (1990). Biochemical response of Norway spruce (*Picea abies* (L) Karst) toward 14-month exposure to ozone and acid mist, effect on amino acid, glutathione and polyamine titers. *Environmental Pollution, 64*, 375–383.

Escobedo, F. J., Wagner, J. E., & Nowak, D. J., (2008). Analyzing the cost effectiveness of Santiago, Chile's policy of using urban forest to improve air quality. *J. Envi. Manag., 86*, 148–157.

Finkelstein, R. R., & Gibson, S. I., (2001). ABA and sugar interactions in regulating development: Cross-talk or voices in a crowd. *Curr. Opin. Plant Biol., 5*, 26–32.

Forooq, M., & Beg, M. U., (1980). Effect of aqueous sulfur dioxide on the membrane permeability of common Indian tree leaves. *New Botanist, 7*, 213–217.

Foyer, C. H., & Lelandais, M., (1996). A comparison of the relative rates of transport of ascorbate and glucose across the thylakoid, chloroplast and plasma membranes of pea leaf mesophyll cells. *J. Plant Physiol.*, *148*, 391–398.

Foyer, C. H., & Noctor, G., (2009). Redox regulation in photosynthetic organisms: signaling, acclimation, and practical implications. *Antioxid Redox Signal*, *11*, 861–905.

Foyer, C. H., Lopez-Delgado, H., Dat, J. F., & Scott, I. M., (1997). Hydrogen peroxide- and glutathione-associated mechanisms of acclamatory stress tolerance and signaling. *Physiologia Plantarum*, *100*(2), 241–254.

Fryer, M. J., (1992). The antioxidant effects of thylakoid vitamin E (α tocopherol). *Plant Cell and Environment*, *15*, 381–392.

Gill, S. S., Khan, N. A., Anjum, N. A., & Tuteja, N., (2011). Amelioration of cadmium stress in crop plants by nutrients management: Morphological, physiological and biochemical aspects. *Plant Stress*, *1*, 1–23.

Godzic, S., & Linskens, H. F., (1974). Concentration of free amino acids in primary bean leaves after continuous and interrupted SO_2 fumigation and recovery. *Environmental Pollution*, *7*, 25–38.

Godzic, S., & Sassen, M. M. A., (1978). A scanning electron microscope examination of *Aesculushippo castanum* L. leaves from control and air polluted areas. *Environ. Pollution*, *17*, 13–18.

Gostin, I., & Ivanescu, L., (2007). Structural and micromorphological changes in leaves of Salix alba under air pollution effect. *Int. J. Energy. Environ.*, *1*, 219–226.

Hernandez-Jimenez, M. J., Lucas, M. M., & DE Felipe, M. R., (2002). Antioxidant defence and damage in senescing Iupin nodules. *Plant Physiol. Biochem.*, *40*, 645–657.

Huddart, L., (1990). *The Use of Vegetation for Traffic Noise Screening* (p. 238). Crowthorne, Berkshire: U. K. Transport and Road Research Laboratory Research Report.

Huttunen, S., Havas, P., & Laine, K., (1981). Effects of air pollutants on the wintertime water economy of the Scots pine *Pinus silvestris*. *Holarct. Ecol.*, *4*, 94–101.

Ingold, C. T., (1971). *Fungal Spores* (p. 302). Clarendon Press, Oxford.

Iqbal, M., Srivastava, P. S., & Siddiqi, T. O., (2000b). Anthropogenic stresses in the environment and their consequences. In: Igbal, M., et al. (eds.), *Environmental Hazards, Plants and People* (pp. 1–40). CBS Publishers, New Delhi.

Janiszowska, W., & Pennock, J. F., (1976). The biochemistry of vitamin E in plants. *Vitamins and Hormones*, *34*, 77–105.

Jones, H. G., (1994). *Plants and Microclimate* (2nd edn, pp. 407.). Cambridge University, Cambridge, UK.

Joshi, N., & Bora, M., (2011). Impact of air quality on physiological attributes of certain plants. *Report and Opinion*, *3*, 42–47.

Jyothi, S. J., & Jaya, D., (2010). Evaluation of air pollution tolerance index of selected plant species along roadsides in Thiruvanthapuram, Kerala. *Journal of Environmental Biology*, *31*, 379–386.

Kapoor, R. K., & Gupta, V. K., (1984). A pollution attenuation coefficient concept for optimization of green belt. *Atmospheric Environment*, *18*(6), 1107–1117.

Keller, T., (1986). The electrical conductivity of Norway spruce needle diffuse at ease affected by air pollutants. *Tree Physiol.*, *1*, 85–94.

Khattab, H., (2007). The defense mechanism of cabbage plant against phloem-stucking aphid (*Brevicoryne brassicae* L.). *Aust. J. Basic Applied Sci.*, *1*, 56–62.

Kolattukudy, P. E., (1980). Cutin, suberin and waxes. In: Stumpf, P. K., & Conn, E. E., (eds.), *The Biochemistry of Plants, Lipids: Structure and Function* (Vol. 4, pp. 571–645). Academic Press, New York.

Kolattukudy, P. E., Espelie, K. E., & Soliday, C. L., (1981). Hydrophobic layers attached to cell walls, cutin, suberin and associated waxes. In: Tanner, W., & Loewus, P. A., (eds.), *Plant Carbohydrates II* (pp. 225–248). Springer Verlag, Berlin-Heidelberg-New York.

Kondo, N., Akiyama, Y., Fujiwara, M., & Sugahara, K., (1980). Sulfite oxidizing activities in plants. In: *Studies on the Effects of Air Pollutants in Plants and Mechanism of Phytotoxicity. Res. Rep. Nat. Environ. Study Jpn.* (Vol. 11, pp. 137–150).

Krishnaveni, M., Chandrasekar, R., Amsavalli, L., Madhaiyan, P., & Durairaj, S., (2013). Air pollution tolerance index of plants at Perumalmalai Hills, Salem, Tamil Nadu, India. *Int. J. Pharm. Sci. Rev. Res.*, *20*(1), 234–239.

Kuddus, M., Kumari, S., & Ramteke, W. P., (2011). Studies on air pollution tolerance of selected plants in Allahabad city, India. *J. Environ. Res. Manag.*, *2*(3), 42–46.

Kumar, G. S., & Dubey, P. S., (1998). Differential response and detoxifying mechanism of *Cassia siamea* Lam. and *Dalbergia sissoo* Roxb. of different ages to SO_2 treatment. *J. Environ. Biol.*, *9*(3), 243–249.

Legge, A. H., & Krupa, S. V., (2002). Effects of sulfur dioxide. In: Bell, J. N. B., & Treshow, M., (eds.), *Air Pollution and Plant Life* (pp. 130–162). John Wiley & Sons, West Sussex, England

Lendzian, K. J., (1984). Permeability of plant cuticles to gaseous air pollutants.' In: *Gaseous Air Pollutants and Plant Metabolism* (pp. 77–81). Butter Worths, London.

Lewin, S., (1976). *Vitamin C: Its Molecular Biology and Medical Potential Literature Cited* (p. 231). Academic Press, New York.

Lewis, N., & Yamamoto, E., (1990). Lignin: Occurrence, biogenesis and biodegradation. *Annual Review of Plant Physiology and Plant Molecular Biology*, *41*, 455–496.

Liang, X., Zhang, L., Natarajan, S. K., & Becker, D. F., (2013). Proline mechanisms of stress survival. *Antioxidants and Redox Signaling*, *19*(9), 998–1011.

Liu, Y. J., & Ding, H., (2008). Variation in air pollution tolerance index of plants near a steel factory: Implication for landscape-plant species selection for industrial areas. *Wseas Trans. Environ. Dev.*, *4*, 24–32.

Lohe, R. N., Tyagi, B., Singh, V., Kumar, T. P., Khannam, D. R., & Bhutiani, R., (2015). A comparative study for air pollution tolerance index of some terrestrial plant species. *Global J. Environ. Sci. Manage.*, *1*(4), 315–324. doi: 10.7508/gjesm.2015.04.006.

Lorenc-Plucinska, G., (1982). Influence of SO_2 on CO_2 assimilation and carbon metabolism in photosynthetic processes in Scotspine. In: *Arboretum Kornickie (in Polish)* (Vol. XXVII, pp. 285–310).

Mansfield, T. A., & Majernik, O., (1970). Can stomata play a part in protecting plants against air pollutants? *Environ. Pollut.*, *1*, 149–154.

Martin, J. T., & Juniper, P. E., (1970). *The Cuticles of Plants* (p. 347). Edward Arnold Ltd. Great Britain.

Masuch, G., Kicinski, H., Kettrup, A., & Boss, K. S., (1988). Single and combined effects of continuous and discontinuous O_3 and SO_2 emission on Norway spruce needle: Histochemical and cytological changes. *Int. J. Environ. Anal. Chem.*, *32*, 213–241.

Mittler, R., Vanderauwera, S., Gollery, M., & Breusegem, F., (2004). Reactive oxygen gene network of plants. *Trends Plant Sci.*, *9*, 490–498.

Moller, I. M., (2001). Plant mitochondria and oxidative stress: Electron transport, NADPH turnover, and metabolism of reactive oxygen species. *Annu. Rev. Plant Physiol. Mol. Biol.*, *52*, 561–591.

Nelson, D. L., & Cox, M. M., (2005). *The Three Dimensional Structure of Proteins, Lehninger Principles of Biochemistry* (4th edn., pp. 75–110). Freeman, W. H., & Co, New York.

Noctor, G., & Foyer, C. H., (1998). A re-evaluation of the ATP: NADPH budget during C3 photosynthesis. A contribution from nitrate assimilation and its associated respiratory activity? *J. Exp. Bot.*, *49*, 1895–1908.

Noland, T. L., & Kozlowski, T. T., (1979). Effect of SO_2 on stomatal aperture and sulfur uptake of woody angiosperm seedlings. *Can. J. For. Res.*, *9*, 52–62.

Nowak, D. J., Kevin, L. C., Rao, S. T., Sistia, G., Luley, C. J., & Crane, D. E., (2000). A modeling study of the impact of urban trees on ozone. *Atmospheric Environment*, *34*, 1601–1603.

Nrusimha, T. K. Kumar, S., & Srinivas, N., (2005). Air pollution tolerance index of tree species growing in industrial and traffic areas of Visakhapatinam, India. *J. Env. Protection*, *25*, 1057–1060.

Ogawa, K., & Iwabuchi, M., (2001). A mechanism for promoting the germination of Zinnia elegans seeds by hydrogen peroxide. *Plant Cell Physiol.*, *42*, 286–291.

Palit, D., Kar, D., Misra, P., & Banerjee, A., (2013). Assessment of air quality using several bio-monitor of selected sites of Duragapur, Burdwan district by Air Pollution Tolerance Index approach. *Indian J. Sci. Res.*, *4*(1), 149–152.

Panda, L. S., & Rai, P. K., (2015). *Roadside Plants – Study on Eco-Sustainability* (pp. 152). Germany: Lambert Publisher.

Pasqualini, S. Batini, P. Ederli, L., Porceddu, A., Piccioni, F., Marchis, D. E., & Antonielli, M., (2001). Effects of short-term ozone fumigation on tobacco plants: Response of the scavenging system and expression of the glutathione reductase. *Plant Cell Environ.*, *24*, 245–252.

Pokhriyal, T. C., & Raturi, A. S., (1984). A note on proline content in eucalyptus hybrid leaves. *Indian For.*, *12*, 1070–1075.

Prado, F. E., Boero, C., Gallarodo, M., & Gonzalez, J. A., (2000). Effect of NaCl on germination, growth and soluble sugar content in *Chenopodium quinoa* wild seeds. *Bot. Bull. Acad. Sin.*, *41*, 27–34.

Prasad, M. N. V., & Inamdar, J. A., (1990). Effect of cement kiln dust pollution on ground nut. *Indian Bot. Cont.*, *7*(4), 159–162.

Prusty, B. A. K., Mishra, P. C., & Azeezb, P. A., (2005). Dust accumulation and leaf pigment content in vegetation near the national highway at Sambalpur, Orissa, India. *Ecotoxicol. Environ. Saf.*, *60*, 1193–1204.

Puckett, K. J., Nieboer, E., Flora, W. P., & Richardson, D. H. S., (1973). Sulfurdioxide: Its effect on photosynthetic14C fixation in lichens and suggested mechanism of phytotoxicity. *New Phytol.*, *72*, 141–154.

Puntarulo, S., Sanchez, R. A., & Boveris, A., (1988). Hydrogen peroxide metabolism in soybean embryonic axes at the onset of germination. *Plant Physiol.*, *86*, 626–630.

Rahmawati, N., Rosmayati, D., & Basyuni, M., (2014). Chlorophyll content of soybean as affected by foliar application of ascorbic acid and inoculation of arbuscular mycorrhizal fungi in saline soil. *Int. J. Sci. Technol. Res.*, *3*(7), 127–131.

Rai, P. K., & Panda, L. S., (2015a). Assessment of air pollution tolerance index (APTI) with road side plants in East and North East India: An eco-sustainable approach. *Journal of Pollution Effects and Control.* (in press).

Rai, P. K., & Panda, L. S., (2015b). Roadside plants as bio indicators of air pollution in an industrial region, Rourkela, India. *International Journal of Advancements in Research and Technology, 4,* 19–41.

Rai, P. K., (2016a). *Biomagnetic Monitoring Through Roadside Plants of an Indo-Burma Hot Spot Region* (p. 196). Elsevier, UK, ISBN: 978–0–12–805135–1.

Rai, P. K., (2016b). Biodiversity of roadside plants and the response to air pollution in an Indo-Burma hotspot region: Implications for urban ecosystem restoration. *J. Asia-Pac. Biodivers., 9*(1), 47–55.

Raina, A. K., & Sharma, A., (2003). Effects of vehicular pollution on the leaf micromorphology, anatomy and chlorophyll contents of *Syzygium cumini* L. *Indian J. Environ. Prot., 23*(8), 897–902.

Rao, M. N., & Rao, H. V. N., (1989). *Air Pollution* (pp. 271–272). Tata McGraw-Hill publishing company limited, New Delhi.

Rao, M. V., (1992). Cellular detoxifying mechanisms determine age dependent injury in tropical plants exposed to SO_2. *Plant Physiology, 140,* 733–740.

Rao, M. V., Jennifer, R. K., & Davis, K. R., (2000). Ozone: A tool for probing programmed cell death in plants. *Plant Molecular Biology, 44,* 345–358.

Ruth, A. E., & William, R. E., (1994). *The Encyclopedia of the Environment.* The Rene Dubor Centre for Human Environments, Houghton Miffin Company, U.S.A.

Sairam, R. K., & Tyagi, A., (2004). Physiology and molecular biology of salinity stress tolerance in plants. *Current Science, 86,* 407–421.

Schuppler, U., He, P. H., John, P. C. I., & Munns, R., (1998). Effect of water stress on cell division and Cdc2-like cell cycle kinase activity in wheat leaves. *Plant Physiol., 117,* 667–678.

Schütt, R., & Schuck, H. J., (1972). Zusammenhangez wischen Rauchhàrte und cuticular wax hesenbei Koniferen. Mitt. Bundesversanst. *Wien., 97*(11), 399–417.

Seyyednejad, S. M., & Koochak, H., (2011). A study on air pollution-induced biochemical alterations in *Eucalyptus camaldulensis. Aust. J. Basic Appl. Sci., 5*(3), 601–606.

Seyyednejad, S. M., Majdian, K., Koochak, H., & Nikneland, M., (2011). Air pollution tolerance indices of some plants around industrial zone in South of Iran. *Asian J. Biol. Sci., 4*(3), 300–305.

Seyyednejad, S. M., Niknejad, M., & Yusefi, M., (2009). The effect of air pollution on some morphological and biochemical factors of *Callistemon citrinus* in petrochemical zone in South of Iran. *Asian J. Plant Sci., 8,* 562–565.

Shannigrahi, A. S., Fukushima, T., & Sharma, R. C., (2004). Anticipated air pollution tolerance of some plant species considered for green belt development in and around an industrial/urban area in India: an overview. *International Journal of Environmental Studies, 61*(2), 125–137.

Sharma, P., Jha, A. B., Dubey, R. S., & Pessarakli, M., (2012). Reactive oxygen species, oxidative damage, and antioxidative defense mechanism in plants under stressful conditions. *Journal of Botany,* 1–26. doi: 10.1155/2012/217037.

Sharma, S. C., & Roy, R. K., (1997). Green belt – an effective means of mitigating industrial pollution. *Indian J. Environ. Prot., 17*(10), 724–727.

Shimazaki, K., Sakaki, T., Kondo, N., & Sugahara, K., (1980). Active oxygen participation in chlorophyll destruction and lipid peroxidation in SO_2-fumigated leaves of spinach. *Plant Cell Physiol.*, *21*, 1193–1204.
Siefermann-Harms, D., (1987). The light harvesting and protective function of carotenoids in photosynthetic membranes. *Physiologia Plantarum*, *69*, 561–568.
Singh, N., & Jothi, S., (1999). Reduction in protein contents in a few plants as indicator of air pollution. *Poll. Res.*, *18*, 281–283.
Singh, S. K., & Rao, D. N., (1983). Evaluation of plants for their tolerance to air pollution. In: *Proceedings of the Symposium on Air Pollution Control* (pp. 218–224).
Singh, S. N., & Verma, (2007). Phytoremediation of air pollutants: A review. In: Singh, S. N., & Tripathi, R. D., (eds.), *Environmental Bioremediation Technology* (Vol. 1, pp. 293–314). Springer, Berlin Heidelberg.
Sisodia, A., & Dutta, S., (2018). Air pollution tolerance index of certain plant species: A study of National Highway no-8, India. *Journal of Environmental Research and Development*, *10*(4), 723.
Slater, T. F., (1976). Biochemical pathology in microtime. In: Dianzani, M. U., et al. (eds.), *Recent Advances in Biochemical Pathology: Toxic Liver Injury* (Vol. 18, 388–390.). Minerva Medica: Torino, Italy.
Slater, T. F., (1984). Free radical mechanisms in tissue injury. *Biochem. J.*, *222*(1), 1–15.
Smirnoff, N., (1996). The function and metabolism of ascorbic acid inplants. *Annals of Botany*, *78*, 661–669.
Smirnoff, N., (2000). Ascorbic acid: Metabolism and functions of a multifaceted molecule. *Current Opinion in Plant Biology*, *3*, 229–235.
Smith, P. A., & Raven, J. A., (1979). Intercellular pH and its regulation. *Annu. Rev. Plant Physiol.*, *30*, 289–311.
Stewart, C. R., & Boggess, S. F., (1978). Metabolism of [5-h] proline by barley leaves and its use in measuring the effects of water stress on proline oxidation. *Plant Physiol.*, *61*(4), 654–657.
Tankha, K., & Gupta, R. K., (1992). Effect of water deficit and Sulfur dioxide on total soluble proteins, nitrate reductase activity and free proline content in sunflower leaves. *Biol. Plant*, *34*(3/4), 305–310.
Tausz, M., Van der Kooij, T. A. W., Muller, M., DeKok, L. J., & Grill, D., (1998). Uptake and metabolism of oxidized and reduced sulfur pollutants by spruce trees,' In: DeKok, L. J., & Stulen, I.. (eds.), *Responses of Plant Metabolism to Air Pollution and Global Change* (pp. 457–460). Backhuys Publishers, Leiden.
Tripathi, A. K., & Gautam, M., (2007). Biochemical parameters of plants as indicators of air pollution. *J. Environ. Biol.*, *28*, 127–132.
Tripathi, A., Tiwari, P. B., & Singh, D., (2009). Assessment of air pollution tolerance index of some trees in Moradabad city, India. *J. Environ Biol.*, *30*(4), 545–550.
Trivedi, M. L., & Singh, R. S., (1995). Reduction in protein contents in a few plants as indicators of air pollution. *Pollut. Res.*, *14*(3), 269–273.
Turk, R., & Wirth, V., (1975). The pH dependence of SO_2 damage to lichens. *Oecologia*, *19*, 285–291.
Tzvetkova, N., & Kolarov, D. C., (1996). Effect of air pollution on carbohydrate and nutrient concentration in some deciduous tree species. *Bulg. J. Plant Physiol.*, *22*, 53–63.

Verma, A., & Singh, S. N., (2006). Biochemical and ultra-structural changes in plant foliage exposed to auto- pollution. *Environmental Monitoring and Assessment, 120*, 585–602.

Wang, F., Zeng, B., Sun, Z., & Zhu, C., (2009). Relationship between proline and Hg2þ-induced oxidative stress in a tolerant rice mutant. *Arch. Environ. Contam. Toxicol., 56*, 723–731.

Woodward, A. J., & Bennett, I. J., (2005). The effect of salt stress and abscisic acid on proline production, chlorophyll content and growth of *in vitro* propagated shoots of *Eucalyptus camaldulensis*. *Plant Cell Tissue Organ Cult., 82*, 189–200.

Yu, S. W., (1988). Plant resistance to sulfur dioxide injury. In: *Perspectives in Environmental Botany* (Vol. 2, pp. 251–282). Today and Tomorrow's Printers and Publishers, New Delhi.

Zaefyzadeh, M., Quliyev, R. A., Babayeva, S. M., & Abbasov, M. A., (2009). The effect of the interaction between genotypes and drought stress on the superoxide dismutase and chlorophyll content in durum wheat landraces. *Turkish Journal of Biology, 33*, 1–7.

Zobel, A. M., (1996). Phenolic compounds against in defense air pollution. In: Yunus, M., & Iqbal, M., (eds.), *Plant Response to Air Pollution* (pp. 241–266). Chichester: John Wiley.

CHAPTER 6

The Uses and Benefits of Sludge and Ways to Protect Against Its Probable Harmful Effects

NAZIR HUSSAIN

Qatar Shell Research and Technology Center, and Ex-Expert Agriculture, Ministry of Environment, Doha, Qatar;
Present Address: Safari Villas 247, Bahria Town, Lahore, Pakistan,
Mobile: +923014117927, E-mail: drnazirhussain@gmail.com

ABSTRACT

An ever-increasing population of the globe, presently exhausted land and water resources, enhanced rate of urbanization, non-stopping changes in versatile lifestyle, innovations in science and technology and modern developments in industries demand a search for newer agricultural resources to meet requirements of the 21st century and the time ahead. Sludge is a heavily renewable source that can be very useful byproduct of the industry for the present and future. It has potential uses in many sectors; agriculture, rangelands, forestry, greening deserts and wastelands, construction industry, biofuel, and energy. It can successfully be used for reclamation of degraded soils, as manures and fertilizers, stabilization of sand dunes and deserts, growing of plantations on degraded lands and deserts, growing of nurseries, establishing rangelands and forests, production of cement, bricks, and grains for the construction industry, and production of biofuel and energy (electricity and biogas). Still, manifold uses are emerging with the development of technology. However, long-term use of sludge without observing regulations and any remediation has potential risks and threats like soil salinization, and piling of heavy metals and harmful microbes. For tackling potential problems and risks, strict enforcement of regulations and implementation and monitoring systems is highly necessary so that the probable harms to humans, animals, soils, and the environment are safeguarded. This will also help to make the use of sludge socially acceptable. The remedial measures and cares during transportation, storage, handling, and application are mandatory whenever sludge is planned to any probable use.

6.1 INTRODUCTION

Industrial wastes are unwanted waste materials but necessarily attached to almost all industries as integral components which cannot be avoided. These are generated from primary or secondary industrial processes and may be in the state of liquid, solids, slurry, or semisolids. As a result of various evolutions and blooming of industries during the past century, industrial wastes became a big problem and disposal of these materials on land and water resources have caused serious environment problems (Rattan et al., 2001, 2004). Resultantly, environmental consideration started strengthening and maximum allowable compositions were fixed, safe limits of various constituents were standardized, and regulations were issued and implemented regarding disposal of industrial wastes so that the environment be safeguarded and protected for the future generations. However, with further development of science and technology, shrinkage of resources, increasing of population and active search of alternatives, the wastes started converting into byproducts which could meet many needs of human beings. However, by, and by, limitations of resources are severing, human needs are spreading, and new technological innovations are emerging whereas the uses of industrial wastes are increasing simultaneously because new techniques and horizons are unlocking every day by current and future innovations.

Despite many in number, there are two broad categories of industrial wastes; treated effluent water and the biosolids/sludge. Both these are formed when industrial water is treated to amend some of its parameters for meeting critical environmental standards fixed by regulations. During the water treatment, chemicals are added to address suspended solids and Biodegradable Organics (e.g., BOD) which can introduce hidden problems like salts, and heavy metals (zinc, iron, copper, manganese, lead, cadmium, and cobalt) or threat of bacterial infection when treated water is used indiscriminately (Salem et al., 2000; Mohammad and Mazahareh, 2003; Al-Omranet al., 2012). Therefore, a vigilant and appropriate use is appreciable which is supported by the local research to match prevailing conditions.

There are tremendous possibilities of using industrial treated effluent to irrigate crops, fruit plantations, forest trees, fodder shrubs and grasses, landscape plants and reclamation of degraded soils, especially in dry regions of the world which are facing severe problems of irrigation water scarcity and still are under more threat of climate changes ahead. Such type of industrial water use can spare good quality water for growing costly cash or food crops and for the utilization in cities and industries, as reported by Abdelrahman et al. (2011). In many parts of the world, treated wastewater is now being used vastly for irrigation (Abd-Elfattah et al., 2002; Levine and Sanot, 2004). The rapid urbanization everywhere in the world and high population growth, results most of the places facing water deficit. Demands search for alternative sources of water while the treated wastewater is the only major supplemental source. This source can not only replace partially the use of fresh good quality water for irrigation of selected plants but can also help in meeting a part of nutrient requirements because of some macro and micronutrient' constituents in wastewater. The sludge can be further helpful in this regard (Usman et al., 2012) because of greater organic matter (OM) constituents (60–70%), N, P, K, and a few microelements. Thus, a significant saving in fertilizer application could be expected.

The use of sludge is not restricted to agriculture only but it may be utilized for land reclamation, re-vegetation of degraded lands, remediation of contaminated soils, mitigation of landfills, topsoil manufacturing, using iron-rich bio-solids for land development, landscaping/horticulture, feedstock as compost, vegetation, and enhancement of parks, sports fields and roadsides, potting, and mulch mixtures, garden fertilizer, sod production and erosion control, etc. The strict precautions required are observing the standard critical values, using the known techniques and continuous monitoring of chemical composition of the substrate and the products being harvested/manufactured. This all will help in winning the public trust which is the key to all biosolids' management and usage program. The necessary component is the transparency and involvement of all the stakeholders in the wastewater and sludge utilization plans. In this way, the shared understanding of facts could be safer, sustainable, and trustable and welcomed by the public but it needs hard work and the way may prove lengthier than it is normally expected.

For sharing the above possibilities, extending the know-how of the subject to the stakeholders and acquainting with the precautions and cares strictly required to follow when using treated industrial effluent and the sludge, the present review is being presented. Though, salient researches have been quoted here but the review paper does not claim to be exhaustive and still very important and useful research works could be left merely due to unavailability or easy access.

6.2 PRODUCTION OF SLUDGE

Sludge can be produced from three major sources: drinking water treatment plants, industrial wastewater and municipal wastewater. The municipal wastewater sludge can contain salts, night soils, washing detergents, night soils, pathogens, and several unidentified and unexpected constituents. The drinking water treatment plants sludge may contain mostly salts but free from contaminants as present in sewage waste. This type of sludge is mostly considered as nonhazardous waste and disposed of into landfill sites. Hidalgo et al. (2017) claimed that it will be wrong to associate sludge from drinking water treatment plants with sludge from wastewater treatment plants because this type of sludge is safer, but only limitation for reuse are its low OM and high silica contents. Industrial sludge is produced during treatment of industrial effluent. Every industry is using larger quantities of water in its various processes, and in turn producing treated industrial effluent. However, this effluent cannot be disposed of directly unless it is treated to contain various constituents matching the fixed standards by regulations. For this purpose, industrial effluent must be treated through waste treatment plants whereby sludge large quantities of sludge are produced as well.

The quantity of industrial sludge depends upon the type and size of industry as well as size of the treatment plant. The precise values of total production of sludge in a country are not mostly available because the reports often total up industrial sludge and sewage sludge. Moreover, some of the reports indicate dry figures while others mention wet values. Therefore, in several cases, total production estimates of different countries are not available accurately. For instance, Cimochowicz-Rybicka (2013) reported that Germany, UK, France, Spain, Italy, Poland, Netherlands, Hungry, Australia, Czech

Republic, and Switzerland were producing respectively 2049, 1771, 1087, 1065, 1056, 563, 353, 260, 254, 220, and 210 thousand ton/year sewage sludge. Industrial sludge is not included in these estimates. Out of these huge quantities; respective percentages used in agriculture are 53, 70, 73, 65, 44, 26, 0, 60, 38, 78, and 10%. Yang et al. (2015) reported that disposal of large quantities of sludge is becoming a huge problem China. On average, an annual growth of 13% from 2007 to 2013 has been recorded in total production of sludge and 6.25 million tons dry solids was produced in 2013 whereas sludge management is poor, and laws and regulations of sludge management are incomplete, rather sometimes unrealistic.

According to Lechtenberg (2018), MENA region is producing big amounts of municipal wastewater which is posing a serious problem because of high treatment costs and risk to environment. The treatment of Municipal wastewater is resulting in production of large quantities of sludge disposal of which is of major concern to authorities. Kuwait is producing around 250 tons of sludge daily in its 6 treatment plants while Tunisia generates around 1 million tons of sewage sludge annually. Current disposal is into landfills. Sewage sludge production is expected to upscale at higher rates in MENA due to increase in urban population and industrial sector growth. The sewage sludge production data for some other countries is presented in Figure 6.1 which indicate the problem of disposal as well as huge potential probable resource that can be put into highly beneficial uses if managed appropriately. It can be observed that as low as 120 metric tons sludge produced in Hungry while as huge as 6.5, 2.0 and 1.5 million metric tons in the USA, Japan/Germany, and the UK/the Netherlands, respectively. These large quantities can be put to highly useful usage solving disposal problem of sludge wastes as well as getting many side benefits. The use of sludge also depends upon the level of treatment given to the sludge material. The modern treatment plants pass preliminary sludge material through three stages: (a) Primary stage; (b) Secondary stage; and (c) Tertiary stage. Demirbas et al. (2017) also identified three stages of sludge treatment but named a little bit differently as- pretreatment, primary treatment, and secondary treatment. Basically, these treatments are applied to refine quality of effluent and remove almost all the suspended material and some of the dissolved constituents.

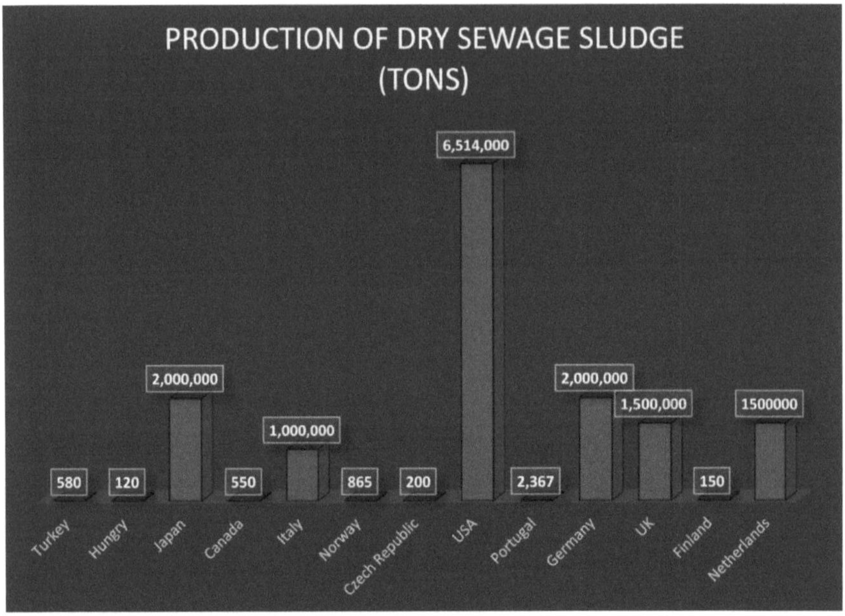

FIGURE 6.1 Production of sewage sludge in various countries (Data taken from UN-Habitat, 2008).

6.3 STANDARDS FIXED FOR UTILIZATION OF SLUDGE

Sludge from any source is containing a few or combination of constitutes like salts, OM residues, heavy metals, contaminants, and microbes the accumulation of which may prove harmful for humans, animals, soil, microbiological population and environment. During the last decades of the previous century, when sludge and wastewater were recognized as potential resources rather than wastes, these were applied indiscriminately without any care and management. However, long-term use of sludge started indicating harmful effects on crops, animals, and the environment as whole. The research urged fixing of different standards for constituents contained in sludge so that harmful effects can be controlled, and sustainable use of wastes becomes possible. However, the minimum values fixed for different salts, heavy metals/constituents vary greatly from location to location and country to country as well sources of sludge. Sometimes the objectives of sludge use is considered for fixing the limits while the quantities to be used will also influence the allowable limits. For example, there will be somewhat liberal limits if sludge is to put into restricted use but if it is to be used continuously, then, the limits will be more restricted and narrower. The sludge is regarded "A class" if it meets restricted and hard limits and its use for longer time is allowed. However, in any case, measures, and cares prescribed for sludge use must be observed for safer use and guarding harmful effects to human beings and other living beings associated with soil and production of crops like animal. The environment must be protected from pollutants as well (US Environmental Protection Agency (EPA), 2004). The limits range very widely for each heavy metal/constituent in different localities of the world even within the country. The national/local governments and international agencies promulgate rules and regulations and issue from time to time for matching the changes taking place in weather, soil, and technologies (EU, 2006; Hossni and El-Tarras, 1997; Hue and Ranjith, 1994; Kumamoto Municipal Government, 1983; Pescod, 1992; Raymond and Okieimen, 2011; U.S. EPA, 1984; WHO/FAO, 2007).

UN-Habitat (2008) reported the standard values adopted by selected countries based on their country reports which are presented in Table 6.1. For example, the fixed values for Arsenic (As) vary from 10 ppm (Russian Federation) to 170 ppm (Ontario) ppm while for Ba these range from 150 ppm (China) to 1300 ppm (Brazil). The values of Cd range from 1.25 (Netherlands) to 40 ppm in Jordan. Similar is the case of other heavy metals/constituents. Soil characteristics and climatic factors play a major role in fixing these criteria because if the soil is sandy with good drainage or the rainfall annual averages are higher, then somewhat liberal standards can be adopted. The site-specific conditions will be playing the major role when such criteria are devised for a certain locality. The local research should support standardization of such values.

6.4 ORGANIC MATTER (OM) AND NUTRIENTS CONTENT IN SLUDGE

Although the sludge varies highly in its constitutions, due to the production source, type of industry, nature of wastewater treatment plant and the stage of treatment imposed, but major common point is very high OM content which normally vary from 65–75% and resultantly the high N content (3.5%). Despite individual site variations, the benchmark analysis suggested by UN-Habitat (2008) is presented in Table 6.2. It is also high in Zn (1000mg/kg), Cu (500mg/kg) and P_2O_5 (3.5%) content whereas K_2O is 0.2% on the average. The composition of sludge is variable in different

TABLE 6.1 Limits of Heavy Metals (ppm) in Sludge Imposed by Various Agencies for Agricultural Use in Different Countries (UN-Habitat, 2008)

Region/ Country	Regulatory Act/ Year	As	Ba	Cd	Co	Cr	Cu	Pb	Hg	Mo	Ni	Se	Zn
South & Central America	Brazil	41	1300	39		1000	1500	300	17	50		100	2800
	Mexico (established 2002)	41		39		1200	1500	300	17		420		2800
North America–USA	U. S. EPA best quality	41		39			1500	300	17	75	420	100	2800
North America, Canada	Ontario, Canada	170		34	340	2800	1700	1100	11		420		4200
	British Columbia	75		20	150	1060	2200	500	5	20	180	14	1850
Europe	European Directiveregulatory limits			20–40			1000 to 1750	750 to 1200	16–25		300to 400		2500 to 4000
	Czech Republic	30		5		200	500	200	4		100		2500
	Germany			10		900	800	900	8		200		2500
	Netherlands	15		1.25		75	75	100	0.75		30		300
	Slovakia	20		10		1000	1000	750	10		300		2500
	Italy			20			1000	750	10		300		2500
Asia	China	75/ 75	150/ 150	20/ 5		1000/600	1500/800	1000/300	15/5		200/ 100		3000/ 2000
	Japan	50		5		500		100	2		300		
	Russian Federation	10		15		500	750	250	7.5		200		1750
Middle East	Jordan	41		40		900	1500	300	17	75	300	100	2800

parts of the world (Table 6.3, UN-Habitat, 2008). Thus, sludge contains major and minor elements of plant nutrition and there is a significant possibility of saving fertilizer application costs partially. The OM content has many benefits in improving soil characteristics and enhancing fertility (Basta, 1995; Chamber et al., 2003; Cogger, 2005). Singh and Agrawal (2008) claimed that because sludge is rich in organic and inorganic plant nutrients, so it can substitute fertilizer application, and may amend the soil physical, chemical, and biological properties. Crop yields are usually adequate in sludge-applied soils while bioavailability of metals increases. The OM can also be big source of energy if incineration is adopted. According to Thomas Nutrient Solutions (2018), incineration is the process of burn of wastes to produce energy. The large treatment plants can use large quantities of as sources of energy for producing steam used as combined heat and power (electricity generation).

TABLE 6.2 Analysis of Benchmark Sludge (W/W) (UN-Habitat, 2008)

Constituents	Values
Organic Matter	75%
Zinc	1000 mg/kg
Copper	500 mg/kg
Nickel	140 mg/kg
Mercury	3 mg/kg
Cadmium	3 mg/kg
Lead	200 mg/kg
Total Nitrogen	3.5%
P_20_5	3.5%
K_2O	0.2%

TABLE 6.3 Major Nutrients in Representative Wastewater Sludge (UN-Habitat, 2008)

Countries	N (%)	P_2O_5 (%)	K_20 (%)
The sludge indicator sample	3.50	3.50	0.20
Australia	7.40	1.80	0.97
Brazil	5.75	1.82	0.36
Canada	2.10	0.50	0.10
Italy	5.20	1.40	–
Turkey	1.68	0.68	0.49
USA	5.80	4.35	0.43

6.5 USES OF SLUDGE

Wastewater sludge and industrial sludge will remain produced as far as the populations are there and industries are running. Therefore, these must be managed because just piling in empty spaces or use in landfills cannot continue indefinitely, keeping in view the quantities being generated. Communities do not want to see sludge staked everywhere as well as it will not be friendly to environment at all. There are, then, only three options for managing sludge:

1. Thermal processing called incineration and subsequently disposing as very less volume-reduced-ash;
2. Recycling into soils, which is presently dominating; and
3. Putting sludge to other uses like construction industries.

For example, in the UK, 159,000 tons sludge is produced out of which 67% is recycled in agricultural lands while 5.2% is used in non-agricultural lands. Disposal to landfills is 1.5% only, 1.8% is put to other uses whereas 19.5% goes to incineration and other thermal destruction processes (Matthews, 2006). Even though the harmful effects may be very lesser, these must be safeguarded. Currently, Sludge has been overwhelmingly put into agricultural uses in most of the countries. Yu-Chi et al. (2012) reported that disposal of sludge because of constraints of landfill space and environmental concern is becoming costly and a huge challenge in Taiwan. They suggested, after performing slump test, that sludge can be used as concrete mix which is a viable and beneficial alternative of this solid waste. Moffat (2006) reported that woodland establishment and subsequent performance can be improved by application of organic materials (composts and biosolids) but depends upon correctly matching material to site type.

6.5.1 AGRICULTURAL USES

The use of sludges started in agriculture which is still a major sector for their usage mainly for rehabilitation of lands, reclamation of salt affected soils, increasing soil fertility and as source of nutrients for growing plants. Sludge is usually an organic rich source, so, when applied and incorporated into the soil surface, it gives all the benefits as of an OM carrier. The soil parameters like water holding capacity, cation exchange capacity and nutrient retention capacity are enhanced due to granulation and improvement of soil structure. The evaporation and leaching of water can be controlled resulting in saving of costly irrigation water. The soil physical characteristics like bulk density and porosity can also be improved/restored. The biological, chemical, and physical, soil characteristics can be improved because of amending effect of the sludge. The yields of crops have been reported to increase due to application of sludge majorly because of increased availability of macro and micronutrients as well as a significant improvement in soil properties. Authors like Singh and Agrawal (2008); Lindsay, and Lagon (1998); Chamber et al. (2003); Gracia-Orenes et al. (2005) and Cogger (2005) expressed their views in favor of these concepts, based upon their respective studies.

Sludge proves especially useful in salt-affected soils because it can help in leaching of accumulated salts resulting in reclamation of salt affected lands that have gone out of cultivation temporarily due to excess of salts. The Ca^{2+} and Mg^{2+} ions contained in sludge or H^{+} ion released during OM decomposition can lower soil pH and release excessive Na^{+} ions from the soil exchange complex, thus behaving like an amendment and ultimately reclaiming sodic soils. Study of Hussain et al. (2006) indicated that combination of treated sludge with gypsum or sulfuric acid proved most successful in rehabilitation of sodic soil (EC = 2.56 d S/m SAR = 38.89 and pH = 10.1 Texture = Sandy clay loam) because SAR and pH decreased significantly and were brought within permissible limits after growing of two crops subsequently. The yield of rice and wheat also increased significantly. Treated sludge alone proved superior to gypsum alone due to plant nutrients contained in it but both were statistically better than control (where no amendment was added). Some of the soil fertility parameters (OM, phosphorus & potassium) were also improved appreciably as compared to control.

As documented earlier (see Tables 6.2 and 6.3), sludge is constituted of many elements (N, P, K, Ca, Mg, Zn, Cu) involved in plant nutrition. The decomposition of OM present in sludge make available these nutrients to plants in addition to those which are already present in the soil but converted to available forms because of the dissolution action of decomposing OM and thus, increasing their availability. For example, research of Mbagwu and Piccolo (1990) proved that when sludge was applied (200 t ha^{-1}), the total nitrogen and available P of soil was increased by 57% and 64.2% respectively. However, these nutrients were released slowly as compared to inorganic fertilizer, therefore, can remain available for prolonged periods and can reduce losses (Ayers et al., 2008). The form of OM in sludge is soluble matter with comparatively lesser portion of lignin and cellulose. Therefore, sludge OM can mineralize more briskly. Hence, sludge supplementation can serve as sources of nutrients and crop fertilization, which can save cost of chemical fertilizer application. Global use of sludge as nutrients source is no more a mystery. The ultimate consequence can be reduction of the cost of crop production and increase in net income of farmers and agricultural industry. Sludge, as an organic fertilizer, is being used for crop husbandry in many developed and developing countries of the world. Jamil et al. (2006) applied 20 ton of sewage sludge combined with half recommended dose of NPK and obtained maximum wheat grains. Above this level, a declining trend was observed for straw and grain yields. In research studies of Manguiat (1997), in Philippine, rice, and wheat yields were increased due to sewage sludge application alone or with N fertilizer. Aerobically digested sludge increased maize biomass whereas undigested sludge revealed reductions in Hawaii, Hue, and Ranjith (1994) obtained increased maize yields in Hawaii, when digested sludge was applied whereas undigested indicated the reduced yields.

Wind and water soil erosion are the significant problems in land degradation. The sludge application can help in reducing these problems and can prove conducive for soil conservation, especially sandy soils eroded during wind erosion or loamy and clayey soils highly eroded during intensive rainfall or very high-intensity floods in monsoon season. The looseness of soil, a basic cause for erosion, can be decreased when OM is incorporated into the soil, even maybe from sludge. The applied organic matter enhances soil coagulation and reduces erosion because chances of eroding comparatively bigger and relatively heavier soil grains are far lesser compared to the loose and lighter particles. Dry soil is more prone to wind erosion compared with wet or moistened soil. The increase in water holding capacity because of OM reduces the probability of soil becoming rapidly dry after irrigations, especially in arid and semiarid regions and desert lands which are mostly sands. The frequent sand storms in these areas are mainly due to dryness and looseness of soils. The application of sludge can play a significant role in reducing sand storms and resultant soil erosion. A study in Spain by Marques et al. (2014) indicated that sludge application at the rate of 20 t ha^{-1} caused a significant reduction in sediments removal from soil, the values were 12.55 mg ml^{-1} m^{-2} in control plots and 2.33 mg ml^{-1}

m^{-2} in sludge applied plots. Thus, there is a clear-cut role of sludge in decreasing potential erosion of lands.

6.5.2 REMEDIATION OF CONTAMINATED SOILS

Soil contamination may occur due to heavy metals, oil, insecticides, and toxins, etc. accumulation beyond the permissible limits that may prove injurious for plants, animals, humans, and other living ecosystems. For example, when wastes from tanneries are disposed of consistently, it will contaminate the costly lands (maybe agricultural or urban). Chromium (Cr), a toxic heavy metal, accumulates in such situation which may hinder plant growth and the land becomes just barren. The oil and other hydrocarbon fumes contaminate the nearby lands of oil refineries. Disposal of wastes of pesticide chemicals near factories or heavy and consistent sprays to kill the crop pests may also contaminate lands to the levels that do not permit growing of plants in lands that have been ever very fertile and productive. The other possible sources of contamination may be wastewater use, metal mining, and milling processes and airborne sources (Raymond and Okieimen, 2011). Remediation of such lands is not easy. Though, remediation may prove; costly, lengthy, and difficult, but it is not impossible. Contaminated soils may prove highly difficult and challenging to remediate, but the soils free from any contamination will become an asset of agriculture, increase production and significantly decrease problems and threats to the environment.

The strategy applied for rehabilitation of contaminated lands is either binding of these contaminants so that these are not available to plants and other living beings or solubilizing and leaching from the soil. The sludge can perform usefully in this regard (McLaren et al., 2005). The sludge can perform both the functions because OM contained in it can bind contaminants by forming complex compounds of long organic chains or it can form chelates that are highly soluble and can be compelled to leave the contaminated soils. For example, Navarro (2012) made efforts to remediate the contaminated soils which were having Copper 250μg/L, Lead 80μg/L and Antimony (Sb) 6μg/L determined in the leachate. However, after restoration by application of sludge, the respective determined values decreased and were found as Cu 20μg/L, Pb 40μg/L and Sb 4μg/L. Thus, there was clear sorption of contaminants by the OM applied in the form of sludge. The author suggested that Pb could form a precipitate of chloropyromorphite while Sb may be precipitated as Sb_2O_3, thus causing partial retention by OM of sludge. Metal and metalloid concentrations such as Fe, Mn, Ni, As, and Se increased in the leachates, which suggested that the organic constituents of sludge could cause solubilization of heavy metals and consequently amending the contaminated soils. The quantities of sludge to be applied are of high importance, considering all factors; climate, soil, and the gravity of the problem, because higher amounts over those allowed under regulations for the area even may become source of further contamination itself. Therefore, all cares must be taken, and local research and regulations must be consulted before moving in this direction.

6.5.3 LANDSCAPING/HORTICULTURE

Sludge can also be used for landscaping and growing of horticultural plants as well as flowers. However, the objective here is different as in the case of agricultural uses. The sludge is used as soil conditioner here and not as organic fertilizer. However, sludge is mostly composted and applied for landscaping and horticultural purposes which can improve soil characteristics and supplying plant nutrients. Generally, the dried sludge or sludge compost is available as small packed bags which may be used to prepare a mixture of clay/soil, sand, and sludge/compost to fill in the pits dug for sowing/transplanting of fruit plants. The sludge/compost is spread over beds for growing of grasses and flowers in fields and greenhouses. The overall objectives are to improve the physical conditions of growing medium, manageability, water retention, aeration, root penetration, and avoiding water and nutrient losses. Such conditions also help in seasonal hardness, especially keeping the growing medium comparatively cooler by maintaining moisture for longer periods. Using sludge/compost is equally beneficial in sandy as well as clayey soils because it will normalize extreme conditions in both the cases. Sludge is highly useful for growing grass on golf courses. Establishment and growth of grasses have been reported to improve significantly after spreading of sludge on the surface of the soil and subsequent incorporation.

Stabnikova et al. (2005) reported from their studies conducted in Singapore that application of compost and sludge into subsoil proves highly helpful to improve physical properties of the soil, nutritional supplying capacity and growth of horticultural plants. The combined application of compost and sludge assessed as more useful. The biomass yield of *Ipomoea aquatica* highly increased when grown in artificially prepared soil by mixing of 4% of horticultural compost, 2% of sewage sludge and remaining (94%) of subsoil each by weight. The heavy metal uptake by plants was determined lower than the levels declared toxic by standards. The conclusion of their study was that artificially prepared soil by the above composition may be used safely for landscape and growing of garden plants.

6.5.4 CONVERSION INTO COMPOST

Treated sludge is being used vastly for composting in many countries of the world. During the composting of sludge, some organic materials like leaves, bark, and straw of plants, residual material, or green waste and kitchen waste with the sludge and anaerobic composting is permitted in composting plants. These supplementary materials serve as bulking agents during composting of dewatered sludge (35%DS). Alexander et al. (2002) regarded this technique highly useful. Due to usefulness, the technology has found vast use in the UK. Such type of compost is claimed to be useful for applications in agriculture and horticulture industry for amending the soils. Thus, sludge waste is ultimately converted into a useful product called compost. During this aerobic decomposition, the volume of waste decreases, pathogens are reduced (Herrmann and Shann, 1997), if not finished in total. A compost is more stable product having a C/N ratio varying from 30:1 to 20:1 (the ideal one). The composting process takes 4–6 weeks. There are many methods of composting like; pile method, heap method and oxygenic composting in reactors (vertical reactor- silo and horizontal reactor- tunnel reactor) as presented by Kosobucki et al. (2000). The compost from sludge can be used as manure, reclaimants of degraded soils and media for growing of nurseries, grasses, landscaping ornamental plants, flowers, etc.

6.5.5 RANGELAND AND FORESTRY

Rangelands and forestry are mostly practiced in lands which are not very fertile and not categorized under suitability class S1 (highly suitable) for growing normal crops and fruit plants. Lands with some problems or stresses are opted to grow range and pasture plants/grasses or forest plants. Sludge can be used effectively as improver of such hard sites for conversion into rangeland and pastures because it contains OM and nutrients. Such marginal lands usually give comparatively lesser returns; therefore, costly manures and fertilizer proves uneconomical to use for rangeland plants. Sludge is available as alternative cheaper source. Rangeland trees, shrubs, and grasses can easily be grown on comparatively hard sites using sludge as soil conditioner and source of nutrients. Greening of deserts by growing forest plants is another useful area of sludge utilization. Instead of disposing sludge into desert areas it can be used systematically for forest fertilization and saving money which is spent on the purchase of conventional manures and fertilizers. Sludge is a good source for growing fiber crops and regenerating of forests. Forest soils are particularly suitable for sludge application because their infiltration rate is high which can help a lot in reducing runoff and ponding of water. Forest soils also have high OM content which has a capacity to immobilize the sludge metals. The perennial root systems of forest trees permit application of sludge whole the year-round (U.S. EPA, 1984). The forest products like mushrooms, wild edible berries, gums, and nuts are taken in minor quantities, therefore, are not significant part of the human food chain. This is an additional benefit and creates chances of sludge application to forest trees. The vast forest lands are extensively spread in many areas and regions, providing good and substantial forest crop-lands for sludge application.

A study of Aguilar et al. (1994) during 1985 to 1989, indicated that when sludge was applied once on the surface of the forest land (22.5 to 45 Mg ha^{-1}), the production of plants and the cover on the land increased significantly. No evidence of accumulation of potentially hazardous elements or heavy metals in soils or plant parts above the permissible values was recorded. Their second research study investigated the hydrological parameters' changes in forest soils after supplementation of sludge (45 Mg ha^{-1} as surface broadcast). The slope gradients of the land were 6% and 10 to 11%. The surface roughness and water absorption were increased by applied sludge which caused decrease in water runoff notably. The increasing of plant cover over on the forest land with passing of time, and root penetration improved infiltration of water which caused further reduction in runoff rain or irrigation water. Kaposts et al. (2000) reported from two five-years studies; one on coastal sand dunes and the other on inland sandy soils that sludge application (3–4 kg per planting spot) increased survival rate of pine plants by 72% while in control, plants died for want of water availability. The sludge fertilizer positively affected not only the survival but also the increment in the height of the young trees. Depending on climatic conditions, the increment was 2–3 times higher as compared with the control. The conclusions of the research of Basta (1995) regarding the use of sludge were that the growth of forest trees and associated vegetation increased on infertile lands, especially the stand of young trees.

Basta (1995) conducted several experiments at Oklahoma Agricultural Experiment Station and Center for Agricultural and the Environment to find effects of soil-applied sludge on forage crops and the forest soils. His findings were that risks from constituents and microbiological population of sludge are safeguarded by barriers created by the natural soil-plant-animal response and organic compounds present in the sludge. He reported these problems as manageable and finally concluded that sludge could be a beneficial source of nutrients and OM for soil applications. The grazing animals on forages or fed on grains which were grown on sludge-amended soils remained safe and healthy and their milk and meat products met standards permitted for consumption of human beings.

The N, P, and K nutrient at the rates of 0-0-0, 180-78-20, 360-156-40 and 540-234-60 kg ha^{-1} from chemical fertilizer and sludge were applied to Bermuda grass (Cynodondactylon) sod in the research investigations after completion of which Robert (1988) reported that mean production of dry matter production was non-significant with respect to nutrient source at the end of a season. The higher applied quantities of nutrients caused significantly increased fodder mass. The forage from sludge treatment compared with chemical fertilizer was having more crude protein content at the 360–156–40 (NPK) rate. There was no metal concentration posing a threat to grazing animals in any case. Sigua et al. (2005) suggested repeated applications of sludge useful and safe for perennial ryegrass (BG) from their experimentation in the USA which revealed no harmful effects on soil and forage quality. The good and economical practice, they found, was regular land application of sludge for at least three years followed by no sludge for at least two years. This practice can boost or maintain forage productivity on sustainable basis. This could cause minimum accumulation of heavy metals in the soil.

6.5.6 USE IN INDUSTRIAL PROCESSES

Many byproducts are being produced through various industrial processes using sludge as raw material. These include; bricks and other building material, glass aggregates used in tiles and pavements and daily landfill cover. For example, Marble Processing Industry in India was generating about 7 million tons of waste material as powder during various industrial processes (mainly sawing and polishing) which were being piled in the open fields as landfills. However, this waste material, in the form of powder, was polluting and damaging the environment, rather raising as big and serious pollution issue in the 16 districts of the state of Rajasthan because almost 4000 marble mines and 1100 marble cutters were working under the SME sector. The conducted research opened the doors of using marble sludge in building industry as various materials (fired clay brick, and concrete products) for constructions of roofs, walls, and floors. Hence, the developed technology turned a 'waste' into 'wealthy product' by using marble sludge in brick-making (TARA, 2018). Using sludge (15–25%), clay, and shale mixture, a bio-brick was developed by Alleman and Berman (1984) which, according to Chin et al. (1998), looked, and have feel and smell like regular and conventional bricks. Later, a novel brick was developed by Tay et al. (2002) who used conventional clay and dried sludge from wastewater. By their suggestion, sludge might make 0–40% of brick weight beyond which compressive strength may decrease. However, uneven texture surface and porosity could arise due to the burning off the OM of sludge during the firing process. To solve this problem, sludge was replaced by dry sludge ash. The assessment showed that 10% of sludge ash constitution of bricks has same strength as of normal bricks made of only clay. Chihpin et al. (2005) reported the sintering temperature required for sludge containing bricks was higher than what the Brick Industry uses normally in brick works because of higher Al_2O_3 and lower SiO_2 content in mixed bricks. The parameters like specific gravity, water absorption and compressive strength of the mixed ceramic bricks met the limits fixed for testing and thus, confirmed their feasibility in construction industry. Milica et al. (2012) emphasized that sludge can be used to produce eco-friendly bricks for using in masonry industry.

Sewage sludge can also be used to produce aggregates for construction industry. Chou et al. (2006) used a combination of sludge and dried ash of sludge ash and produced aggregates synthetically. The results of their study indicated that normal-weight aggregate could be produced when sewage sludge ash and clay were combined. However, a mixture of 20–30% of sewage sludge and the rest of clay can be advantageous if lightweight aggregates are targeted. Sewage sludge has also been investigated for preparing cement-like material and partial replacement in the construction activities. Some researchers produced successfully cement-like specimens from a mixture of dried sludge and limestone powder. For instance, Tay et al. (2002) used digested and dewatered sludge in the production of cement-like material which was grounded and incinerated subsequently. Their test revealed that sludge cement was demanding high water quantity and has quick setting time in comparison to ordinary cement. The mortar cube strength assessment test clarified that most possibly masonry binder made of cement sludge might be produced that could meet the strength requirements of the ASTM standard of cement being used for masonry purposes currently. However, still more research is required for assessing long-term parameters like durability before the sludge cement is accepted for masonry purposes. Clay was replaced partly by sludge to produce a ceramic body by Martínez-García et al. (2012). The tests of mechanical properties of the products showed that ceramic bodies having 5% sludge were possessing good values of water absorption, compressive strength and water suction, making favorable of the preparation of sludge cement ceramic products.

6.5.7 USE IN ENERGY/BIOFUEL PRODUCTION

Sludge can also be used for energy and biofuel production. Biogas can be produced from the digestion of sludge. Heat and electricity can also be produced by incineration of this byproduct. It can be a dry product supplement for coal. Gasification, pyrolysis, and hi-tech energy development techniques can be used to produce energy from sludge. The production of energy is possible by burning/digestion/incineration of organic compounds present in the sludge. Rulkens (2008) identified the following nine groups of different processes/options employed for energy/biofuel delivery from sludge:

1. Anaerobic digestion of sludge;
2. Production of biofuels from sludge;
3. Direct production of electricity from sludge in microbial fuel cells;
4. Incineration of sludge with energy recovery;
5. Co-incineration of sludge in coal-fired power plants;
6. Gasification and pyrolysis of sludge;
7. Supercritical wet oxidation of sewage sludge;

8. Use of sludge as an energy and raw material source in the production of Portland cement and building materials; and
9. Hydrothermal treatment of sewage sludge.

According to Rulkens (2008), some of these options are already in practice in different industries whereas the rest are still under research studies. Cassiday (2012) reported a research study of Eilhann et al. (2012) conducted in South Korea regarding a thermochemical process in which lipids contained in sewage sludge can be converted into biodiesel. According to these authors, this new source of biofuel is economically feasible due to its low cost of preparation and very high quantities of production of the sludge in industrial processes and municipal wastes. At present, biofuel is produced from vegetable oils or animal fats which provide a mixture of fatty acid methyl esters for this industrial process. The biodiesel preliminarily produced from such procedures has been regarded as compatible with present diesel engines. Burning of biodiesel produces comparatively less pollution to the environment than the diesel derived from raw petroleum. However, the high cost of biodiesel production has limited its widespread usage and popularity because the cost of biodiesel is in proportion to the price of source raw materials (such as refined soybean or rapeseed oil) used in producing biodiesel. Lipids, the source material for biodiesel, are contained in sample quantities in the sludge. The team of Eilhann et al. (2012) used *n*-hexane for extracting lipids from sludge converted into pellets from a wastewater treatment plant. The results of their study indicated that the sludge produced 2,200 times more lipids per gram of feedstock when compared with yields of lipids from soybeans. According to authors of this work, sewage sludge is also highly cheaper lipid source than soybeans and other organic sources coming from agricultural crops. Prices calculated at the time of study (2012) revealed that each liter of lipids extracted from sludge costed $0.03, that from soybeans costed $0.80 per liter.

Stolarek and Ledakowicz (2001) identified two processes; pyrolysis and gasification, for conversion of sludge into combustible gases and energy. In Pyrolysis, a thermal process, sludge is heated under pressure and in the absence of oxygen to a temperature of 350–500°C. Thus, sludge is converted into char, ash, oils, water vapors, gases. Some part of the solid and gaseous products of the pyrolysis process is used as heating energy in the pyrolysis process itself. Gasification comprises of breaking down of dried sludge (or biomass) into ash and combustible gases at temperatures at 1000°C with a decreased quantity of oxygen. However, some new thermal techniques have been identified which combine pyrolysis and gasification (Jaeger and Mayer, 2000). But, this subject is under a lot of research (The pyrolysis and gasification of the biomass). These two processes sewage sludge have some advantages when compared to incineration; the conversion of the combustible gases of both systems into electrical power more efficiently while valuable gases can be saved as chemicals and fuel.

There are two processes viz., pyrolysis, and gasification, for conversion of sludge into combustible gases and energy. Pyrolysis is a thermal treatment process in which the sludge is heated under pressure to a temperature of 350–500°C in the absence of oxygen. In this process, the sludge is converted into char, ash, pyrolysis oils, water vapor, and combustible gases. Part of the solid and/or gaseous products of the pyrolysis process are incinerated and used as heating energy in the pyrolysis process itself (Stolarek and Ledakowicz, 2001). Gasification involves the breakdown of dried sludge (or biomass) in ash and combustible gases at temperatures usually about 1000°C in an atmosphere with a reduced amount of oxygen. However, thermal treatment techniques also exist that combine the specific aspects of both pyrolysis and gasification (Jaeger and Mayer, 2000). A lot of research is going on regarding the pyrolysis and gasification of the biomass. Pyrolysis and gasification of sewage sludge have some potential advantages compared to incineration. One advantage is that the conversion of the combustible gases of both systems into electrical power can be achieved more efficiently. In addition, valuable gases can be produced as basic chemicals or as fuel.

Incineration is a modern process aiming at a complete oxidation of the organic compounds of sludge at high temperature, including the toxic organic compounds (Braguglia et al., 2003; Guibelin, 2004). Some types of sludge cannot be put to any use due to various technical and social issues and just burned in incineration process at very high temperature. The sludge must be dried or mechanically dewatered before incineration. However, the emissions of pollutants with the evolution of gases to the atmosphere and quality of the ashes are some potential environmental problems attached to sludge incineration. But, a lot of new technologies available to cope the problem of gaseous emissions very efficiently. The ash quality (heavy metals in the ash) is not a real environmental problem because of the high temperatures applied in the incineration process and the composition of the inorganic compounds in the sludge. The heavy metals are very well-immobilized and become resistant to leaching. This ash must be disposed of or used as a source to produce building materials. The sludge incineration is expensive due to high costs of gas treatment system. Currently, sludge incineration processes are focusing on the recovery of energy and heat from sludge, as steam and electricity (Guibelin, 2004; Simonsen and Bruus, 2003). The amount of energy obtained depends upon the water content of the sludge and the performance of the incineration processes. Incineration of sludge is now being applied worldwide on a large scale.

6.6 CARES AND TECHNIQUES REQUIRED FOR UTILIZATION OF SLUDGE

Widely variable uses of sludge have been discussed in earlier sections, but the question is still there that sludge cannot be used like a blank bank cheque. It has many hidden threats that must be looked at and dealt with, to keep the land and water resources unpolluted, human, and animals health guarded from harmful effects and make the uses socially acceptable. It is also important to use and manage sludge applications in such a manner that encourages sustainable use and making long term use possible. To obtain this target, following cares must be adapted when sludge of various types is going to put into any use, especially relevant to human and animals as well as land and water resources.

6.6.1 STRICTLY PRACTICING REGULATIONS (INTERNATIONAL AND LOCAL)

Because of research conducted in various sectors relevant to the finding use of sludge, certain regulations; both global and local, have been imposed. These are guarantees to guard human and animal health, land, and water resources and the environment. Some local modifications of global regulations from country to country have been made, keeping in view the locally prevailing conditions of land, water, climate, and social customs. These regulations must be abided by fully and observed strictly while practicing use of any type of sludge. At no cost, deviations should neither be allowed nor practiced. This is the only way for making sludge use sustainable without any danger, threat, harm or a hidden loss. In the countries where no such regulations have still constituted and implemented, appropriate regulations should immediately be drafted, in the light of international code with local modifications if required and implemented. It is very important to mention that mere issuance of regulations is not enough but its implementation in the letter and spirit is highly essential in developing countries. Such actions will help in utilizations of resources appropriately on the one hand, and levy guarantees, on the other hand, so that threats do not exist and subsequent incidents may not occur.

6.6.2 REGULAR MONITORING OF SLUDGE QUALITY AND COMPARISON WITH STANDARD PARAMETERS

A double system of quality assessment of sludge should be imposed. There should be regular and continuous sampling and analysis of each batch of sludge preparation by the treating and preparation agencies. So, this analysis will be at source. Then, whenever a user plans to use certain quantity of sludge for a specified utilization, he should get a sample of the proposed bulk and get it analyzed from national agencies while providing information of the use of sludge being planned. After getting analysis and report, the user should satisfy him after discussion with the national lab personnel that the sample meets the standards fixed for the use under consideration and it is safe to use the sampled sludge. The adaptation of this type of procedure will safeguard the possible harms and threats as well as satisfy the sludge users in all respects.

6.6.3 CARES AND PREPARATIONS DURING TRANSPORTATION AND APPLICATION OF SLUDGE

The transportation, storage, and application are the most delicate steps of the whole process starting from sludge preparation to monitoring because here the sludge will be handled by non-professionals who are mostly less educated and not trained. Therefore, all cares must be undertaken and communicated pre-hand by professionals to non-professional involved in this mechanism. Some of the general cares are:

- Sludge should never be touched and handled without gloves.
- When loading, unloading, and applying, all persons should stand pro-direction of wind but behind the application front even if sludge is being applied through machines.

- They should always bear glasses/spectacles/ goggles covering delicate portions of eyes.
- They should always use the masks for inhaling.
- The hands and bodies of the persons should be washed well when the sludge touches apart of the body as well as after completion of the process of sludge application, even if machines were used for this purpose.
- Never store the sludge in moist places that are not safe from rains and other water sources.
- The storage should remain locked and not remain in the access of animals and irrelevant persons.
- The sludge should be uniformly applied avoiding over application at certain points of the fields.
- The quantity of application should never be applied more than that permitted under regulations.
- Food or feed crops nearby the location, especially pro-wind direction should not be consumed immediately after sludge application. These should be washed well after harvesting/ plucking before use.
- A time should be provided for decomposition of sludge in the soil after application and mixing of sludge and seeds/seedlings should not be planted immediately after application of sludge.

6.6.4 REGULAR MONITORING OF FOOD AND FEED QUALITY

The major thinking of communities about use of sludge is suspicious quality food and feed and contamination of these, especially with heavy metals and pathogens. To have a regular check on quality changes and contamination as well as satisfaction of the end consumers of commodities grown after sludge application, a regular monitoring system should be established by responsible agencies. A complete record of monitoring should be kept and made readily available whenever any person wants to investigate it. Such a mechanism is in the interest of all the stakeholders.

6.6.5 MONITORING OF SOIL REGULARLY

Soil is the seat where sludge is mostly applied during agricultural uses; as manure/fertilizer/soil conditioner. Thus, this is medium directly receiving sludge and will be containing its components even after decomposition. The non-consumed ingredients not undertaken by plants are going to sustain in the soil for longer time. If sludge is being applied after regular intervals, its cumulative effect will be there definitely, and the remaining constituents are going to accumulate slowly and steadily. Therefore, it is highly imperative that a regular monitoring system of soil assessment regarding sludge constituents including those that are expected after decomposition, should be devised. There should be periodic soil analysis for soil EC, pH, SAR, heavy metals and pathogens. Whenever, any contamination or values higher than the permitted limit are found, growing of crops should be abandoned, remedial measure taken and use of sludge should be stopped for many years to come.

6.6.6 MONITORING OF GROUND WATER QUALITY

Another important consideration is regarding groundwater quality. It may be affected if a significant leaching of heavy metals and pathogens occurs. This may happen when the texture of the soil is light, even in lower profiles, and there is no hard or clay layer in between aquifers and the soil surface/subsurface. Deeper and frequent irrigations can also support leaching of salts, heavy metals and pathogens if clay and OM content in the soil are low because no binding places exist which can check leaching of these materials. Application of sludge itself may help to increase OM content in the soil, help in binding against leaching and decrease chances of leaching of heavy metals and pathogens. Nearby water bodies, if any, may also be affected possibly due to lateral seepage or drainage. Therefore, a continuous monitoring of groundwater quality as well as quality of water in nearby water bodies, if present, will highly be desirable. In this a way, an eye may be kept for overcoming such probabilities.

6.7 CONCLUSIONS

1. Sludge is a renewable source that can be very useful byproduct of industry in current and future times. It has potential uses in many sectors; agriculture, rangelands, forestry, and greening deserts and wastelands, construction industry, biofuel, and energy.

2. It can successfully be used for reclamation of degraded soils, as manures and fertilizers, stabilization of sand dunes and deserts, greening of wastelands/deserts, growing of nurseries, establishing rangelands and forests, production of cement, bricks, and grains for construction industry, and production of biofuel and energy (electricity and biogas). Still manifold uses are emerging with development of technology.
3. Long term use of sludge without observing regulations and any remediation has potential risks and threats like accumulation of salts, heavy metals and pathogens, if values are higher than those fixed.
4. Strict enforcement of regulations, implementation, and monitoring systems are required to safeguard the probable harms to humans, animals, soils, and the environment as well as making the use of sludge socially acceptable.
5. The remedial measures and cares are mandatory whenever sludge is put to any probable use.

KEYWORDS

- **agriculture**
- **cares and management**
- **energy source**
- **sewage sludge**
- **sludge utilization**

REFERENCES

Abd-Elfattah, A., Shehata, S. M., & Talab, A. S., (2002). Evaluation of irrigation with either raw municipal treated waste or river water on elements up take and yield of lettuce and potato plants. *Egypt J. Soil Sci.*, *42*, 705–714.

Abdelrahman, H. A., Alkhamisi, S. A., Ahmed, M., & Ali, H., (2011). *Effects of Treated Wastewater Irrigation on Element Concentrations in Soil and Maize Plants*. Conference Proceeding. Sultan Qaboos University, Oman. *42*(17), 18. Doi: 10.1080/00103624.2011.596236

Aguilar, R., Loftin, S. R., & Fresquez, P. R., (1994). Chapter 29: Rangeland restoration with treated municipal sewage sludge. In: Clapp, C. E., Dowdy, R. H., & Larson, W. E., (eds.),

Sewage Sludge: Land Utilization and the Environment (p. 211). SSSA Miscellaneous Publication, ASA-CSSA-SSSA, 677 South Segoe Road, Madison, WI 53711, USA.

Alexander, R., Szmidt, R., & Cruz, P., (2002). *Factsheet on Composting Biosolids*. Remade Scotland, Glasgow.

Alleman, J. E., & Berman, N. A., (1984). Constructive sludge management: Bio-brick. *Journal of Environmental Engineering, 110*(2), 301–311.

Al-Omran, M. A., El-Maghraby, S. E., Nadeem, M. E. A., El-Eterand, A. M., & Al-Mohani, H., (2012). Long term effect of irrigation with the treated sewage effluent on some soil properties of Al-Hassa Governorate, Saudi Arabia. *Journal of Saudi Society Agricultural Sciences, 11*, 15–18.

Ayres, J. G., Dick, S., Semple, S., Ogden, I., & Fowler, P., (2008). *Human Health and the Environmental Impacts of Using Sewage Sludge on Forestry and for Restoration of Derelict Land: Task 1 – Desk-Based Literature Review of the Human Health Impacts of Spreading Sewage Sludge on Non-Agricultural Land*. Edinburgh, SNIFFER. https://abdn.pure.elsevier.com/en/publications/human-health-and-the-environmental-impacts-of-using-sewage-sludge (Accessed on 4 August 2019).

Basta, N. T., (1995). *Land Application of Biosolids: A Review of Research Concerning Benefits*. Environmental impacts, and regulations of applying treated sewage sludge, B-808. Oklahoma Agricultural Experiment Station and Center for Agriculture and the Environment, Division of Agricultural Sciences and Natural Resources, Oklahoma State University. http://pods.dasnr.okstate.edu/docushare/dsweb/Get/Document-3109/B-808.pdf (Accessed on 4 August 2019).

Braguglia, C. M., Mininni, G., Marani, D., & Lotito, V., (2003). *Sludge Incineration: Good Practice and Environmental Aspects, Biosolids* (pp. 523–530). Wastewater sludge as a resource, Trondheim, Norway, International Water Association (IWA), Norwegian University of Science and Technology (NTNU).

Cassiday, L., (2012). *Biodiesel from Sewage Sludge*. Biofuels: New method may be cheaper than using conventional biodiesel feedstocks. Chemical, & Engineering News. Retrieved from: https://cen.acs.org/articles/90/web/2012/08/Biodiesel-Sewage-Sludge.html (Accessed on 4 August 2019).

Chamber, B. J., Nicholson, F. A., Aitken, M., Cartmell, E., & Rowlands, C., (2003). Benefits of biosolids to soil quality and fertility. *Water Env. Journal, 17*(3), 162–166.

Chihpin, H., Jill, R. P., & Yaorey, L., (2005). Mixing water treatment residual with excavation waste soil in brick and artificial aggregate making. *Journal of Environmental Engineering, 101*, 272–277.

Chin, T. L., Hui, L. C., Wen-Ching, H., & Chi, R. H., (1998).A novel method to reuse paper sludge and co-generation ashes from paper mill. *Journal of Hazard Matter,58*, 93–102.

Chou, I. J., Wang, K. S., Chen, C. H., & Lin, Y. T., (2006). Lightweight aggregate made from sewage sludge and incinerated ash. *Waste Management, 26*, 1453–1461.

Cimochowicz-Rybicka, M., (2012). *Minimization of Sewage Sludge Production-European Trends and Selected Technologies*. Retrieved from: https://www.kth.se/polopoly_fs/1.651120!/JPSU18P12.pdf (Accessed on 4 August 2019).

Cogger, C. G., (2005). Potential compost benefits for restoration of soils disturbed by urban development. *Compost Science Utilization, 13*(4), 243–251.

Demirbas, A., Edris, G., & Alalayah, W. M., (2017). Sludge production from municipal wastewater treatment in sewage treatment plant. *Energy Sources, Part A:*

Recovery, Utilization, and Environmental Effects, 39(10), 999–1006. doi: 10.1016/j.watres.2015.04.002.
Eilhann, E. K., Sungpyo, K., Young, J. J., & Haakrho, Y., (2012). Biodiesel production from sewage sludge: New paradigm for mining energy from municipal hazardous material. *Environmental Science and Technology, 46*(18), 10222–10228. doi: 10.1021/es3019435.
EU, (2006). Commission regulation (EC) No. 1881/2006 of 19 December 2006. Setting maximum levels for certain contaminants in foodstuffs. *Official Journal of Euro Union,* L364/5.
Garcia-Orenes, F., Guerrero, C., Mataix-Solera, J., Nava, P. J., Gomez, I., & Mataix-Beneyto, J., (2005). Factors controlling the aggregate stability and bulk density in two different degraded soils amended with Biosolids. *Soil Tillage Research, 82*, 65–76.
Guibelin, E., (2004). Sludge thermal oxidation processes: Mineral recycling, energy impact, and greenhouse effect gases release. *Water Science Technology, 49*(10), 209–216.
Herrmann, R. F., & Shann, J. F., (1997). Microbial community changes during the composting of municipal solid waste. *Microbial Ecology, 33,* 78.
Hidalgo, A. M., Murcia, M. D., Gómez, M., Gómez, E., García-Izquierdo, C., & Solano, C., (2017). Possible uses for sludge from drinking water treatment plants. *Journal of Environmental Engineering, 143*(3). https://doi.org/10.1061/(ASCE)EE.1943-7870.0001176.
Hossni, Y. A., & El-Tarras, A., (1997). Safety of treated wastewater in the irrigation of plants. *Egypt Journal of Horticulture, 24*, 261–270. doi: 10.5402/2011/402647.
Hue, N. V., & Ranjith, S. A., (1994). Sewage sludge in Hawaii: Chemical composition reactions with soils and plants. *Water, Air and Soil Pollution, 72*(1), 265–283.
Hussain, N., Rashid, S., Farooqi, J., & Ahmed, A., (2006). Control of environmental pollution by using treated solid sludge waste as soil amendment. *Science International, 18*(4), 321–325.
Jaeger, M., & Mayer, M., (2000).The noel conversion process-a gasification process for the pollutant-free disposal of sewage sludge and the recovery of energy and materials. *Water Science and Technology, 41*(8), 37–44.
Jamil, M., Qasim, M., & Umar, M., (2006). Utilization of sewage sludge on organic fertilizer in sustainable agriculture. *Journal of Applied Science, 6*(3), 531–535.
Kaposts, V., Karins, Z., & Lazdins, A., (2000). Use of sewage sludge in forest cultivation. *Baltic Forestry, 6*(2), 24–28.
Kosobucki, P., Chmarzyński, A., & Buszewski, B., (2000). Sewage sludge composting. *Polish Journal of Environment Studies, 9*(4), 243–248.
Kumamoto Municipal Government, (1983). *Report on Public Sewage Treated Water for Agriculture Use.*
Lechtenberg, D., (2018). *Use of Sewage Sludge in Cement Industry.* Retrieved from: EcoMENA, https://www.ecomena.org/sewage-cement/ (Accessed on 4 August 2019).
Lindsay, B. J., & Logan, T. J., (1998). Field response of soil physical properties to sewage sludge. *Journal of Environmental Quality, 27*(3), 534–542.
Manguiat, I. J., (1997). *Sewage Sludge: Turning an Environmental Pollutant into an Agriculture Resource,* 2nd Professional Lecture Delivered in Department of Soil Science. University of Philippine, LOSBanos.
Marques, M. J., Jimenez, L., & Pérez-Rodríguez, R., (2014). Chapter: Influence of shrub re-vegetation and sewage sludge in erosion control of gypsic soils in the south of Madrid (Spain). In: Ángel, F., & Roque, O., (eds.), *Book: Fourth International Conference on Land Degradation.* (Extended Abstracts) Gregorio García.

Martínez-García, C., Eliche-Quesada, D., Pérez-Villarejo, L., Iglesias-Godino, F. J., & Corpas-Iglesias, F. J., (2012). Sludge valorization from wastewater treatment plant to its application on the ceramic industry. *Journal of Environmental Engineering, 95*, 5343–5348.

Matthews, P., (2006). In: Ronald, J., LeBlanc, Peter, M., and Roland, P. R., (eds.), *England and Wales in Global Atlas of Excreta, Wastewater Sludge, and Biosolids Management: Moving Forward the Sustainable and Welcome Uses of a Global Resource, Greater Moncton Sewerage Commission UN Habitat* (p. 285). ISBN: 978–92–1–132009–1.

Mbagwu, J. S. C., & Piccolo, A., (1990). Carbon, nitrogen and phosphorus concentration in aggregates of organic waste-amended soils. *Biological Wastes, 31*(2), 97–111.

McLaren, R. G., Clucas, L. M., & Taylor, M. D., (2005). Leaching of macronutrients and metals from undisturbed soils treated with metal-spiked sewage sludge. 3. Distribution of residual metals. *Australian Journal of Soil Research, 43*(2), 159–170.

Milica, A., Zagorka, R., & Slavka, S., (2012). Removal of toxic metals from industrial sludge by fixing in brick structure. *Journal of Construction, and Building Material, 37*, 7–14.

Moffat, A. J. (2006). The use of sewage sludges and composts in forestry. Forestry Commission Information Note 79, Forestry Commission, Edinburgh.

Mohammed, M. J., & Mazahareh, N., (2003). Changes in soil fertility parameters in response to irrigation of forage crops with secondary treated waste water. *Communications in Soil Science and Plant Analysis, 34*, 1281–1294.

Navarro, A., (2012). Effect of sludge amendment on remediation of metal contaminated soils. *Minerals, 2*(4), 473–492.

Pescod, M., (1992). *Waste Water Treatment and Use in Agriculture* (No. 47, p. 125). Bull. FAO, Rome.

Rattan, R. K., Datta, S. P., Chhankar, P. K., Suribabu, K., & Singh, A. K., (2004). Long term impact of irrigation with sewage effluents on heavy metal contents in soils, crops and groundwater – a case study. *Agriculture, Ecosystems, & Environment, 109*, 310–322.

Rattan, R. K., Datta, S. P., Singh, A. K., Chhonkar, P. K., & Suribabu, K., (2001). Effects of long-term application of sewage effluents on available nutrient and available water status in soils under Keshopur effluent irrigation scheme in Delhi. *Journal of Water Management, 9*, 21–26.

Raymond, A. W., & Okieimen, F. E., (2011). Review article: Heavy metals in contaminated soils: A review of sources, chemistry, risks and best available strategies for remediation. *International Scholarly Research Notices of Ecology*, p. 20.

Robert, A. L., (1988). The effect of sewage sludge application to Bermuda grass on forage quality production, and metal accumulation. *Agriculture, Ecosystem and Environment, 20*(3), 209–219.

Rulkens, W., (2008). Sewage sludge as a biomass resource for the production of energy: Overview and assessment of the various options. *Energy & Fuels, 22*, 9–15. Available online at https://pdfs.semanticscholar.org/4c26/a86de99ba118fdea398d214b40f57d16c3e4.pdf (Accessed on 4 August 2019).

Salem, M. M., Amir, S., Abdel-Aziz, S. M., Kandil, M. F., & Mansour, S. F., (2000). Effect of irrigation with treated waste water on some chemical characteristics of soils and plants. *Egypt Journal of Soil Science, 40*, 49–59.

Sigua, G., Adjei, M., & Rechcigl, J., (2005). Cumulative and residual effects of repeated sewage sludge applications: Forage productivity and soil quality implications in South Florida, USA. *Environmental Science and Pollution Research, 12*(2), 80–88.

Simonsen, N., & Bruus, J., (2003). Drying and incineration of wastewater sludge. Experience and perspectives based on the development in Denmark. Biosolids 2003, wastewater sludge as a resource, Trondheim, Norway. *International Water Association (IWA), Norwegian University of Science and Technology (NTNU)*, pp. 399–406.

Singh, R. P., & Agrawal, M., (2008). Potential benefits and risks of land application of sewage sludge. *Waste Management*, *28*(2), 347–358.

Stabnikova, O., Goh, W. K., Ding, H. B., Tay, J. H., & Wang, J. Y., (2005). The use of sewage sludge and horticultural waste to develop artificial soil for plant cultivation in Singapore. *Bioresource Technology*, *96*(9), 1073–1080.

Stolarek, P., & Ledakowicz, S., (2001). Thermal processing of sewage sludge by drying, pyrolysis, gasification and combustion. *Water Science Technology*, *44*(10), 333–340.

TARA, (2018). *Scalable Solutions for People and Our Planet: Use of Marble Sludge Waste in Building Materials*. Retrieved from: https://www.tara.in/marble_sludge.aspx (Accessed on 4 August 2019).

Tay, J. H., Show, K. Y., Hong, S. Y., Chien, C. Y., & Lee, D. J., (2002). Potential reuse of wastewater sludge for innovative applications in construction aggregates. *Water Science Technology*, *50*(9), 189–196.

Thomas Nutrient Solutions, (2018). *What are the Advantages and Disadvantages of Sewage Sludge Incineration*. Retrieved from: http://www.thomasnutrientsolutions.com/sewage-sludge-incineration/ (Accessed on 4 August 2019).

U.S. EPA, (1984). *Environmental Regulations and Technology: Use and Disposal of Municipal Wastewater Sludge*. EPA/625/10–84/003. Washington, DC.

UN Habitat, (2008). *Global Atlas of Excreta, Wastewater, Sludge and Biosolids Management: Moving Forward the Sustainable and Welcome Uses of a Global Resource*. Greater Moncton Sewerage Commission, UN Habitat, ISBN: 978–92–1–132009–1.

US Environmental Protection Agency, (2004). *Guidelines for Waste Water Reuse*. Municipal Support Office of wastewater management, Washington DC, EPA/625/R-04/108.

Usman, K., Khan, S., Ghulam, S., Khan, M. U., Khan, N., Khan, M. A., & Khalil, S. K., (2012). Sewage sludge: An important biological resource for sustainable agriculture and its environmental implications. *American Journal of Plant Sciences*, *3*, 1708–1721. http://dx.doi.org/10.4236/ajps.2012.312209 Published Online December 2012 http://www. SciRP.org/journal/ajps (Accessed on 4 August 2019).

WHO, FAO, (2007). *Joint FAO/WHO Food Standard Program Codex Alimentarius Commission 13th Session*. Report of the 38 Session of the Codex Committee on Food Hygiene, Houston, United States of America, ALINORM 07/30/13. www.remade.org.uk/Organics/organics_documents/20020713_BiosolidsFactsSheet.pdf (Accessed on 4 August 2019).

Yadar, R. K., Goyal, R., Sharma, R. K., Dubey, S. K., & Minchas, R. S., (2002). Post irrigation impact of domestic sewage effluent on composition of soils, crops and ground water – a case study. *Journal of Environment International*, *28*, 481–486.

Yang, G., Zhang, G., & Wang, H., (2015). Current state of sludge production, management, treatment and disposal in China. *Water Research*, *78*, 60–73.

Yu-Chi, L., Shang-Lien, L., Kuo, J., & Chung-Chou, T., (2012). Beneficial uses of sludge from water purification plants in concrete mix. *Environmental Engineering Science*, *29*(4), doi: 10.1089/ees.2010.0479.

CHAPTER 7

Contributions of Pit Lakes to the Socioeconomic Status: A Case Study of the Raniganj Coal Field Area

DEBNATH PALIT[1] and ANJAN RAY CHAUDHURY[2]

[1]*Department of Botany, Durgapur Government College, Durgapur, West Bengal, India, Mobile: +919832175737, E-mail: debnath_palit@yahoo.com*

[2]*Department of Economics, Durgapur Government College, Durgapur, West Bengal, India*

ABSTRACT

In this chapter, we look into the contribution of the pit lakes to the society by using the primary data collected from the neighboring villages of the pit lakes of the Raniganj coalfield areas (henceforth RCF), in the state of West Bengal, India. It has been found that pit lake resources are helping the poor rather than the relative rich households of the adjacent areas of the pit lakes. In some cases, the poor people get some advantages through the use of the resources of the pit lakes. They can raise their earnings by using these resources and maintain their livelihoods in a better way, which can reduce the gap between the rich and poor people in the adjacent areas of the pit lakes.

7.1 INTRODUCTION

Large-scale open-cut mining activity has left a legacy of many thousands of mine pit lakes across the world (Klapper and Geller, 2002). The ecosystems of these pit lakes are the most varied and productive (Convention on Wetlands, 1971; Jha, 2008; Gurung et al., 2009). These lakes are considered as the sources of enormous water resources as well as these are important from the point of view of the conservation and sustainable management of their rich diversity of flora and fauna. These pit lakes are actually similar in all aspects to the wetlands. According to wide recognition, these lakes like other wetlands also provide several ecosystem services that contribute to human well-being. The major ecosystem services that these lakes provide include fish, fiber, water supply, water purification, climate regulation, flood regulation, coastal protection, recreational opportunities and eco-tourism (Millennium Ecosystem Assessment, 2005). The loss of Pit Lake breaks the ecosystem services, which leads to the damages of the health and well-being of individuals and local communities and diminishes their development prospects (Millennium Ecosystem Assessment, 2005).

Actually, the people living in the mining areas are suffering from the scarcity of water resources. Managed and efficient utilization of this huge water resource may be beneficial to the people living in the adjacent areas of these lakes (Castro and Moore, 2000). Members of the households living in the neighboring places of the pit lakes can use huge water of pit lakes for domestic purposes, such as for cooking, washing clothes, washing utensils and so forth. As these lakes are huge water bodies, huge amount with various kind of fishes are automatically produced in these lakes. The individuals living in the neighboring places can also collect these fishes for their self-consumption as well as these collected fish can be marketed which will raise household incomes and standard of living of the neighboring regions of these lakes.

Likewise, the use of the water resources of the wetlands, the water resources of the pit lakes can also be rigorously used for irrigation purpose. It is noteworthy that the cost of irrigation, in this case, is significantly low compared to the irrigation through deep tube well. As a result, pit lakes are raising agricultural productivity indirectly (Jones and McCullough, 2011; McCullough and Lund, 2006). The availability of huge water resources enables the farmers of the neighboring areas to initiate the diversified farming method. Moreover, even in summer, these lakes can hold adequate water due to their deepness. Therefore, the farmers farming in the lands of the neighboring areas of these lakes are usually able to cultivate throughout the year.

Due to the availability of water resources in the neighboring places of the pit lakes natural forests are developed. It is frequently observed that a significant part of the poor residents of the neighboring villages are using the fuelwoods collected from these forests as fuel for cooking in their own households. A large section of the population has taken collection of fuelwood from the natural forests as their livelihood, i.e., the poor people are earning money by selling the collected fuelwoods to the distant areas. In addition, the individuals living in the neighboring areas earn money and raise their household incomes by providing various services to the tourists, who usually visit the areas of the pit lakes to get pleasure from the scenic beauty of these areas (Miller and Lund, 1996; McCullough and Lund, 2006).

Despite all of the aforementioned beneficial services provided by the pit lakes and resources, the water bodies and the adjacent areas of these lakes are often polluted and contaminated by the activities of the local residents either deliberately or in their ignorance (Younger and Wolkersdorfer, 2004; Lund et al., 2012). Our main objective of this study is to assess the beneficial effects provided by the pit lakes to the local residents across diverse socio-economic backgrounds. In addition, we have also looked at the activities of the local residents from different economic status, different educational groups and different age groups, which have been polluting and contaminating the pit lakes and the surrounding areas of these lakes.

For the execution of the aforementioned objectives, we have conducted a primary survey and collected data from the Raniganj coalfield areas, West Bengal, India. The collected information has several dimensions and several variables which are related with the benefits and services provided by the pit lakes to the local residents and the activities of the local residents. Further, we describe the primary data collection methodology and discuss the contributions of pit lakes to households from different economic status along with attitudes of the local residents towards the pit lakes.

7.2 PRIMARY SURVEY: THE DATA COLLECTION METHODOLOGY

For this study, we have used the data collected through the primary survey in neighboring two hundred and fifteen villages surrounding the pit lakes of RCF areas. We have followed a multi-stage stratified random sampling method in this context. Based upon the report of the Indian Census, 2011, first we have listed all villages in the earlier mentioned areas, and then selected the villages. There are forty pit lakes in the RCF. From the neighboring place of each pit lake, we have selected five villages by the method of random sampling. Therefore, the first stage units are villages, and second stage or final stage units are the households and individuals.

The selected villages are stratified into two to five strata based upon the size of the population in the villages. If the population size of the village is less than 500, then we don't stratify the village. If the population size of the village is 501 to 1000, then we have stratified the village into two strata; 1001 to 1500, then we stratify the village into three strata; 1501 to 2000, then we stratify the village into four strata; and finally we classify the village into five strata if the size of the population is greater than 2000.

After classifying the sample villages into strata, we select one to two strata from total number of strata by using the probability proportional to size method of sampling. These selected strata are second stage units of the sample design. If the population size of the village is less than 500, then these is only one stratum of the village, and we select that single stratum. If the number of strata of the villages is 2 to 4, then two strata are selected randomly, and if the number of available strata is 5 then three strata are selected randomly.

After selecting the strata from the villages, the strata are re-stratified into some household groups on the basis of monthly per capita consumer expenditure. Each of the strata is sub-divided into three sub-strata, and from each sub-stratum we select the sample households in the following proportion 1:2:1. Here, the sub-stratification is done in the following way: (i) Rich households, i.e., mpce ≥ Rs. 10,000; (ii) Households having mpce lies between Rs. 10,000 and Rs. 1000; and (iii) Poor households, i.e., mpce lies below Rs. 1000. From the first strata we select one household, from the second sub-strata we select two households and from the third sub-strata we select only one household. If the population size of the village is small and there is only one stratum within the village then we select the households from three sub-strata on the basis of the following ratio –2:4:2.

So, on the whole, 1795 households and 8518 individuals were selected as respondents. We have collected of the selected households and individual's information by interviewing the persons based upon the well-framed questionnaire containing questions on different personal, familial, and societal characteristics of the respondents and their households. The response period is last 365 days from the date of survey.

We classify the population of the samples of into some categories according to the mpce, such as rich households, middle-class households and poor households. It is observed that in the sample around 25% of the households are rich, 50% of the households are middle class and 25% of the households are poor. It has also been found that 23% of the individuals are from rich households, 46% of the individuals are from middle-class households and 31% of the sampled individuals are from poor households.

According to the ownership of assets, such as different types of vehicles and different types of livestock we again classify the sample of the households into rich and poor groups. There are four kinds of vehicles which are available in the villages, such as bi-cycle, motorbike, four-wheeler, and tractors. Tractors are usually used for cultivation purpose. On the other hand, the livestock can also be classified into two groups, one is large/big live stocks, and the other is small livestock. The big livestock are cow and bullock, and the small livestock are hen and duck.

In the case of rich households in the village usually these is an ownership of tractors and in some case, there is some ownership of four-wheelers. A significant percentage of middle-class households own motorbike and bicycle. It has been observed from the collected data that a significant percentage of households don't have any type of vehicle.

Therefore, it can be assumed that according to the ownership of the vehicles, we can classify the sample into three classes-one is rich who own four-wheelers and tractors, the middle-class households own motorbike. Motorbikes are also owned by rich households, but for the middle-class households, the ownership of bike is common.

In the case of owning livestock, it is observed that the rich and middle-class households usually own big livestock, but the poor households usually own small livestock. In some cases, the rich and middle-class households own the small livestock.

According to the sample, it is observed that around 90% of the so-called rich households defined by *mpce* own tractors, 70% of the rich households own four-wheeler, 98% of the rich households own motorbike, and all rich households own bi-cycle. Around 30% of the middle-class households own tractors, 22% of the middle-class households own four-wheeler, 87% of the middle-class households own motorbike, and all middle-class households own bicycle. However, none of the poor households own tractor, four-wheeler, and motorbike, but around 34% of the poor households own bicycle.

Likewise, the ownership of vehicles, the ownership of livestock can also be used as an indicator of the economic status of the households. The observations collected from the primary survey reveals that around

27% of the poor households are having small livestock, and 71% and 79% of the middle-class and rich village households are having small livestock. In case of the ownership of big livestock, a sharp disparity between rich and poor can also be identified. Around 3% of poor household living in the adjacent villages of the pit lakes is owning big livestock, but in case of rich and middle-class households, these figures are around 94%.

7.3 CONTRIBUTION OF PIT LAKES TO HOUSEHOLDS FROM DIFFERENT ECONOMIC STATUS

Initially, for this analysis, we classify the sample into three classes-rich, middle-class, and poor according to the monthly per capita consumer expenditure. Later on, we invoke the household asset ownership for the classification of the households (Table 7.1).

TABLE 7.1 Percentage of the Households from Different Classes Using Pit Lakes Resources for Their Daily Needs

	Percentages of Households		
	Rich	Middle-Class	Poor
Using the water of pit lakes for domestic use	10	59	87
Taking the fishes of pit lakes as food	7	61	83
Using woods of forest nearest of the pit lakes as fuel in their households	2	40	70
Bathing in the pit lake water	0.8	26	65
Taking wild-fruits collected from the forests near the pit lakes	0	21	54

(Source: Primary survey done by the investigators).
Note: Total number of households is the number of households in the nearby 200 villages of forty pit lakes as described in Census India, 2011.

7.3.1 CLASSIFICATION OF THE SAMPLE ACCORDING TO MONTHLY PER CAPITA CONSUMER EXPENDITURE

i. **Using Pit Lakes Resources for Daily Needs:** There are several contributions of the pit lakes to the lives of the individuals living in the neighboring places of these lakes. However, the dependence on the pit lakes is attributed to the household's economic status assessed by the monthly per capita consumer expenditure of the households. The percentages of the households across the economic classes who are depending on the pit lake resources for their daily needs are reported in Table 7.1. It can be observed that 10% of rich households use the water resource of the pit lakes for their, while around 60% and 90% of the middle-class and poor households are using the pit lake water for their domestic use respectively. Likewise the first case it is observed that only 7% of the rich households are taking the fishes of the pit lakes as their food, but around 60% and middle-class and poor households are taking the fishes collected from pit lakes as food respectively. In the case of using woods of the forest nearby the pit lakes as fuel, bathing in the pit lake water, and taking wild-fruits collected from the forests near the pit lakes the middle-class and poor households are getting more benefits rather than the rich households. Actually, with the rise in economic status household's dependence on pit lake resources declines as the rich households can afford other sources of water, fuel, fishes, and so on. The relatively poor households sometimes are compelled to avail the resources provided by the pit lakes as the costs in these cases are low. Figure 7.1 depicts the same fact as described by Table 7.1.

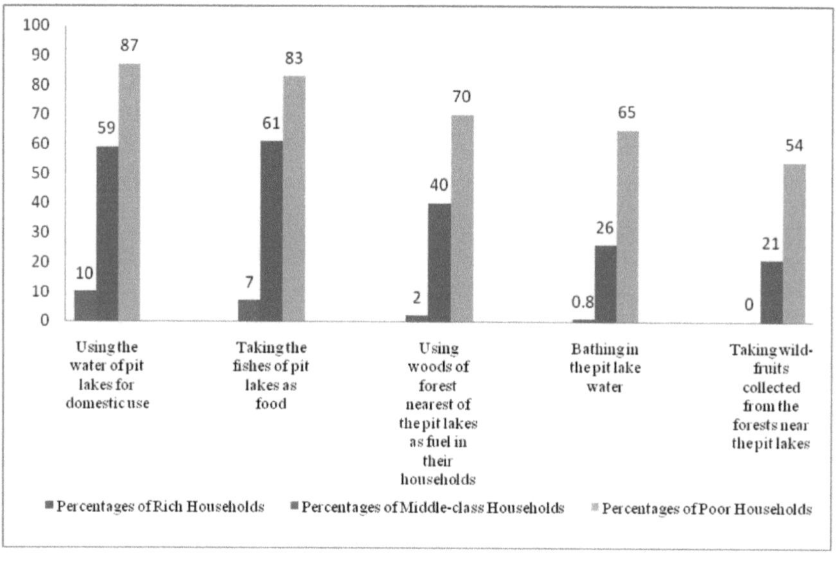

FIGURE 7.1 (See color insert.) Percentage of households from different economic classes using pit lakes resources for their daily needs.

ii. Maintenance of the Livelihood Through the Use of the Pit Lake Resources: Not only the goods and services provided by the pit lakes are used by the household living in the neighboring places of the pit lakes, but those resources are often used to raise the earnings of the households and it will ease the lives of the individuals living in those households. Table 7.2 reports of the percentage figures of the rich, middle-class, and poor households using different resources of the pit lakes in RCF for the maintenance of their livelihoods. Likewise, the previous case in this case also it can be easily identified that the relatively poor households are depending on the resources of the pit lakes in RCF more than that of the rich households.

In the case of using pit water for irrigating the cultivated lands, it can be observed that only 20% and 23% of the rich and poor households are depending on the water resources of the pit lake. However, almost 80% of the poor households who are depending on cultivation using the water resources of the pit lakes for irrigation purpose.

It can also be observed that around 10% of the middle-class and 50% of the poor households are earning by marketing the fishes caught from the pit lakes. However, none of the rich households are reaping these benefits from the pit lakes, as they have several other sources of income, they don't depend on the pit lake water for maintaining their livelihoods.

Apart from the use of the woods collected from the nearby forests developed from the water resources of the pit lakes, it can also be observed that the individuals of the households are selling the collected fuelwoods in the market for earning, and a considerable fraction of earnings of poor households living in the neighboring places of the pit lakes comes from this source. Even though none of the members of the middle-class and rich households don't try to raise their earnings in this way.

It is can be found according to the reported figures in Table 7.2 that a major section of poor households are using the grasses grown in nearby fields of the pit lakes to use it as fodder for their cattle. It is noteworthy to state that negligible percentage of the poor households are owning cattle. If they own the cattle then almost all of them depend on the pit lake resources for feeding their cattle. Dependence of rich and middle-class households on the pit lake resources in this is very insignificant. For rich households in villages nearby the pit lakes most of which persons own cattle only 7% of these rich persons are using the grasses grown in the adjacent lands of the pit lakes as the fodder for their cattle. For the middle-class households this figure is greater than the rich households; however, this figure very negligible (only 10%).

It is also found that a significant proportion of the poor population earns by marketing the collected wild fruits grown in the forests nearby the pit lakes (38%). However, none of the rich and middle-class households take the fruits grown in the forests of the pit lakes as their fruits. In the case of diversified farming also dependence of the poor households pit lake water resources is quite high, where there is no dependence of the rich and middle-class households on the water resources of the pit lakes in this manner. As almost of the adult persons of the rich and middle-class households are earners and they have different types of livelihoods, these households do not want to earn by providing different types of services to the visitors of the pit lakes. However, almost one-third of the poor households living in the neighboring areas of the pit lakes depends on the pit lakes in this manner.

Based upon the aforementioned findings, it can be perceived that pit lakes are inevitable for the maintenance of the lives of the poor households living in the neighboring villages of the pit lakes. Therefore, pit lakes have an immense contribution for the maintaining socio-economic equality. Figure 7.2 is the diagrammatic representation of the reported figures in Table 7.2.

TABLE 7.2 Percentage of the Households from Different Classes Maintaining Their Livelihoods by Using Pit Lake Resources

	Percentages of Households		
	Rich	**Middle-Class**	**Poor**
Using pit water for irrigating their cultivated lands	20	23	78
Depending on the fish farming in pit lake water	0	10	46
Earning from selling the fuelwoods	0	0	74
Using the grasses nearby the pit lakes as the fodder for cattle	7	10	95
Using pit water for their diversified farming	17	31	52
Earning from wild fruits collected from the forests nearby of the pit lakes	0	0	38
Earning from the visitors of the pit lakes by providing different services	0	0	33

(Source: Primary survey done by the investigators).

Note: Total number of households is the number of households in the nearby 200 villages of forty pit lakes as described in Census India, 2011.

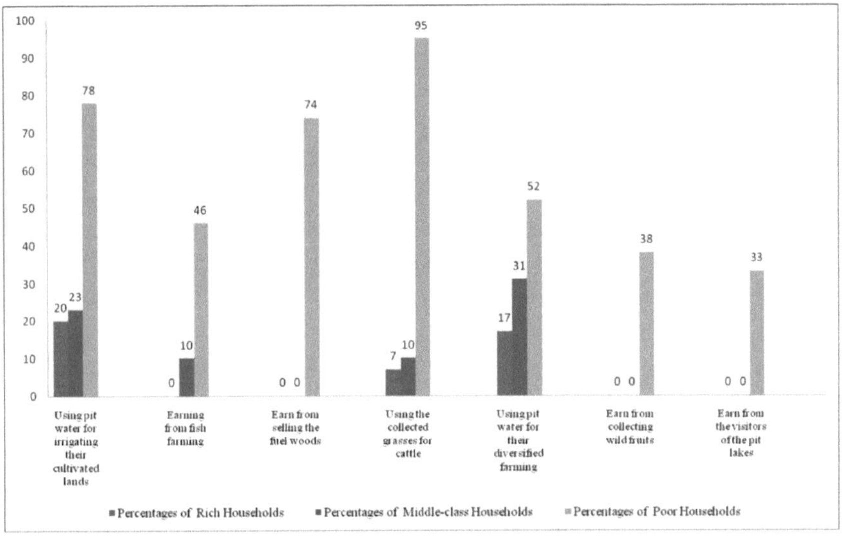

FIGURE 7.2 (See color insert.) Percentage of the households from different classes maintaining their livelihoods by using pit lake resources.

7.3.2 CLASSIFICATION OF THE SAMPLE BY THE OWNERSHIP OF VEHICLES AND LIVESTOCK

7.3.2.1 DISPARITY BETWEEN THE HOUSEHOLDS IN THE USE OF THE PIT LAKE RESOURCES FOR DAILY NEEDS

In this section, we describe the benefit received by the local households. Here we classify the sample of households into rich and poor groups according to the ownership of vehicles. There are four types of vehicles, such as bicycle, motorbike, tractor, and four-wheeler. It is found that domestic use of pit water declines with the improvement in the economic status of the households. For example, around 89% of the households having bicycle are using pit water for domestic use, whereas only 5% of the households having a personal car are using pit water for domestic use. This indicates that with the improvement of economic status while household members are aware on the polluted and contaminated water of the pit lakes they avoid to use the water for domestic use. In case of taking the fishes of pit lakes as foods household economic status plays a vital role, since 72% of the households having bicycle are taking fishes of pit lakes as food and the figure declines while the household own bike, tractor, and /or car. Household economic status has also important bearing on the household use of fuelwoods from pit forests. In addition, it can also be stated on the basis of the figures reported on Table 7.3 that rising economic status brings down the use of pit-water for bathing and taking wild fruits as foods.

Another way of explaining household economic status is the ownership of livestock of the households. Table 7.4 reports the dependence of the households on pit water and resources for their daily needs, where the households are classified by the small and big livestock. The small livestock comprises hens, duck, etc., big livestock comprises cow, buffalo, and so on. Reported values in Table 7.4 corroborates the findings of Table 7.3, where household economic status has been defines quite differently. It is observed that relatively poor households, i.e., the households having small livestock are more likely to use pit water for domestic purpose: taking fishes collected from pit lakes as foods; fuelwoods collected from pit forests as fuel; using pit water for bathing and taking wild-fruits collected from pit forests as foods, compare to the households having the big livestock. For instance, only 50% of the households having big livestock are using pit water for domestic purpose, whereas around 90% of the households having small livestock using pit water for domestic purpose.

Therefore, all these computed percentages derived from the collected data corroborate the previous findings, i.e., the higher dependence of the poor households on the pit lake resources for meeting their daily needs (Figures 7.3–7.6).

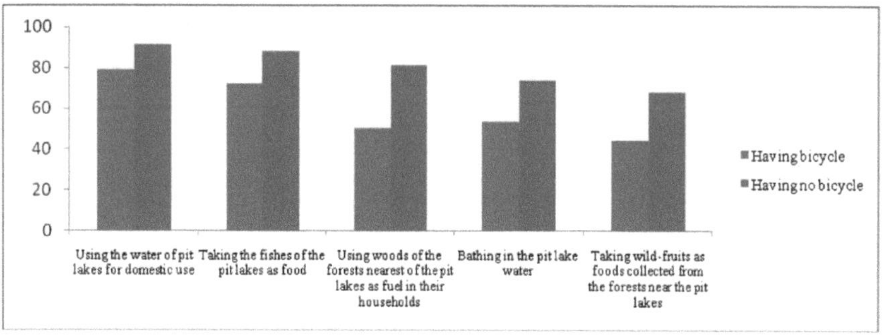

FIGURE 7.3 Use of the pit lakes for daily needs by the household members having bicycle.

TABLE 7.3 Dependence of Different Types of Households on the Pit Lake Resources for Daily Needs

	Percentage of		Percentage of		Percentage of		Percentage of	
	Households Having Bicycle	**Households Having no Bicycle**	**Households Having Bike**	**Households Having no Bike**	**Households Having Tractor**	**Households Having no Tractor**	**Households Having Car**	**Households Having no Car**
Using the water of pit lakes for domestic use	79.2	91.37	61.15	94.12	54.34	96.47	5.23	99.13
Taking the fishes of the pit lakes as food	72.1	88.27	67.94	83.61	62.91	86.11	7.19	91.72
Using woods of the forests nearest of the pit lakes as fuel in their households	50.3	81.29	23.78	70.55	19.89	78.16	7.41	70.44
Bathing in the pit lake water	53.55	74.19	37.27	65.83	28.29	71.86	4.59	61.89
Taking wild-fruits as foods collected from the forests near the pit lakes	44.43	68.36	20.32	70.11	11.32	85.27	3.73	44.15

(Source: Primary survey done by the investigators).

Note:

(i) Total number of households is the number of households in the nearby 200 villages of forty pit lakes as described in Census India, 2011.

(ii) Sum-total of each cell entry and 100 minus cell entry will always be 100.

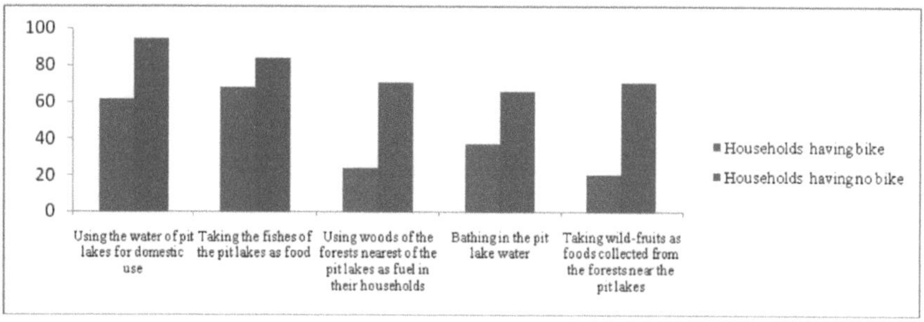

FIGURE 7.4 Use of the pit lakes for daily needs by the household members having bike.

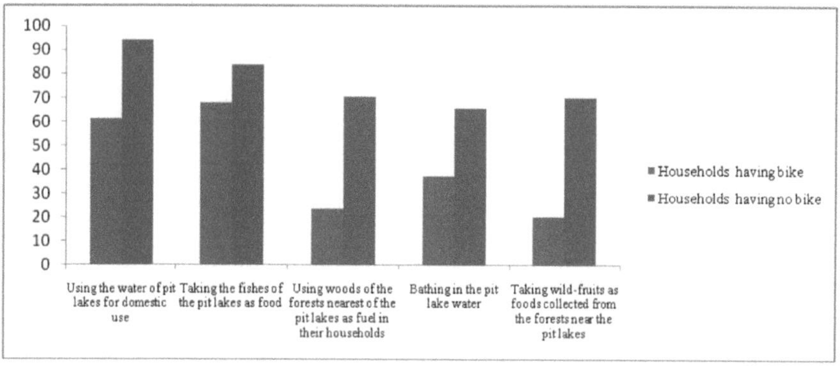

FIGURE 7.5 Use of the pit lakes for daily needs by the household members having tractor.

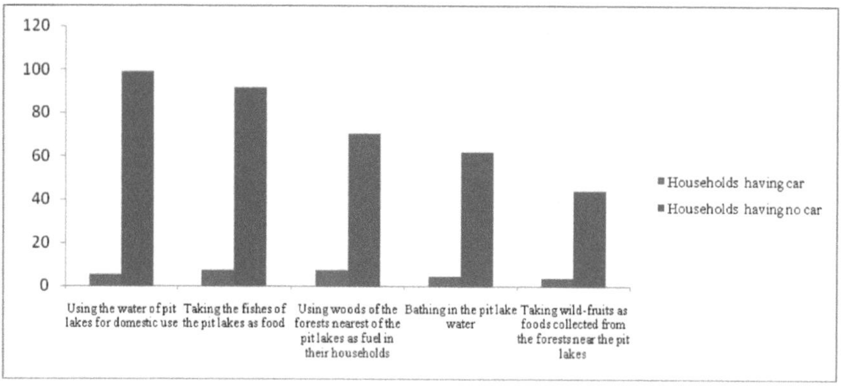

FIGURE 7.6 Use of the pit lakes for daily needs by the household members having personal car.

TABLE 7.4 Household Ownership of Livestock and Dependence on Pit Lakes for Daily Needs

	Households		Households	
	With Small Livestock	**With no Small Livestock**	**With Big Livestock**	**With no Big Livestock**
Using the water of pit lakes for domestic use	85.36	91.87	49.28	92.63
Taking the fishes of the pit lakes as food	70.19	81.37	59.77	90.05
Using woods of the forests nearest of the pit lakes as fuel in their households	51.44	78.91	40.18	79.56
Bathing in the pit lake water	50.02	54.98	40.12	50.87
Taking wild-fruits as foods collected from the forests near the pit lakes	35.71	69.88	9.55	80.98

(Source: Primary survey done by the investigators).

Note:

(i) Total number of households is the number of households in the nearby 200 villages of forty pit lakes as described in Census India, 2011.

(ii) Sum-total of each cell entry and 100 minus cell entry will always be 100.

(iii) Small livestock group comprises hen and duck, and big livestock comprises cow and buffalo.

7.3.2.2 DISPARITY BETWEEN THE HOUSEHOLDS IN THE USE OF THE PIT LAKE RESOURCES FOR LIVELIHOODS

In this section, we discuss the contribution of the pit lakes on the livelihoods of the households residing in the adjacent areas of the lakes. However, here we take the responses of the individuals across different economic status assessed by the ownership of assets, such as vehicles and livestock (Figures 7.7 and 7.8).

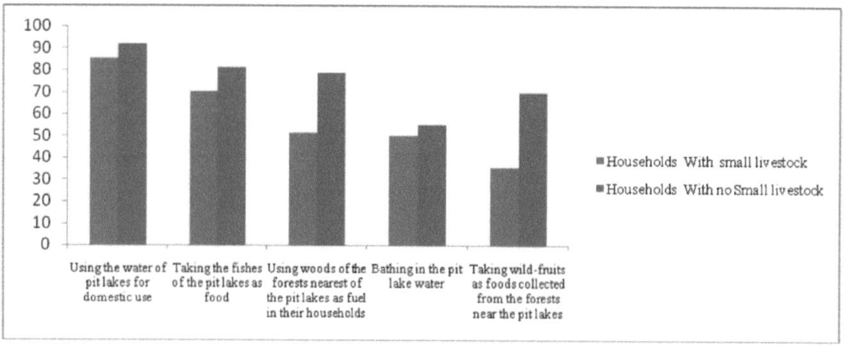

FIGURE 7.7 Use of the pit lakes for daily needs by the household members having small livestock.

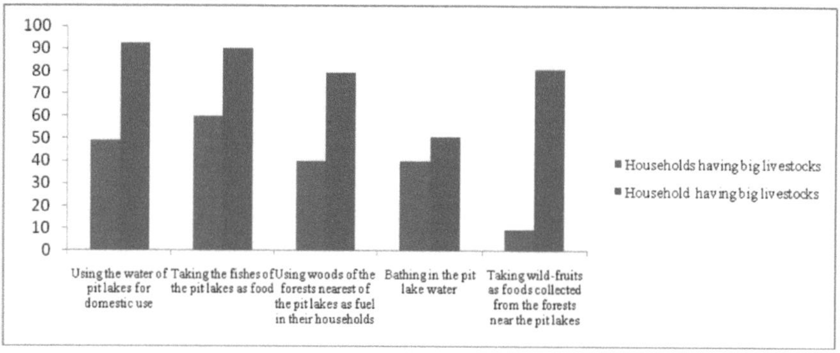

FIGURE 7.8 Use of the pit lakes for daily needs by the household members having big livestock.

It is observed from figures can explain these facts, where we classify the samples by the ownership of vehicles and livestock. There are four vehicles regarding which we collect information. These vehicles are bicycle, motorbike, tractor, and car, and the livestock are classified into two groups, such as small livestock comprises duck and hen, and big livestock, such as buffalo and cow. The households having only bicycle are the relatively poor households, and households owning cars are relatively rich households. If a household has no bicycle, then the household is considered as poorest household. Likewise, it is considered that the households having small livestock are relatively poor households and households having big livestock are relatively rich households.

It is observed from Table 7.5 that in all cases of using the pit lakes and pit water is greater by the households having no bicycle. In case of the households having bicycle around 66% of the households are using the pit lakes water for irrigation purpose for cultivation; 46% of the households are depending on the fish farming; 32% of the households are selling the fuelwood collected from the forests to earn money; 79% of the households are using the grasses collected from the adjacent lands of the pit lakes for their cattle; 39% of the households are earning from the visitors by providing different services to the visitors of the pit lakes. The difference between the households having bicycle and no bicycle is not significant for all of these cases.

It can be observed from the reported figures in Table 7.5 that 62%, 18%, 24%, 76%, 24% 7%, and 5% of the households having bike are using pit lake water for irrigation, fish farming, collecting, and selling fuelwood collected from the forests nearby the pit lakes, using the grasses collected from the adjacent lands of the pit lakes for their cattle, using pit water for diversified farming, using wild fruits collected from the nearby forests of the pit lakes, and providing services to the visitors of the pit lakes for enjoying scenic beauties of these lakes.

Likewise the households having bicycles the percentages of households having tractor and personal car are also low in case of getting the benefits of pit lakes except in the first and fourth cases compared to the percentage of households having bicycle. The percentages of households having tractor and car are significantly low compared to the households bicycle in case of collecting wild fruits and providing services to the visitors of the pit lakes. It has already v=been mentioned that irrespective economic status a major section of the households are using the collected grasses from the nearby fields as the fodder for their cattle. The reason behind this is that these grasses are the cheapest source of fodder for the cattle. Same is true for using the pit water for irrigation purpose. Figures 7.9–7.12 are the diagrammatic representation of Table 7.5.

TABLE 7.5 Disparity in the Dependence of the Households on Pit Lakes for the Maintenance of Livelihoods

	Having Bicycle		Having Bike		Having Tractor		Having personal car	
	Yes	**No**	**Yes**	**No**	**Yes**	**No**	**Yes**	**No**
Using pit water for irrigating their cultivated lands	65.78	73.21	62.18	74.22	66.34	78.27	17.86	74.11
Depending on fish farming	46.18	58.29	17.56	58.99	11.05	63.23	1.27	60.26
Earning from selling the fuelwoods	31.98	32.66	24.31	40.27	14.79	50.11	3.02	37.33
Using the grasses nearby the pit lakes using as the fodder for cattle	78.62	81.21	76.05	85.49	58.13	90.32	43.66	87.15
Using pit water for their diversified farming	37.78	40.13	23.99	44.65	19.87	49.55	11.09	54.77
Earning from collecting wild fruits from the forests nearby of the pit lakes	33.48	41.72	7.35	20.46	3.22	24.69	0.87	19.88
Earning from the visitors of the pit lakes by providing different services	38.92	67.77	4.55	18.18	3.41	19.14	0.31	20.36

(Source: Primary survey done by the investigators).

Note:

(i) Total number of households is the number of households in the nearby 200 villages of forty pit lakes as described in Census India, 2011.

(ii) Sum-total of each cell entry and 100 minus cell entry will always be 100.

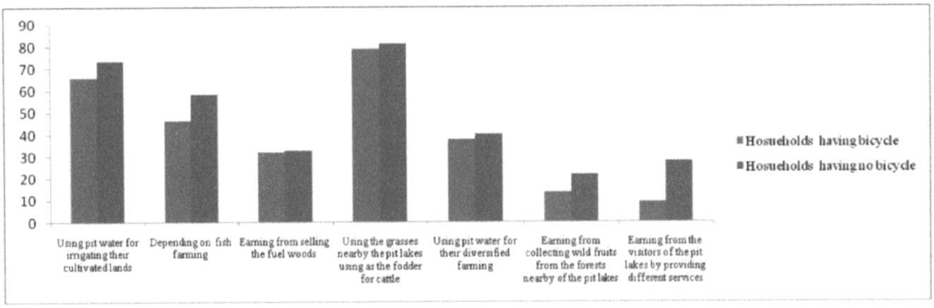

FIGURE 7.9 Use of the pit lakes for the maintenance of livelihoods by the households having bicycle.

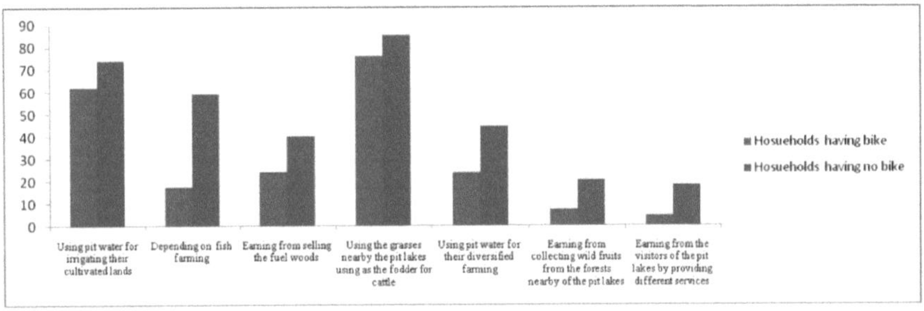

FIGURE 7.10 Use of the pit lakes for the maintenance of livelihoods by the households having bike.

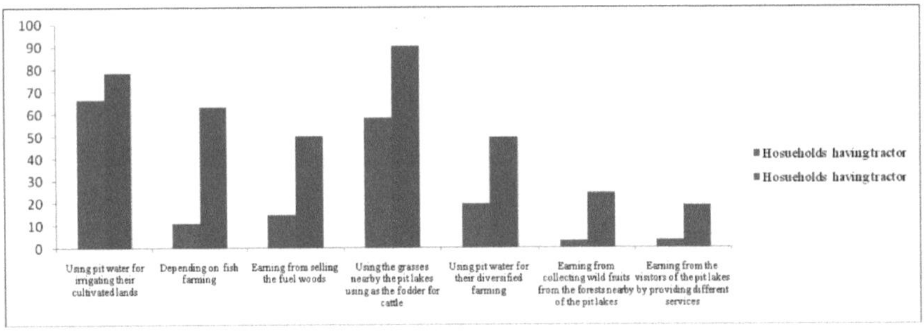

FIGURE 7.11 Use of the pit lakes for the maintenance of livelihoods by the households having tractor.

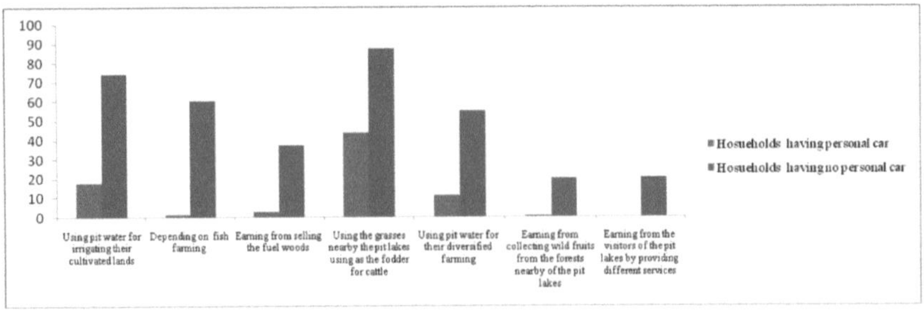

FIGURE 7.12 Use of the pit lakes for the maintenance of livelihoods by the households having personal car (four wheeler).

Table 7.6 reports the percentages of the classified households on the basis of the ownership of small and big livestock, according to their benefits received from the pit lakes for their livelihoods. It can be observed from the reported percentages in this table that except the first case, i.e., except the use of the pit lake water for irrigation purpose in all other cases the percentages of households getting benefits from the pit lakes are low while they are owning big livestock compared to the percentages of the households owning small livestock.

TABLE 7.6 Household Ownership of Vehicles and Dependence on Pit Lakes for Livelihood

	Having Small Livestock		Having Big Livestock	
	Yes	No	Yes	No
Using pit water for irrigating their cultivated lands	60.99	83.52	61.48	80.61
Depend on fish farming	44.06	60.18	10.72	72.49
Earning from selling the fuelwoods collected from the forest nearby of the pit lakes	33.33	38.82	12.38	52.72
Using the grasses nearby the pit lakes using as the fodder for cattle	-	-	52.13	-
Using pit water for their diversified farming	40.33	41.27	16.68	50.23
Earning from collecting wild fruits from the forests nearby of the pit lakes	23.25	37.11	8.47	29.76
Earning from the visitors of the pit lakes by providing different services	17.82	34.54	9.51	27.46

(Source: Primary survey done by the investigators).

Note:

(i) Total number of households is the number of households in the nearby 200 villages of forty pit lakes as described in Census India, 2011.

(ii) Sum-total of each cell entry and 100 minus cell entry will always be 100.

(iii) Small livestock group comprises hen and duck, and big livestock comprises cow and buffalo.

Actually we have mentioned and explained in the earlier analysis that usually rich household share owning big livestock, and small livestock are owned exclusively by the poor households. Therefore, based upon this view this finding corroborates the previous finding. In other words it can be recognized that the poor households are usually the main beneficiaries of the pit lakes resources. Figures 7.13 and 7.14 arte the diagrammatic representation of Table 7.6.

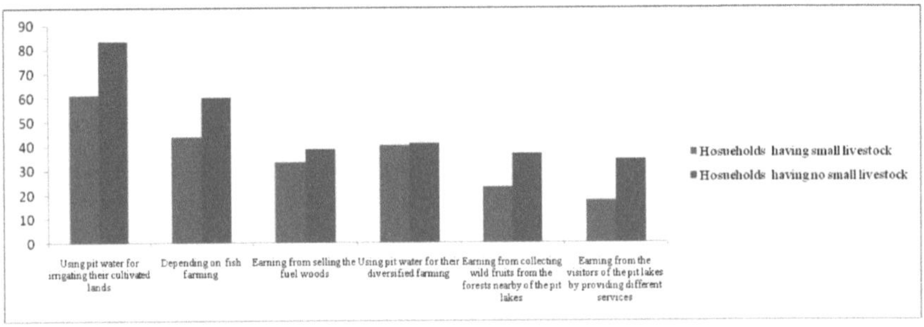

FIGURE 7.13 Use of the pit lakes for the maintenance of livelihoods by the households having small livestock.

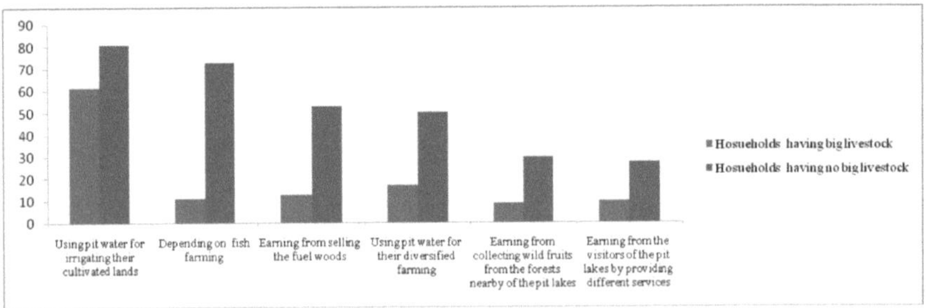

FIGURE 7.14 Use of the pit lakes for the maintenance of livelihoods by the households having big livestock.

7.4 ATTITUDES OF THE LOCAL RESIDENTS TOWARDS THE PIT LAKES

7.4.1 ECONOMIC STATUS CAN CHANGE THE ATTITUDE TOWARDS PIT LAKES

Based upon the earlier analysis it can be perceived that the pit lakes in the RCF areas have significant beneficial impacts on the lives of the residents of the neighboring villages of these lakes. Now, the question is-what is the attitude of the people living in the neighboring places of the pit lakes towards these lakes. For this analysis, we classify the sample into some categories by three ways, such as economic status of the households, educational qualification of the individuals and age of the individuals. We have identified four activities by which the pit lake water and the adjacent areas of the pit lakes may be polluted and contaminated. These activities are-using the pit lake areas as dustbin, using the pit lake as lavatory, using pit lakes for idol immersion and sustainable us of the pit lakes.

Firstly, we classify the sample of the population by the economic status which we have already done in the previous analysis. Table 7.7 reports the percentage distribution of population across the groups defined by economic status and the different types of uses of the pit lakes. It can be observed that except idol immersion, in all cases pit lakes are being polluted and contaminated more by the relatively by the persons from poor and middle-class families. In the case of idol immersion, there is no difference between rich and poor.

TABLE 7.7 Economic status can Change the Attitudes Towards Pit Lakes

Different Types of Use	Percentage of Individuals
1. Use the pit lake areas as dustbin:	
(a) Rich	5.14
(b) Middle-class	17.22
(c) Poor	78.09
2. Use the pit lake areas as lavatory:	
(a) Rich	0
(b) Middle-class	9.88
(c) Poor	52.03
3. Use the pit lakes for idol immersion:	
(a) Rich	87.45
(b) Middle-class	83.99
(c) Poor	90.77
4. Sustainable use of the pit lakes:	
(a) Rich	79.11
(b) Middle-class	71.23
(c) Poor	20.91

(Source: Primary survey done by the investigators).

Contrary is true in case of the sustainable use of the pit lake resources. In this case, the percentage of people from the rich households is greater than that of poor households. This indicates self contradictory behavior of the individuals from poor households. These people are getting more benefits from pit lakes but they are not taking any step with responsibility for the conservation of the pit lake resources. Figure 7.15 depicts the reported figures in Table 7.7 diagrammatically.

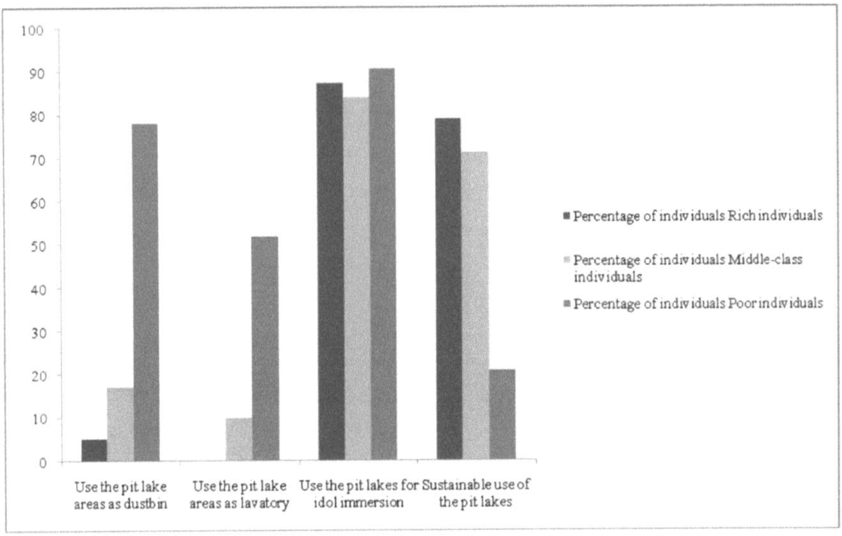

FIGURE 7.15 Use of the pit lakes by the individuals from the households with different economic status.

7.4.2 *EDUCATION CAN CHANGE THE ATTITUDE*

The difference in educational qualification can also influence the mind of the people for using the pit lakes. Uneducated persons have very little awareness of pollution and contamination of any kind of common property resource. We classify the people living in the neighboring areas of the pit lakes of the RCF areas into three categories, such as illiterate and low educated implies without any formal schooling and formal schooling up to 5 years; medium educated implies formal schooling up to 12 years, and higher educated implies educated more than higher secondary level of education. It is found that 93% of the illiterate persons, 62% of the moderately educated persons and 20% of the higher educated persons are using pit water and pit lake areas as dustbin. Likewise, 78% and 98% of the illiterates, 31% and 88% of the moderately educated persons, and 6% and 70% of the higher educated persons are either using pit water and pit lake areas as lavatory or supporting idol immersion in the pit lake water (Table 7.8).

TABLE 7.8 Education can Change the Attitudes Towards Pit Lakes

Different Types of Use	Percentage of Individuals
1. Use the pit lake areas as dustbin:	
(a) Illiterate and low educated	92.79
(b) Medium educated	61.82
(c) Higher educated	20.34
2. Use the pit lake areas as lavatory:	
(a) Illiterate and low educated	78.22
(b) Medium educated	31.49
(c) Higher educated	5.66
3. Use the pit lakes for idol immersion:	
(a) Illiterate and low educated	98.03
(b) Medium educated	88.17
(c) Higher educated	70.33
4. Sustainable use of the pit lakes:	
(a) Illiterate and low educated	34.67
(b) Medium educated	65.99
(c) Higher educated	81.43

(Source: Primary survey done by the investigators).

Note:

(i) Illiterate and low educated implies without any formal schooling and formal schooling upto 5 years; medium educated implies formal schooling up to 12 years, and higher educated implies educated more than higher secondary level of education.

(ii) Sustainable use of the pit lakes means that the individuals residing in the adjacent areas of the pit lakes do not do any work which can pollute the water of the pit lakes.

When the sustainable use of the pit lake water and the areas of the pit lake is concerned, higher educated persons occupies better place compared to the persons included into the first two categories of education. These results indicate that the rise in educational achievement not only improves the socio-economic condition of the individuals, but it also improves their mentality, attitudes, and knowledge on the contribution of the pit lakes towards the development of the society. This averts them to pollute the pit water and pit lake areas. Figure 7.16 is the diagrammatic representation of Table 7.8.

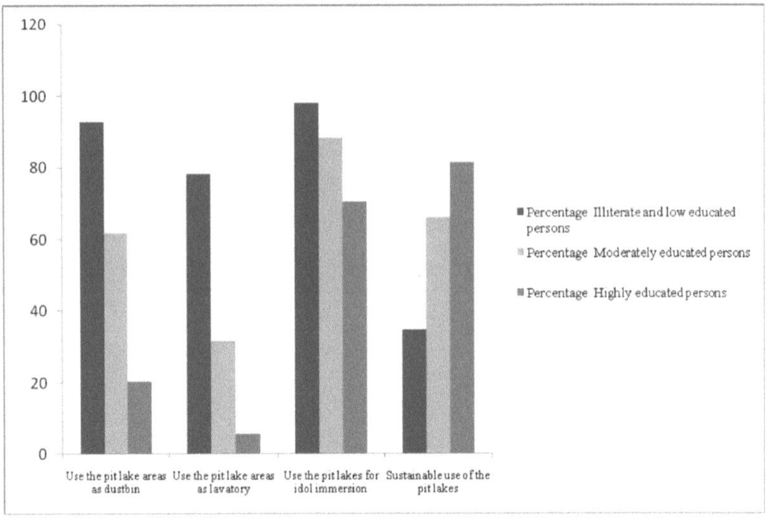

FIGURE 7.16 Use of the pit lakes by the individuals with different educational attainment.

7.4.3 AGE MAY BE AN IMPORTANT DETERMINANT FOR THE ATTITUDES OF THE INDIVIDUALS

If we look at the reported figures in Table 7.9, and the Figure 7.17, it can be found that the rise in age raises the awareness of the individuals regarding worst effect of the pit water and pit areas as dustbin, as this use can pollute the water and adjacent areas. It has been found that around 80% of youngest age group uses the pit water and pit lake areas as dustbin, the figures for medium and oldest age groups are 50% and 55% respectively. When the question was regarding the use of the pit lake areas as lavatory, then the responses of the age groups reversed, since around 20% of the individuals from the youngest age group, 24% of the respondents from medium age group and 45% of the respondents of the oldest age group assured that they usually use the adjacent areas of the pit lakes as lavatory. Likewise, the first case in the third case, i.e., in the case of idol immersion denial of the respondents from the youngest age group is greater. Similarly, in the context of the sustainability youngest age group occupies a better place than the medium and oldest age groups. Figures 7.17 is the pictorial representations of the last three case reported in Table 7.8, which enable us to understand the condition more clearly and quickly.

TABLE 7.9 Attitudes of the Local Residents Towards the Pit Lakes May Varied by Age

Different Types of Use	Percentage of Individuals
1. Use the pit lake areas as dustbin:	
(a) Age group one	78.42
(b) Age group two	48.35
(c) Age group three	53.34
2. Use the pit lake areas as lavatory:	
(a) Age group one	19.64
(b) Age group two	23.77
(c) Age group three	45.13
3. Use the pit lakes for idol immersion:	
(a) Age group one	27.88.
(b) Age group two	54.29
(c) Age group three	60.52
4. Sustainable use of the pit lakes:	
(a) Age group one	94.78
(b) Age group two	81.32
(c) Age group three	56.59

(Source: Primary survey done by the investigators).

Note:

(i) Age group one comprises the individuals aged between 15 years to 30 years; age group two contains the individuals aged between 36 years to 60 years, and age group three includes the individuals aged from 61 years and above.

(ii) Sustainable use of the pit lakes means that the individuals residing in the adjacent areas of the pit lakes do not do any work which can pollute the water of the pit lakes.

The study revealed that the rise in educational achievement changes the value judgment of the individuals completely. Household's income has a positive influence on the educational achievements of the individuals, as it raises the opportunity to achieve better education. Therefore, developing some developmental projects based upon the water and other resources of the pit lakes will have a direct stimulus on the socio-economic status of the surrounding areas of the pit lakes from one side and from other it will enable the society to enter into the track of the virtuous circle of development. Rise in literacy and educational achievement raises the environmental awareness of the people which can change the attitudes of individuals towards the resources of the pit lake like large water bodies. This will lead to the minimization of the risks of pollution and contamination of the pit lake water resources and surrounding areas. As a result, the health-related problems, as well as their problems, originated from low income and low education of the local residents will decline reduced.

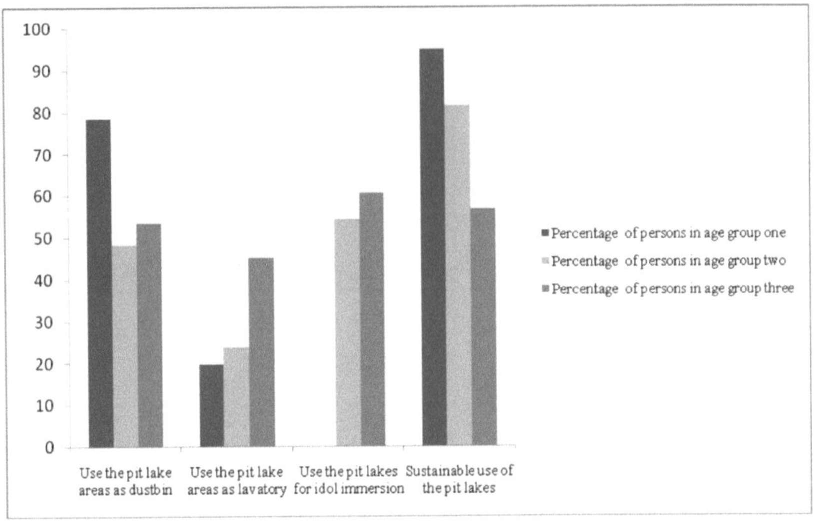

FIGURE 7.17 Use of the pit lakes by the people from different age groups.

7.5 CONCLUSION

This paper attempt to look at the contribution of the pit lakes and their water resources to the society vis-à-vis the attitudes of the individuals living in the neighboring areas towards these lakes, i.e., whether the activities of the individual pollute the water resources and surrounding land surfaces of the pit lakes. This study is based on the sample collected from the neighboring areas of the pit lakes of RCF areas. The prima facie evidence by analyzing the sample drawn from the earlier mentioned areas, that the water resources of the pit lakes and other resources available in the surrounding areas of these lakes have immense value for the society. These resources are usually used by the local residents for different purposes to maintain their livelihoods. In addition, people from distant regions also get different types of benefits from these lakes. If the local people change their attitude towards these lakes and utilize the water and other resources of these lakes, a prudent manner and by proper management instead of creating any pollution and contamination, then the potentiality of these resources to help the society will rise. This will also enable the society to achieve the trajectory of sustainable development.

KEYWORDS

- **asset**
- **livelihood**
- **mining**
- **pit lakes**
- **resources**
- **socio-economic status**

REFERENCES

Castro, J. M., & Moore, J. N., (2000). Pit lakes: Their characteristics and the potential for their remediation. *Environmental Geology*, *39*(11), 1254–1260.

Convention on Wetlands, (1971). *Convention on Wetlands of International Importance Especially as Waterfowl Habitat 1971.* Adopted by the International Conference on the Wetlands and Waterfowl at Ramsar, Iran. UNESCO, Paris, France.

Gurung, S., Sharma, S., & Sharma, C. M., (2009). A brief review on limnological status of high altitude lakes in Nepal. *Journal of Wetlands Ecology*,*3*, 12–22.

Jha, S., (2008). Status and conservation of lowland Terai wetlands in Nepal. *Our Nature*,*6*(1), 67–77.

Jones, H., & Mccullough, C., (2011). Regulator guidance and legislation relevant to pit lakes. In: McCullough, C. D., (ed.), *Mine Pit Lakes: Closure and Management* (p. 183). Australian Centre for Geomechanics, Perth.

Klapper, H., & Geller, W., (2002). Water quality management of mining lakes – a new field of applied hydrobiology. *Acta Hydrochimica et Hydrobiologica*, *29*, 363–374.

Lund, M. A., McCullough, C. D., & Kumar, R. N., (2012). In: McCullough, C. D., Lund, M. A., & Wyse, L., (eds.), *The Collie Pit Lake District, Western Australia: An Overview* (pp. 287–294). International Mine Water Association Symposium, IMWA.

McCullough, C. D., & Lund, M. A., (2006). Opportunities for sustainable mining pit lakes in Australia. *Mine Water and the Environment*, *25*(4), 220–226.

Millennium Ecosystem Assessment, (2005). *Ecosystems and Human Well-Being: Wetlands and Water Synthesis.* World Resources Institute, Washington, D. C., USA. [online] URL: http://www.unep.org/maweb/documents/document.358.aspx.pdf (Accessed on 4 August 2019).

Miller, G. C., Lyons, W. B., & Davis, A., (1996). Peer REVIEWED: Understanding the water quality of pit lakes. *Environmental Science & Technology*, *30*(3), 118–123.

Younger, P., & Wolkersdorfer, C., (2004). Mining impacts on the fresh water environment: Technical and managerial guidelines for catchments scale management. *Mine Water Environment*, *23*, 2–80.

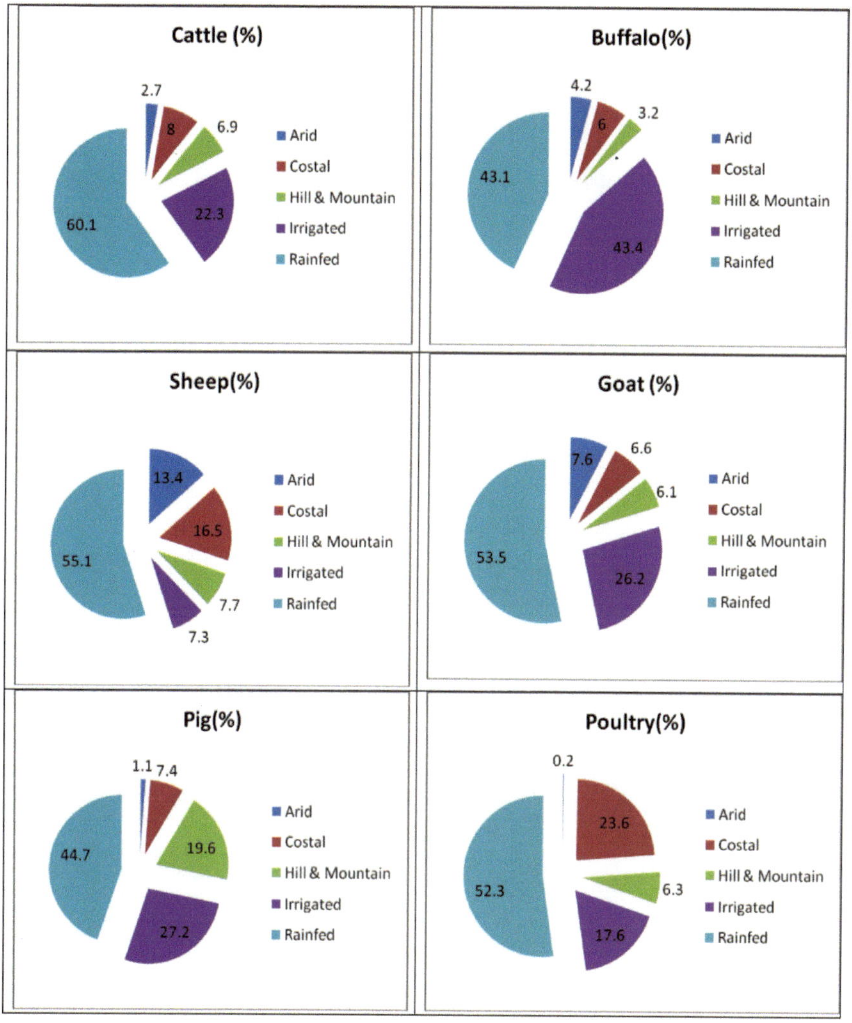

FIGURE 4.1 Distribution of major livestock's across different agro-ecological regions in the year 2003.

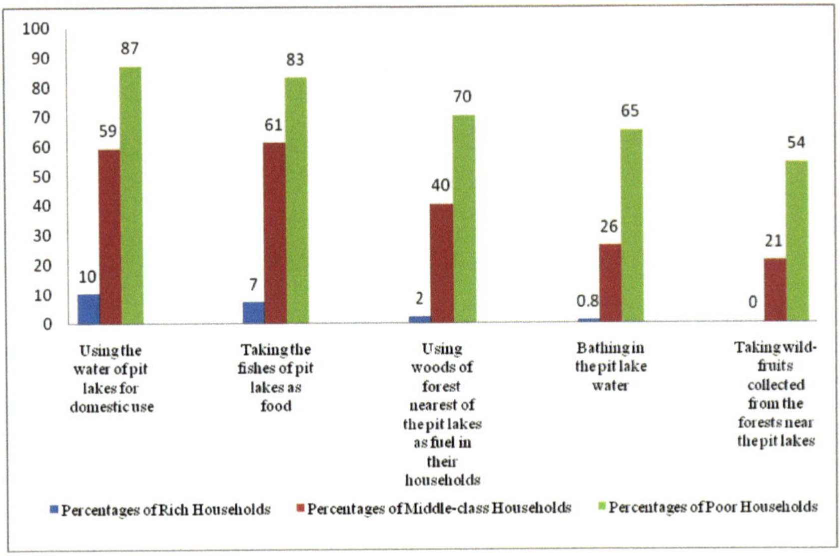

FIGURE 7.1 Percentage of households from different economic classes using pit lakes resources for their daily needs.

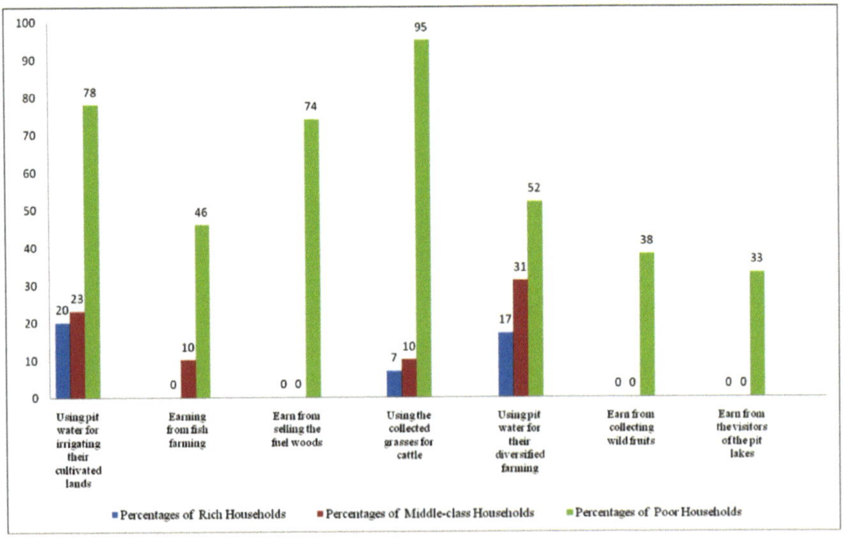

FIGURE 7.2 Percentage of the households from different classes maintaining their livelihoods by using pit lake resources.

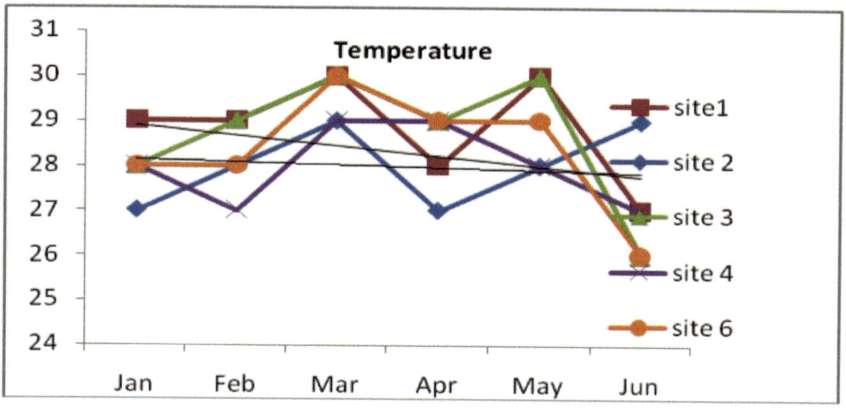

FIGURE 8.2 Temperature variations of six stations of public water supply.

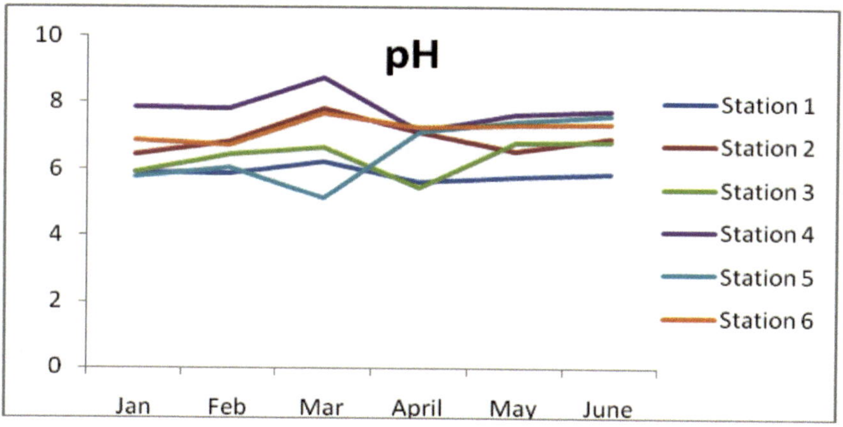

FIGURE 8.3 pH variation of six stations of public water supply.

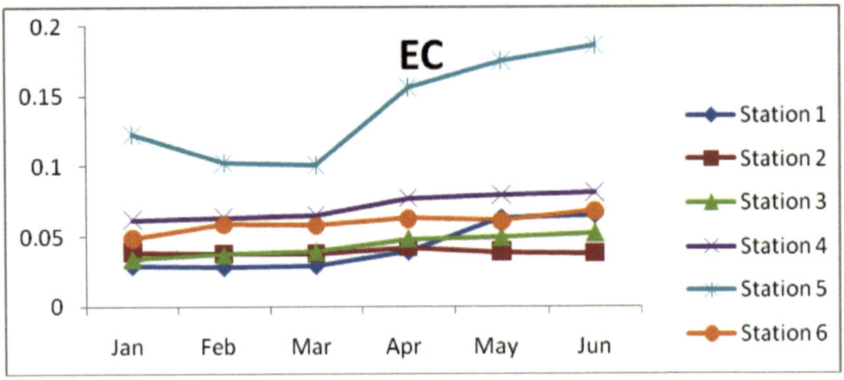

FIGURE 8.4 EC variations of six stations of public water supply.

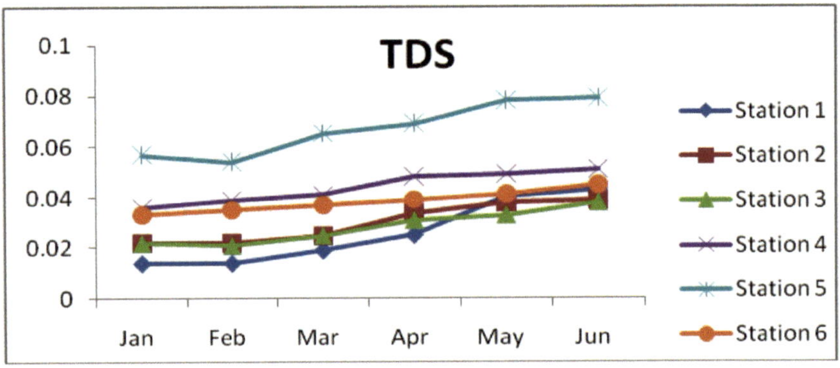

FIGURE 8.5 TDS variations of six stations of public water supply.

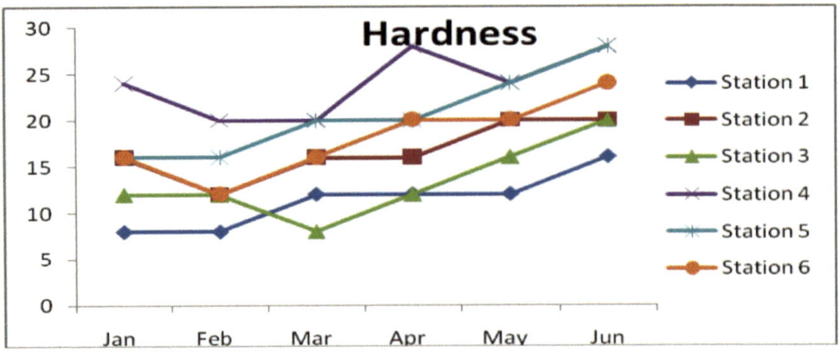

FIGURE 8.6 Hardness variations of six stations of public water supply.

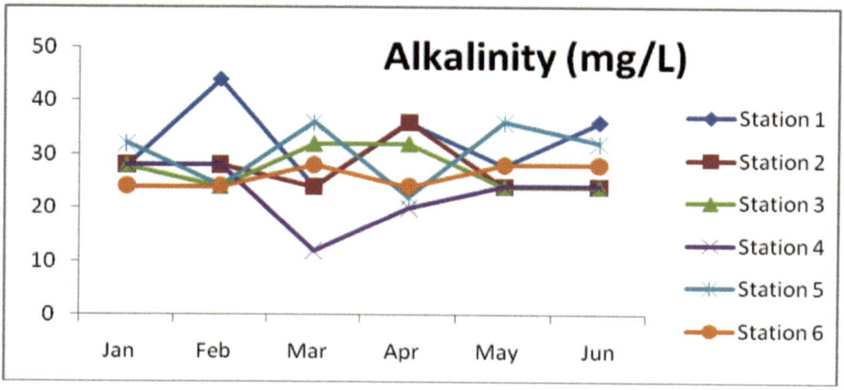

FIGURE 8.7 Alkalinity variations of six stations of public water supply.

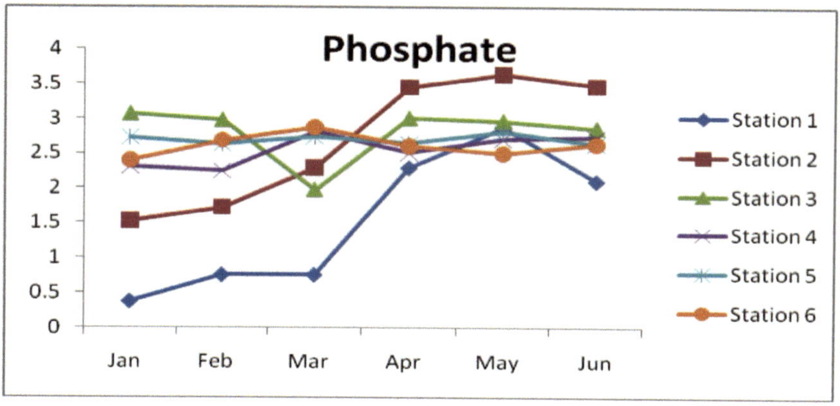

FIGURE 8.8 Phosphate variations of six stations of public water supply.

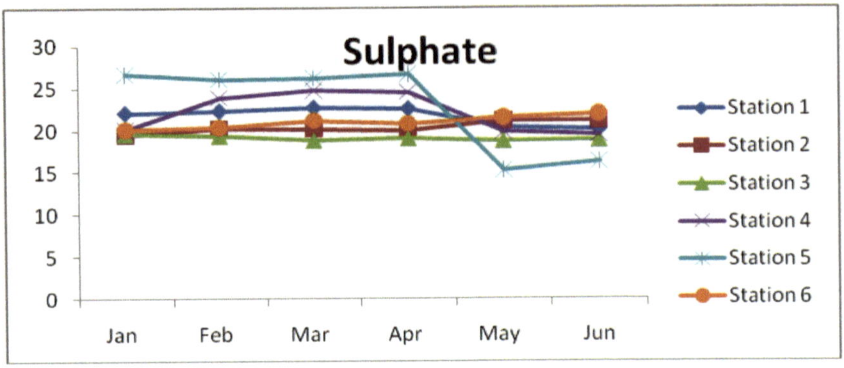

FIGURE 8.9 Sulfate variations of six stations of public water supply.

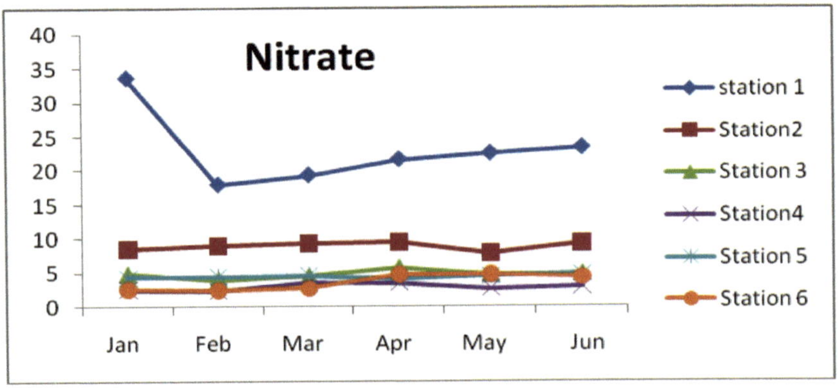

FIGURE 8.10 Nitrate variations of six stations of public water supply.

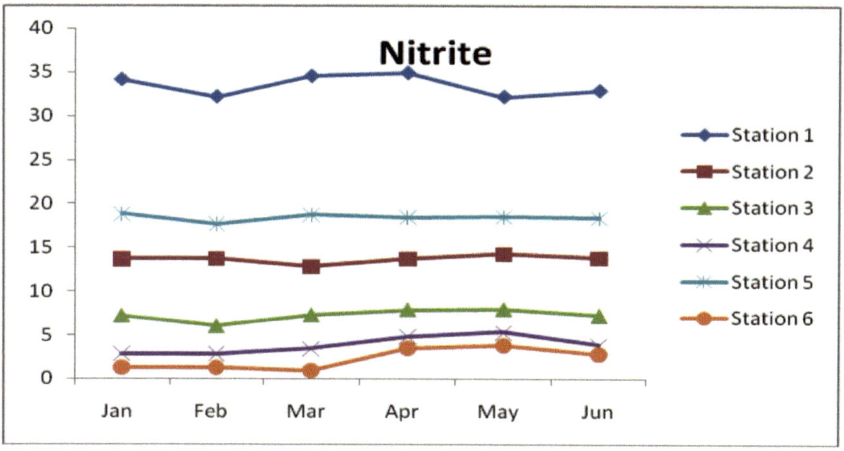

FIGURE 8.11 Nitrite variations of six stations of public water supply.

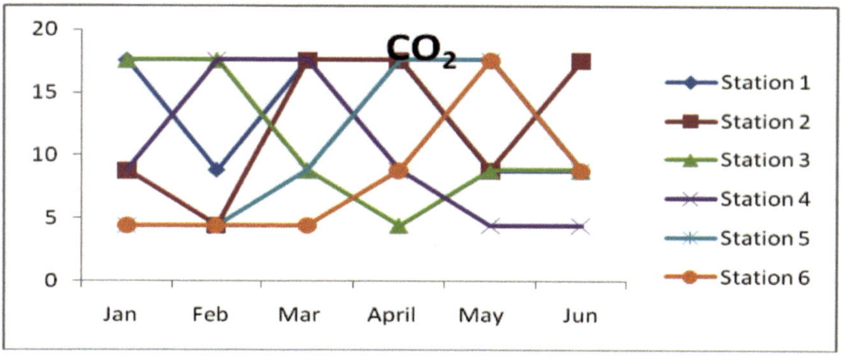

FIGURE 8.12 CO_2 variations of six stations of public water supply.

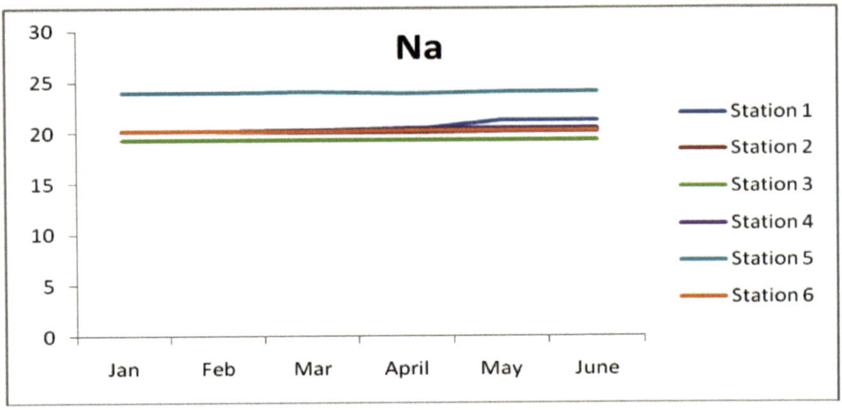

FIGURE 8.13 Sodium variations of six stations of public water supply.

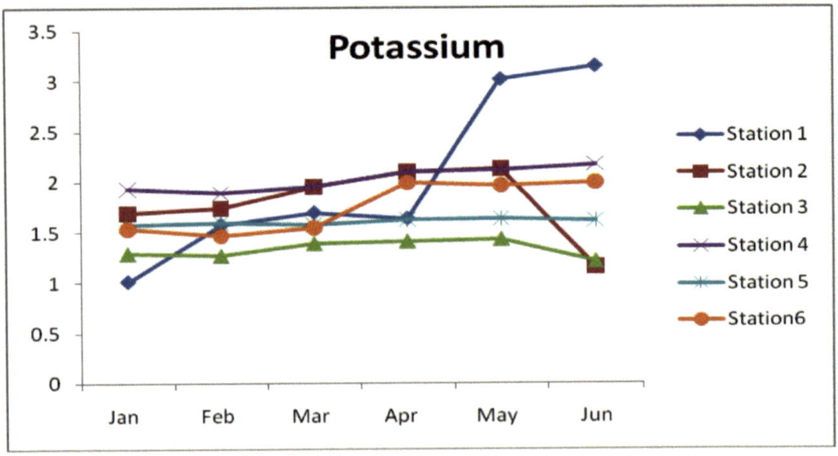

FIGURE 8.14 Potassium variations of six stations of public water supply.

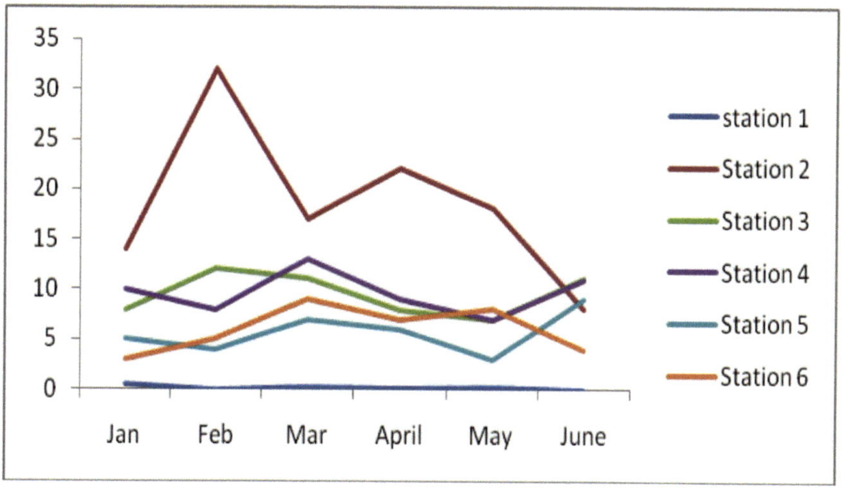

FIGURE 8.15 Residual chlorine variations of six stations of public water supply.

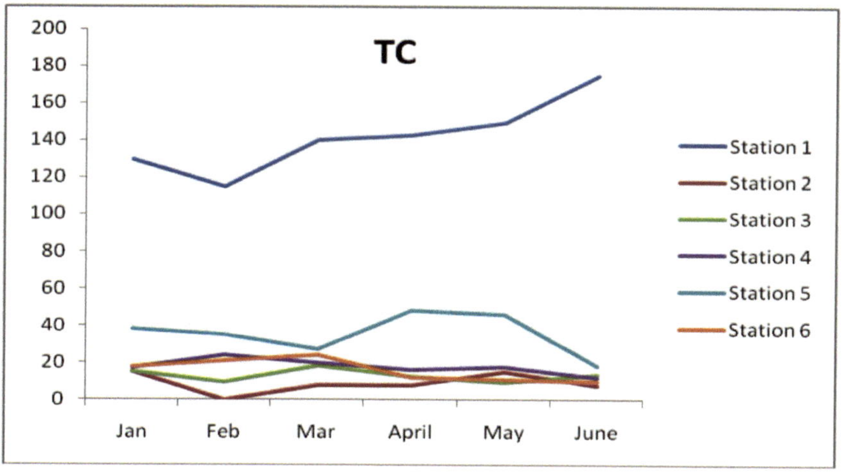

FIGURE 8.16 Total coliforms variations of six stations of public water supply.

FIGURE 9.2 Cow dung and crop residues used in agriculture at horticulture centre in Barjora, West Bengal.

FIGURE 9.3 Organic farming in horticulture centre at Barjora, West Bengal.

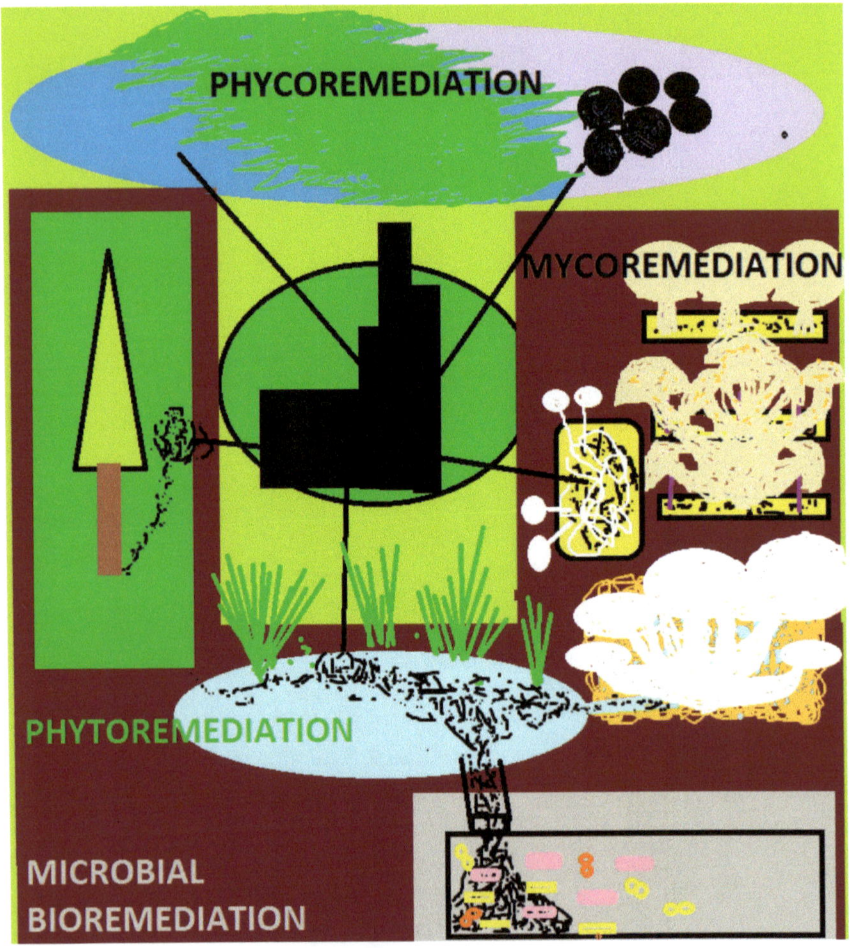

FIGURE 10.1 Different forms of bioremediation.

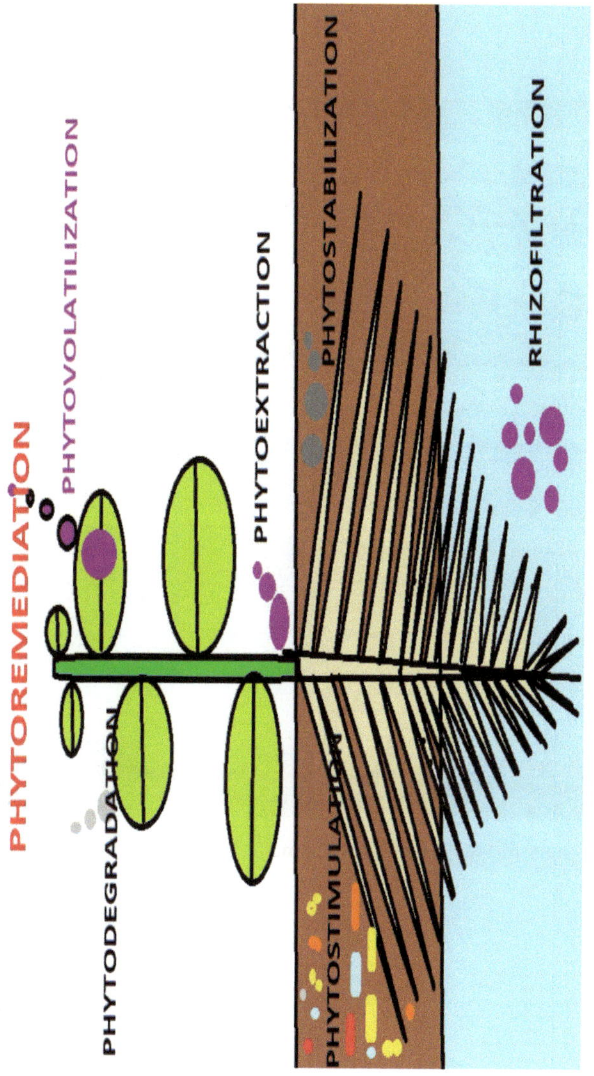

FIGURE 10.2 Different forms of phytoremediation.

FIGURE 10.3 Idea for sustainable development, biodiversity conservation, and profit making.

FIGURE 10.4 Inter-linkage and dependency of organism in ecosystem.

FIGURE 10.6 Spiral gardening foe conservation of biodiversity and to increase green plants in a polluted city with utilizing very less amount of space but can fight with air pollution in the polluted cities, where there is no land for forest building or another green belt or green cover.

CHAPTER 8

Chlorine Status and Drinking Water Quality Monitoring Within a Public Distribution System of Kollam District

ANILA GEORGE

PG Department of Environmental Science, St. John's College, Anchal, Kollam, Kerala, India, Mobile: +919496195458, E-mail: george44@gmail.com

ABSTRACT

Water is an essential commodity for human consumption, and it is one of the renewable resources which must be prevented from deterioration in quality. A reliable supply of clean and safe water is the first and most critical municipal service that people require. The urban and rural populationfaced a big problem of the scarcity of drinking water. In urban areas, the municipal supply of water is available only for a short time. Therefore, people are compelled to use pipeline water for their domestic purposes. The present study was carried out to assess the quality of public water supply distributed through a distribution network of various panchayats of Kollam district. The paper focuses on the Japan drinking water project, which has successfully installed in Kerala. The study was conducted at six stations which include four Panchayats with Japan drinking water household pipeline connection, the source (Kallada River) and treatment plant from January to June in 2016–2017. Water samples collected were analyzed for various physicochemical parameters following the standard methods, and it includes water temperature pH, conductivity, TDS, alkalinity, free carbon dioxide, hardness, free chlorine, nitrate, nitrite, inorganic phosphate, sulfate, and total coliforms (TC). Statistical analysis was done to find the correlation among the parameters using Pearson's Correlation matrix. The results revealed that the chlorine content and TC were present in significant number of water samples indicating microbial contamination of water passing through the distribution network. The overall water quality index (WQI) of Japan drinking water is 0.305, indicated that the water was partially clean on the basis of the desirable limits for physicochemical parameters.

8.1 INTRODUCTION

Water is a vital commodity for the survival of all living things on the earth. Safe and adequate supply of clean water is essential for the human development, health, and well being of living things. The demand for water for domestic, irrigational, and industrial uses also increased as the population increases. Safe drinking water for all is one of the most crucial challenges of present age. The declining water availability of public distribution system is one of the environmental issues which are facing now a day.

Kerala is one of the states blessed with high rainfall and large number of water bodies. Kerala has 41 west-flowing rivers and 3 east-flowing rivers. Kerala receives an average rainfall of 3000 mm (Economic review, 2016). From the point of view of water resources, Kerala is having both abundance and scarcity. In Kerala, water scarcity is one of the main problems during summer months. Encroachment, destruction of forest, wetlands, paddy fields, ponds, and changes in the cropping pattern has increased the problem of water scarcity. The water scarcity leads to lowering of water table and drying of rivers (Yesodharan et al., 2007). A special investigation carried out in villages of Kerala based on quality and accessibility of water identified that 99% panchayats, faced drinking water problems. Water scarcity was more severe in coastal and hilly regions. In the thickly populated coastal stretch of Kerala, freshwater is becoming more and more contaminated that locals are abandoning their domestic wells and depending on public water supply. The scarcity of drinking water is a big problem faced both by the urban and rural population. In urban areas, the municipal supply of water is available only for a short time. Therefore, people are compelled to use pipeline water for drinking and cooking purposes (Chaudhary et al., 2005). The provision of drinking water supply is an imperative need of the society. Water can be safe only if we care and consider during the catchment, treatment, storage, and distribution. Water suppliers and governing authority has a public health responsibility to provide safe water for the public.

8.2 PUBLIC DISTRIBUTION SYSTEMS

Public distribution system is the distribution of the sufficient and uniform quantity of water through the designed network of pipes. The aim of the distribution system is to provide adequate quantity of water at sufficient pressure at their points. Water reaches the consumers through their house connection pipes from a vast network of pipelines from the main branch. The expense of the water is collected by the water distribution agencies.

Water supply system consists of the following components (Figure 8.1):

i. Source of water supply;
ii. Treatment plant;
iii. Transmission and distribution system.

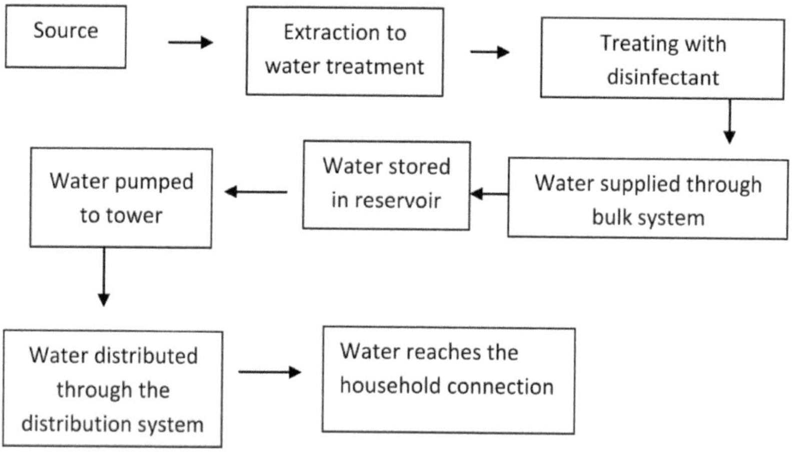

FIGURE 8.1 A small illustration of a typical water supply system.

8.3 DETERIORATION OF PUBLIC DISTRIBUTION

Water leaving after treatment may not be stable and can undergo changes in its properties during distribution. The water quality deterioration in a public distribution system can be caused by various factors. Factors which deteriorate water quality include internal corrosion, leaching of chemicals, accumulation, and release of contaminants, biofilm formation, and bacterial regrowth and physicochemical processes that occur in the system pipe

(EPA, 2006). Due to these factors, there will be a cross-contamination of water supply.

8.3.1 DISTRIBUTION SYSTEM INADEQUACIES

A number of failures will be there in a public distribution system, namely, inadequate pressure, disinfection, intermittent water supply, leaks, corrosion, etc.

1. **Inadequate Disinfection**: Chlorine is a well-known disinfection agent which is used to kill all the microbes present in drinking water, and it is one of the important steps in the distribution system to protect public health. It is good, but its odor, taste, and reactivity may badly influence the people to choice chlorine as a disinfectant (Besner et al.,2002). Like the appropriate treatment in the water treatment plant, the maintenance of chlorine concentration (residual chlorine) in the distribution system is also very important. Chloramines which have a greater persistent capacity are also used in the distribution system. The WHO recommends residual chlorine of 0.2 to 0.5 mg/L in a public distribution system (WHO, 1997). Any failure in the disinfection of public distribution system results in disease outbreaks. Residual chlorine level declines as the distance between the treatment plant and household tap connections increases. The rate of depletion of residual chlorine may be affected by other factors such as water flow velocity, residence time, age, and material of pipes, water pressure, etc. (Egorov et al., 2002).
2. **Inadequate Pressure:** In addition to disinfection, the next important one is maintaining adequate pressure throughout the entire network. Reduction in pressure in the pipe may cause vacuum effect, and these can draw contaminants through leaks into the pipes. When scarcity of water occurs, the pressure in the pipeline is reduced their cause the reversal flow of water.
3. **Intermittent Water Supply:** Intermittent supply provides consumers with water for only certain periods of the day. Factors influencing intermittency are population growth, power shortages, non-revenue, pipe breaks, or leaks. It is common in developing countries. Intermittent supply influences the water usage and wastage than the continual water supply (Kumar, 1998). This is

because during supply hours each household draws a maximum quantity due to unawareness of the arrival of the next water supply. They also construct storage tanks and their own pumps and store a large quantity of water and used for other purposes. Leaks cause a large quantity of water wastage and loss of pressure and thus the contamination due to the entry of pathogenic organisms and making the water quality poor. This problem can be reduced by our awareness of water quality deterioration between the point of entry to a distribution system and to the consumers tap.

8.4 PUBLIC DISTRIBUTION IN KERALA

In the Kerala context, it is found that majority of the population depend on open wells for potable water, i.e., 86% in rural Kerala and 55% in urban Kerala. Piped water supply covers 64.02% of the population in Kerala and 59.19% in Kollam district. Kerala Water Authority has 116 schemes in operation in Kollam district. Drinking water is supplied through 20395 taps and 1685049 (Domestic) and 6691 (Non-domestic) and 219 (Industrial) connections (Economic Review, 2017).

8.4.1 WATER SUPPLY OF KOLLAM DISTRICT

Sasthamcotta Lake serves as one of the source of water supply for Kollam Corporation and a number of Gram Panchayats. Water from the intake well located at Sasthamcotta is treated in the 37.5mld treatment plant at Sasthamcottta and supplied to Kollam Corporation and the Gram Panchayats of Neendakara, Sasthamcotta, Sooranad South and Chavara. Water tapped from Adikkadu and Pottakkannanmukku is supplied after disinfection to West Kallada and Mynagappally, respectively. The rivers flowing through Kollam district includes Kallada River, Ithikkara River, Pallikal River, and Achenkovil River and these rivers are utilized for the water supply in Kollam district. Kallada River serves as the source of water supply schemes to Punalur Municipality and Kundara and adjoining Grama Panchayats with tapping points near Punalur Town and treatment plants at Punalur and Elampal respectively. The Japan drinking water project also utilizes the water from Kallada River. The Achenkovil River serves as a source of water supply scheme for Chithara and adjoining Gram panchayat with intake at Mylamood and treatment plant at Madathara. Ithikkara River also serves as a source of major part of water supply in Kollam district. It includes panchayat such as Kalluvathukkal, Parippally (tapping point at Aduthala), WSS to Adichanalloor and Chathannoor (tapping point at Kattachal), RWSS to Ayoor (tapping point at Peringalloor), WSS to Elavoor Anakkulam (tapping point at Puthar, Channapetta), WSS to Anchal and Puthiyam Kuzhiyathadam (tapping point at Kuzhiyathadam), WSS to Kallumala Harijan colony (tapping point at Kuzhiyam) and WSS to Chadayamangalam (tapping point at Poomkode). Pallikkal River, which is a tributary of Ithikkara River, serves as a source of water supply scheme of Nedumpana and to Pallimon Harijan colony with tapping point at Kundamon. Achenkovil River supply water to Achenkovil grama panchayat.

The present investigation was carried out with the following objectives:

- To analyze the water quality of Japan drinking water project from the source to the distribution system.
- To find out the chlorination level.
- To identify the availability of water during the scarcity period.
- To find out the water quality index (WQI).

8.5 MATERIALS AND METHODS

The Japan drinking water project was launched at Kerala on August 17, 2003. The Japan International Cooperation Agency (JICA) aided drinking water scheme of the Kerala Water Authority has been commissioned successfully. Around 25 lakh people spread out across 40 villages of Kerala will be benefited by the availability of high quality and uninterrupted supply of clean drinking water. There are five projects running under the scheme.

1. WSS to Sherthalai and adjoining panchayats in Alapuzha district;
2. WSS to Meenad and adjoining panchayats in Kollam district;
3. WSS to Trivandrum and adjoining panchayats;
4. WSS to Pattauvam and adjoining panchayats in Kannur district; and
5. WSS to Kozhikode and adjoining panchayats.

A package of 5 water supply schemes was approved by the Japan Bank for International Cooperation (JBIC). Every project has four components viz; intake (pump house) water treatment plant (transmission main line) reservoirs and water distribution lines.

8.5.1 STUDY AREA

Meenad drinking water project in the Kollam district was one of the schemes. Kallada River is the primary source of water for the scheme. Using two 970 HP pumps, water is pumped to the treatment plant at Panamkuttymala, which is about 120 m above the pumping station. It has the capacity to treat and distribute 73 million liters of water per day. Each grama panchayat will have its own reservoir. Water will enter these reservoirs through gravitational force from the balancing reservoir attached to the water treatment plant at Panamkuttymala, thereby saving electricity for the purpose of pumping the water up the reservoir in the panchayats. With the latest technology for the setting up of pumping units and transmission lines, the Supervisory and data acquisition system (SCADA) for monitoring of water supply is employed. The grama panchayats covered by the scheme are Karavaloor, Anchal, Edamulackal, Ummanur, Elamad, Pooyapally, Veliyam, Mayyanad, Adichanallur, Chirakkara, Poothakulam, and Chathanoor.

The supply of water in this area is determined by the availability of water in the Kallada river which is the source of water supply. Kallada River is one of the rivers that flow through the Kollam district of Kerala. The river is formed by the three rivers viz. Kulathupuzha, Chenthurini, and Kazhuthurutty. The length of the river is 121 km and drainage area is 1996 Km^2. Kallada river basin is bounded by Achenkovil basin on the North and Ithikkara basin on the South. Most of the population depends upon the river water for irrigation and drinking water projects.

8.5.2 SAMPLING STATIONS

- Station 1: Source (Kallada River).
- Station 2: Treatment plant (Panamkuttimala).
- Station 3: Karavaloor panchayath.
- Station 4: Edamulakal panchayath.

- Station 5: Pooyapally panchayath.
- Station 6: Chathanoor panchayath.

The samples were collected covering six stations from January to June 2016–2017. Samples for analysis were taken directly from these six stations and stored in well labeled and tightly capped sterilized bottles. Samples for bacteriological analysis were collected separately. Water samples collected were analyzed for various physical, chemical, and microbial parameters followed as per standard methods of APHA (1998) and Trivedi and Goel (1986) as given in Table 8.1.

TABLE 8.1 Methods Applied for Analysis of Various Parameters

Parameter/Attributes	Methods
Temperature	Thermometer
pH	pH meter
Electrical conductivity	Conductivity meter
TDS	Gravimetric method
Total hardness	EDTA method.
Free chlorine	Titrimetric method
Total Alkalinity	Titrimetric method.
Phosphate	Stannous chloride method.
Sulfate	Turbidimetric method
Nitrate	Brucine method
Sodium	Flame photometric method
Potassium	Flame photometric method
Coliforms	MPN method

8.5.3 FIELD SURVEY

There is also a customer's satisfaction survey had been carried out to obtain primary data and information directly from customers using a structured questionnaire. The main aim of the survey is to obtain information to read the pulse of people regarding water supply problems.

8.5.3.1 WATER QUALITY INDEX (WQI)

It is valuable and unique rating to expresses overall water quality at a certain location and time based on several water quality parameters. It was calculated following the guidelines and formula given by Kaur et al. (2001).

$$W.Q.I = \Sigma wi \times Vr$$

where, Wi = unit weight of parameters; Vr = rating value of parameters.

$$Wi = \frac{K}{V_i}$$

where, Wi = Unit weight of the parameters; Vi = desirable limit of the parameters; and K = Constant of proportionality.

Value of K is calculated as $K = 1 \Big/ \sum_{i=1}^{11} \frac{1}{Vi}$

The desirable limits set by BIS (1991) and the mean of estimated parameters formed the basis for index calculation.

8.6 RESULTS AND DISCUSSION

8.6.1 TEMPERATURE

The average temperature values of water samples in public distribution system ranges from 27.5°C to 28.83°C (Table 8.2). The maximum 28.83°C temperature was recorded in the Station 1 (Kallada) and minimum 27°C in Station 5 (Pooyapally). The temperature varies between 27°C to 30°C in Kallada River which is the source of drinking water project (Figure 8.2).

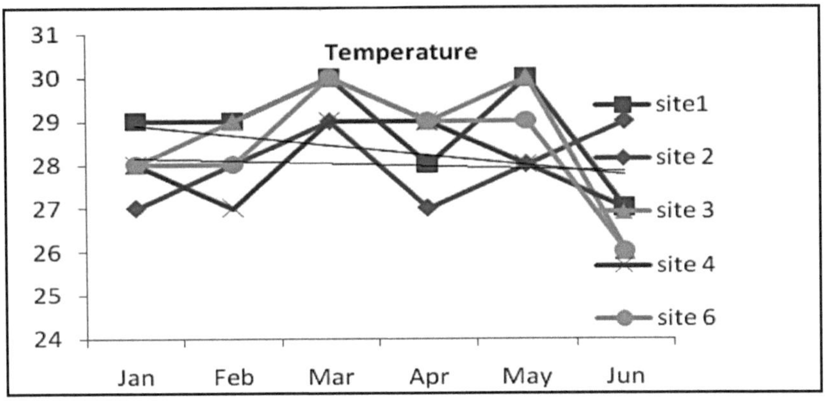

FIGURE 8.2 **(See color insert.)** Temperature variations of six stations of public water supply.

8.6.2 PH

The pH indicates the acidic or alkaline material present in the water. pH balance in an ecosystem is maintained when it is within the range of 5.5 to 8.5. Among the pipeline water samples Station 4 shows the highest pH value (7.83) and the Station 1 shows the minimum (5.89) pH value (Table 8.2). The BIS value of pH for drinking water is 6.5–8.5 while that of the WHO standards in between 7.0–8.5. The pH values observed in the present finding was within the safe limit set by BIS, ICMR (1975) and WHO (1985). The corrosion in pipes starts when pH is less than 6.5. Presence of toxic metals in water increases the pH (Patelet al., 2016) (Figure 8.3).

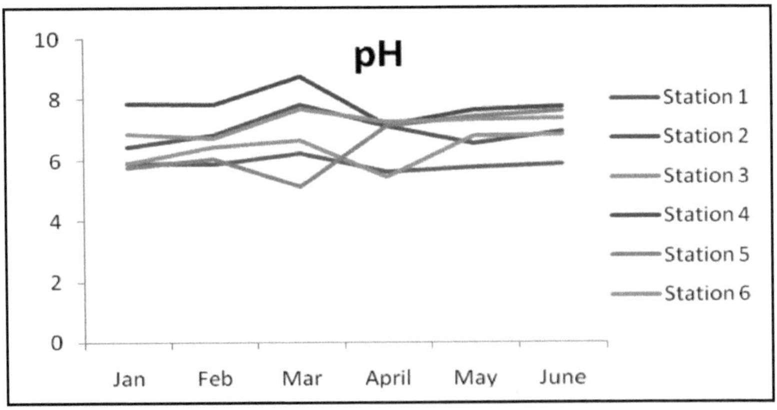

FIGURE 8.3 **(See color insert.)** pH variation of six stations of public water supply.

8.6.3 ELECTRICAL CONDUCTIVITY

The values of electrical conductivity of water samples of Station 5 (Pooyapally) shows the maximum 0.146 µs/cm^{-1} and Station 1(treatment plant) shows the minimum 0.039 µs/cm^{-1} conductivity (Table 8.2). Conductivity of water is attributed to the presence of dissolved solids, especially chloride contained in it (APHA, 1998). According to BIS, the permissible limit is 300µs/cm^{-1}. This indicates that here, no adverse effects associated with the electrical conductivity of the water were expected (Figure 8.4).

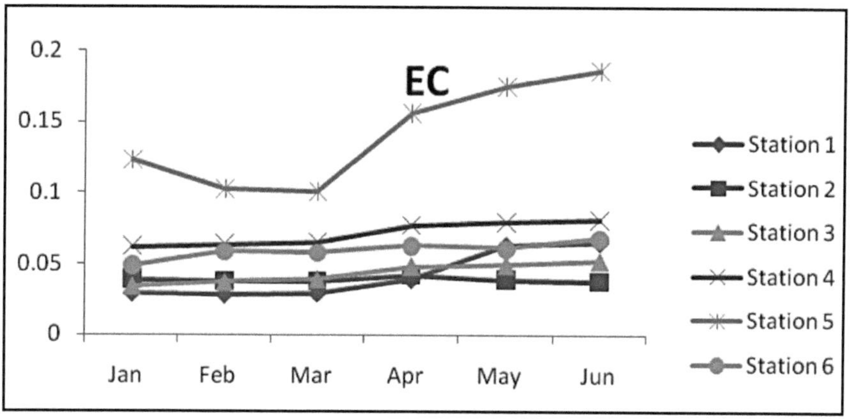

FIGURE 8.4 **(See color insert.)** EC variations of six stations of public water supply.

8.6.4 TOTAL DISSOLVED SOLIDS

Total dissolved solids ranged from 0.026 mg/L to 0.067 mg/L. Highest TDS (0.067) were found in Station 5 (Pooyapally) whereas lowest TDS (0.025) were recorded in samples of Station 1 (source). Cemented storage tank may also contribute to increase the level of TDS during storage. The high amount of TDS adversely affects the quality of running water and it is unsuitable for irrigation and drinking. The permissible limit of TDS in surface water is 500 mg/L (BIS, 1991). In the present study, all the samples are within the limit (Figure 8.5).

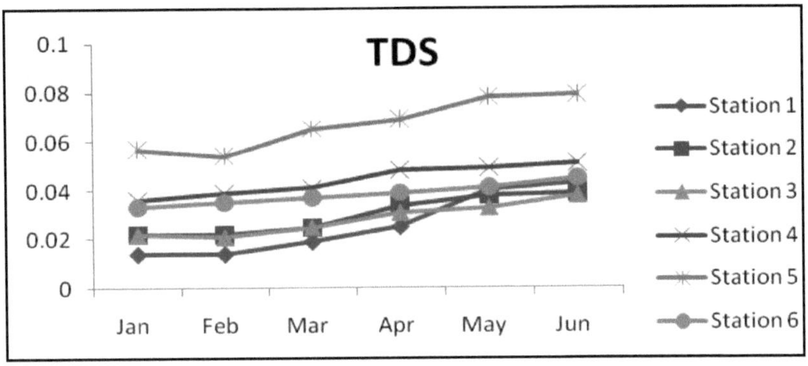

FIGURE 8.5 (See color insert.) TDS variations of six stations of public water supply.

8.6.5 TOTAL HARDNESS

Total hardness is caused primarily by the presence of cations such as calcium and magnesium and anions such as carbonate, bicarbonate, and chloride in water. Total hardness detected minimum 11.3mg/L in Station 1 and maximum value 24 mg/L was recorded in Station 4. According to BIS the maximum permissible limit of hardness for drinking water is 500 mg/L and all the samples were within the limit (Figure 8.6).

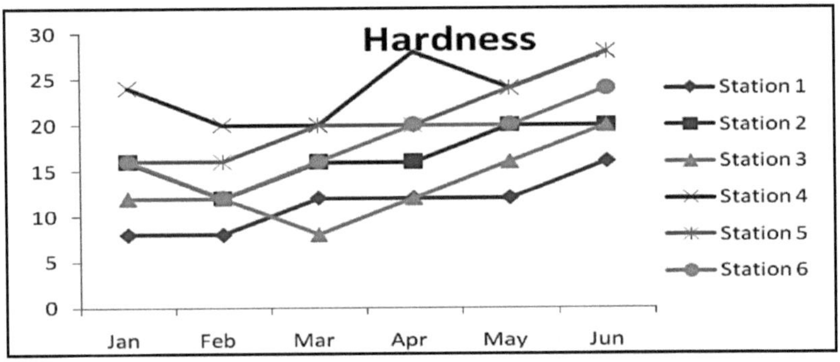

FIGURE 8.6 (See color insert.) Hardness variations of six stations of public water supply.

8.6.6 ALKALINITY

Alkalinity of any water is mainly due to carbonates, bicarbonates, and hydroxides. The value of alkalinity in water samples ranges between 22 mg/L to 32.67 mg/L. Maximum in Station 1 (source) and minimum in

Station 4. The desirable limit of alkalinity is 200 mg/L and the permissible limit is 600 mg/L (WHO, 1997). Here all the samples are within the limit (Figure 8.7).

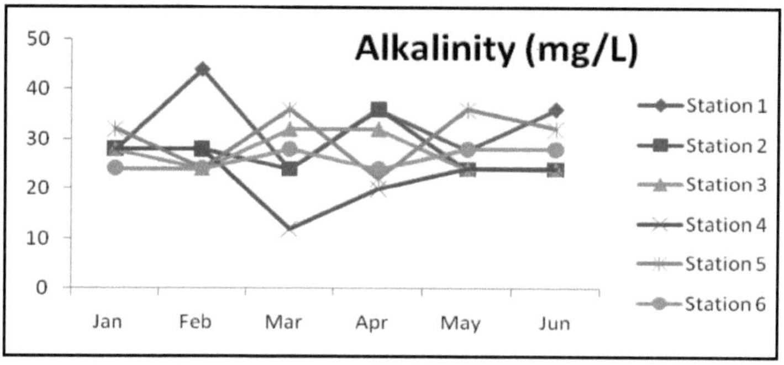

FIGURE 8.7 (See color insert.) Alkalinity variations of six stations of public water supply.

8.6.7 PHOSPHATE

The concentration of phosphate varies from 1.53 mg/L to 2.81 mg/L and the maximum value 2.81 mg/L was recorded in Station 3 and minimum value (1.0mg/L) in Station 1. The high values in Station 2 are mainly due to surface run off and agricultural runoff. According to the United States Public Health Standard, the permissible limit of phosphate is 0.1 mg/L and all the samples have phosphate concentration higher than the prescribed limit. George and Prakasam (2011) also reported the phosphorous value ranged from 0.01 to 0.29 mg/L in Edamulakal panchayat (Figure 8.8).

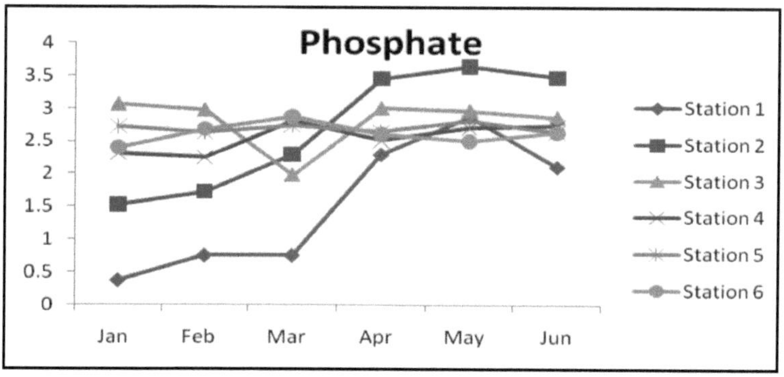

FIGURE 8.8 (See color insert.) Phosphate variations of six stations of public water supply.

8.6.8 SULFATE

Sulfate exist in natural waters and it is one of the least toxic and major anions that present in water. The maximum value 22.84 mg/L was observed in Station 5 while lowest value 20.34 mg/L was recorded in Station 2. High concentration of sulfate may decrease pH of the water and may increase bacterial load (John et al., 2008). The desirable limit of sulfate according to the BIS is 200 mg/L and the water samples were under acceptable limit (Figure 8.9).

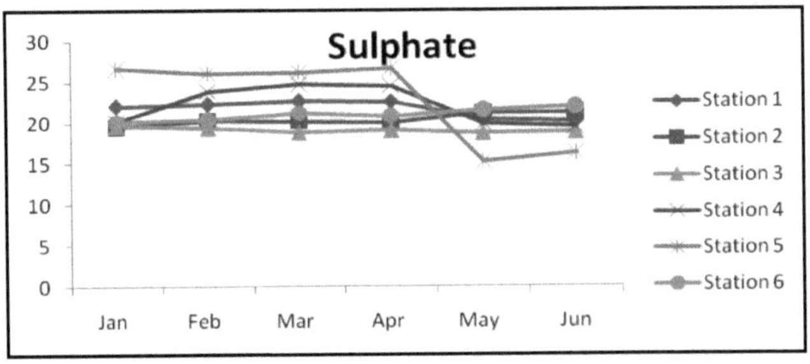

FIGURE 8.9 (See color insert.) Sulfate variations of six stations of public water supply.

8.6.9 NITRATE

Nitrate values ranged from 2.89 mg/L to 23.01 mg/L, and the highest concentration 23.01 mg/L was observed in Station 1 (source), and lowest concentration 2.89 mg/L was found in Station 4 (Edamulackal). Nitrate is the highest oxidized form of nitrogen. The likely reasons for their higher values in Kallada River (source) could be runoff from agricultural fields. The prescribed limit for nitrate, according to BIS is 45 mg/L (Figure 8.10).

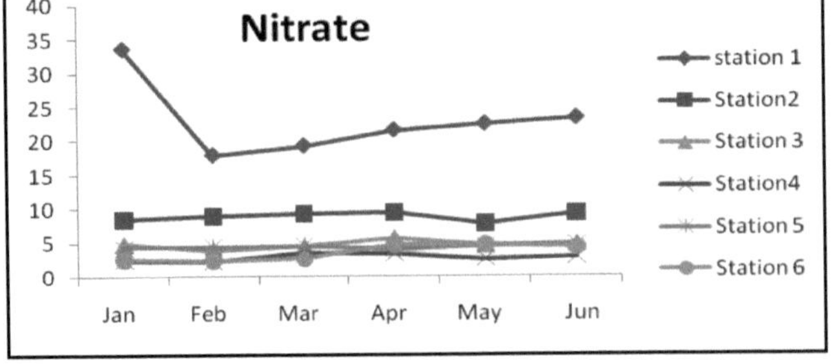

FIGURE 8.10 (See color insert.) Nitrate variations of six stations of public water supply.

8.6.10 NITRITE

The nitrite value of six sample sites ranges from 2.28 mg/L to 33.46 mg/L and the maximum value was recorded in source (Kallada river) and minimum value 2.28 mg/L was in Station 6 (Chathanoor) (Figure 8.11).

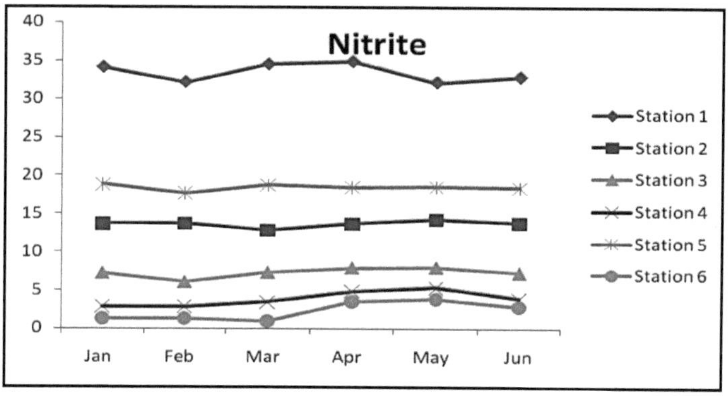

FIGURE 8.11 (See color insert.) Nitrite variations of six stations of public water supply.

8.6.11 CARBON DIOXIDE (CO_2)

The value of free carbon dioxide ranges from 8.07 mg/L to 13.2 mg/L and the maximum value (13.2 mg/L) were recorded in source (Kallada River) and minimum value (8.07 mg/L) in Station 5. According to ISI standards the surface water contains less than 10 mg/L free CO_2. The high acidity is also impacted by the dissolution of CO_2 in groundwater. The processing of organic matter (OM) might be the reason for the high CO_2 content in the water (Figure 8.12).

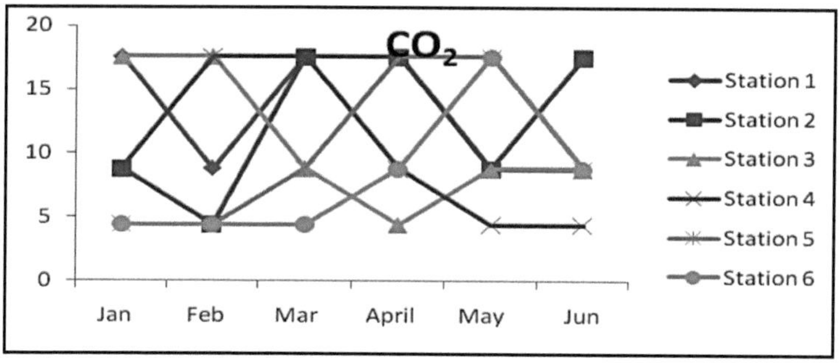

FIGURE 8.12 (See color insert.) CO_2 variations of six stations of public water supply.

8.6.12 SODIUM

The maximum Na (23.94 mg/L) was found in sampling Station 5 while minimum Na (19.3 mg/l) was recorded in Station 3. According to BIS standard the permissible limit of sodium in drinking water permissible limit is 50 mg/L and all water samples were under the safe limits. Concentration of sodium in natural surface water vary considerably depending on local geologic conditions, wastewater discharges, etc. (Figure 8.13). Potassium is a naturally occurring element and the concentration remains quite lower than the sodium and calcium. Maximum concentration (2.02 mg/L) potassium was found in Station 4 (Edamulackkal) whereas the lowest value (1.59 mg/L) was detected in Station 5 (Pooyapally). The BIS value for potassium in drinking water is 100 mg/L. All the samples are within the limit as prescribed by BIS (Figure 8.14).

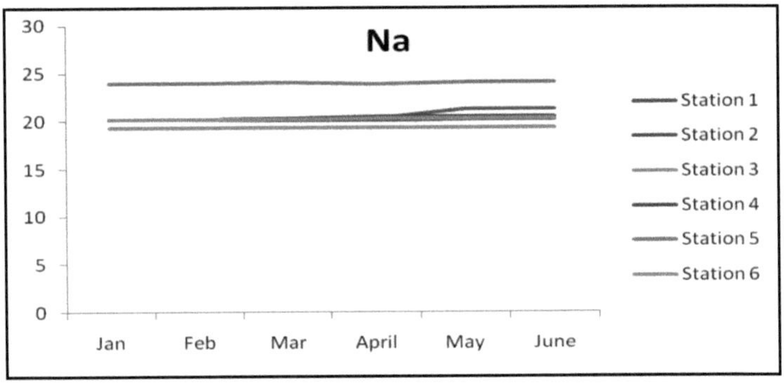

FIGURE 8.13 **(See color insert.)** Sodium variations of six stations of public water supply.

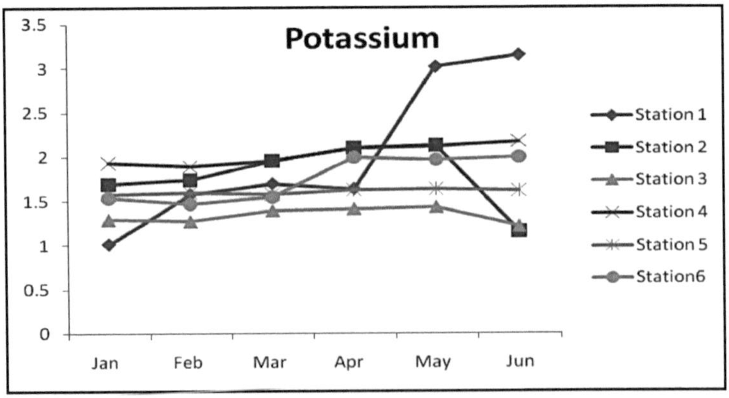

FIGURE 8.14 **(See color insert.)** Potassium variations of six stations of public water supply.

8.6.13 FREE CHLORINE

Free chlorine and chloramines are used in drinking water to provide primary disinfection. Dichloramines and trichloramines are relatively unstable and quickly decompose at pH values above 7. The excessive addition of free chlorine causes some health problems.

In the present study, the free chlorine ranged from 5.66 mg/L to 18.5mg/L. The free chlorine concentration is low in the river water (source), and it is high in the treatment plant, and this is due to the treatment process for the inactivation of pathogens. The residual chlorine values of the water at the treatment plants are higher than those of sampling stations. Chlorine concentration decreases from different time duration. As a matter of fact, some of the samples have no chlorine at all, which is alarming since chlorination is essential to maintain the quality of water. When the concentration of chlorine is about 2 to 3 mg/L, consumers can smell an irritant odor. Concentration below the optimum level for disinfection can cause the bacterial growth in distribution system. In the present study, there is a gradual decrease in chlorine as the water moves in the distribution system. The result showed that the free chlorine at inlet and outlet of the storage tank dropped from 18.5 mg/L to 3 mg/L. Here the chlorine level is high in some sample. If the chlorine is high in treated water, it is dangerous to human health and causes allergy in the skin, irritation to their nose and eyes, stomach discomfort, etc. (USEPA, 2012) (Figure 8.15).

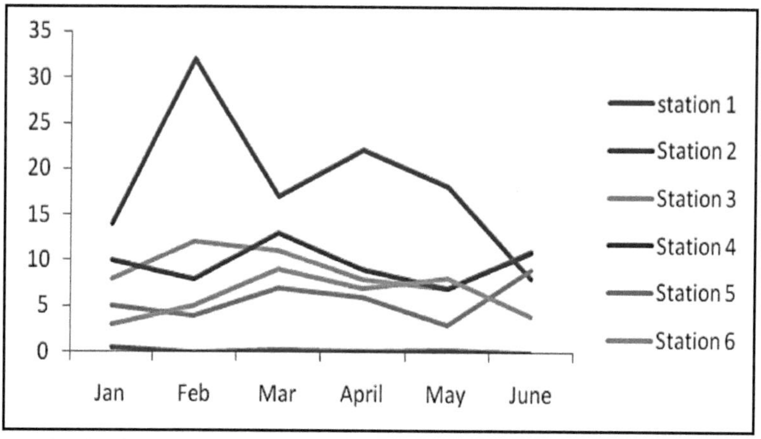

FIGURE 8.15 (See color insert.) Residual chlorine variations of six stations of public water supply.

8.6.14 *TOTAL COLIFORMS (MPN)*

Coliform group has been used extensively as an indicator of water quality which leads to public health protection. Majority of coliform are seen in human intestinal flora and other warm-blooded animals and thus found in fecal waste. Maximum count was found in Station 1 (Kallada river), while the minimum count is 9 MPN/100 ml in sample Station 2 (treatment plant) (Table 8.2). The total count detected in all the samples were above the permissible limit set by WHO (1997). Kallada River (source) is highly polluted with coliforms, and this may be due to anthropogenic activities. According to WHO (World Health Organization) , there should not be a single bacteria per 100 ml of water. The coliform count showed an increasing trend from the treatment plant towards household connections. Presence of coliform in treated water may reveal that treatment is ineffective or the intrusion of contaminated water into the distribution system. Drinking water with coliforms can cause stomach and intestinal illness, including diarrhea and nausea and even lead to death (Figure 8.16).

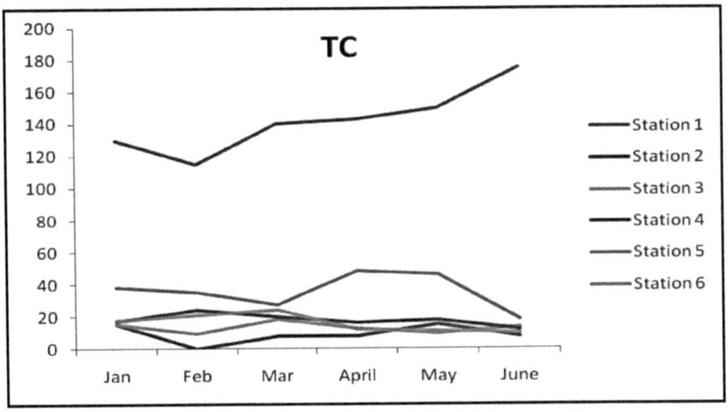

FIGURE 8.16 (See color insert.) Total coliforms variations of six stations of public water supply.

TABLE 8.2 Average Values of Various Parameters at Six Sampling Station with Standard Deviation

Parameters	Station 1	Station 2	Station 3	Station 4	Station 5	Station 6
Temperature (°C)	28.83 ± 1.1	28.0 ± 0.8	28.66 ± 1	28 ± 0.89	27.5 ± 1.2	28.33 ± 17
pH	5.89 ± 0.19	6.94 ± 0.5	6.34 ± 0.1	7.83 ± 1.4	6.51 ± 1.0	7.19 ± 0.3

TABLE 8.2 *(Continued)*

Parameters	Station 1	Station 2	Station 3	Station 4	Station 5	Station 6
TDS(mg/L)	0.026 ± 0.0	0.03 ± 0.0	0.027 ± 0	0.04 ± 0.1	0.06 ± 0.1	0.03 ± 0.1
Conductivity (μs/cm^{-1})	0.042 ± 0.0	0.039 ± 0	0.043 ± 1	0.07 ± 0.0	0.14 ± 0.04	0.06 ± 0.0
Hardness (mg/L)	11.33 ± 3.0	16.6 ± 3.0	13.33 ± 4.	24.0 ± 3.5	20.6 ± 4.6	18.0 ± 4.2
Alkalinity (mg/L)	32.67 ± 7.3	27.3 ± 4.6	27.33 ± 3.	22.67 ± 6.2	30.33 ± 5.9	26.0 ± 2.1
Phosphate (mg/L)	1.53 ± 1.02	2.69 ± 0.9	2.81 ± 0.4	2.55 ± 0.2	2.70 ± 0.0	2.62 ± 0.1
Sulfate (mg/L)	21.69 ± 1.1	20.34 ± 0.	19.08 ± 0.	22.1 ± 2.5	22.84 ± 5.	21.02 ± 7
Nitrate (mg/L)	23.01 ± 5.5	8.9 ± 1.28	4.76 ± 0.5	2.89 ± 0.5	4.47 ± 0.2	3.55 ± 1.1
Nitrite (mg/L)	33.46 ± 1.2	13.6 ± 0.4	7.28 ± 0.7	3.8 ± 1.03	18.4 ± 0.4	2.28 ± 1.2
Chlorine (mg/L)	0.43 ± 0.30	18.5 ± 8.0	9.5 ± 2.07	9.67 ± 1.9	5.67 ± 2.5	6.0 ± 2.37
Carbon dioxide (mg/L)	13.2 ± 3.88	12.5 ± 5.8	11.0 ± 5.3	10.27 ± 6.0	10.27 ± 6.0	8.07 ± 5.1
Sodium (mg/L)	20.49 ± 0.5	20.1 ± 0.0	19.3 ± 0.0	20.34 ± 0.1	23.94 ± 0.0	20.2 ± 0.0
Potassium (mg/L)	2.00 ± 0.86	1.8 ± 0.36	1.33 ± 0.0	2.02 ± 0.12	1.59 ± 0.03	1.75 ± 0.2
TC(MPN/ 100 ml)	142.16 ± 2	9 ± 5.5	12.66 ± 3.	17.83 ± 4.	35.33 ± 11.	16 ± 5.8

Values are in mg/L except for temp, pH, EC, and, and MPN (Mean ± SD).

8.6.15 CORRELATION BETWEEN DIFFERENT PARAMETERS

Correlation studies were done to find the correlation between different parameters by calculating Pearson's correlation coefficient for the parameters like temperature, pH, TDS, EC, total hardness, alkalinity, phosphate, sulfate, nitrate, nitrite, residual chlorine, carbon dioxide, sodium, potassium, and total coliforms (TC) of Japan drinking water distribution system. Tables 8.3–8.8 shows the degree of correlation between parameters for Station 1 to Station 6 respectively. The pH shows positive correlation with TDS (r = 0.918 at Station 3), EC (r = 0.948), TH (r = 0.733), K (r = 0.921) at Station 6 and with alkalinity (r = 0.780), SO_4 (r = 0.722), Cl (r = 0.695) and negatively correlated with nitrite (r = –0.839) at Station 2, SO4 (r = –0.541) at Station 3, TH (r = –0.727 at Station 4, CO_2 (r = 0.749) at Station 5. TDS is positively with EC (r = 0.988) total hardness (r = 0.845 at Station 1 (r = 0.838 at Station 2), phosphorous (r = 0.859 at Station 1 and (r = 0.972) at Station 2, Na (r = 0.955 at Station 1, (r = 0.983), K (r = 0.96), nitrite (r = 0.815) at Station 3 and negative correlation with sulfur (r = –0.89) at Station 5. EC is highly correlated with TDS (r = 0.988 at Station 1, temp (r = 0.722 at Station 2, TDS (r = 0.098 at Station 4, pH (r = 0.948 at Station 5, TDS (r = 0.899) at Station 6.

Total hardness is positively correlated with TDS (r = 0.988 at Station 1, r = 0.838 at Station 2) EC (r = 0.722) at Station 3 pH (r = 0.735) at Station 5. Alkalinity is positively correlated with EC (r = 0.88 at Station 2) pH (r = 0.780 at Station 6) and negatively correlated with total hardness (r = –0.780 at Station 6). Phosphate is positively correlated with TDS (r = 0.857, EC r–0.84 at Station 1). Sulfate is positively correlated with TDS (r = 0788, pH r = 0.722), EC (r = 0.80), TH (r = 0.776), alkalinity (r = 0.844) at Station 6 and negatively correlated with TDS (r = –0.89), EC (r = –0.94) at Station 1.

Chlorine is positively correlated with alkalinity (r = 0.738) at Station 1 and negatively with nitrite(r = –0.765) at Station 3. Sodium is positively correlated with TDS (r = 0.953) EC (r = 0.98) TH (r = 0.713), PO_4 (r = 0.731) and negatively correlated with sulfate (r = –0.977) at Station 1.

8.6.16 WATER QUALITY INDEX (WQI)

In the WQI calculation, only nine physical and chemical parameters as given in the table was chosen. The desirable limits set by BIS (1991) and the annual means of estimated parameters formed the basis for index

TABLE 8.3 Correlation Coefficient Values Among the Physic-Chemical Parameters of Water Samples in Station 1(source)

	Temp	pH	TDS	EC	TH	Alkalinity	PO_4	SO_4	NO_3	Nitrite	Cl	CO_2	Na	K	TC
Temp	1														
pH	0.438	1													
TDS	–0.35	–0.33	1												
EC	–0.34	–0.38	0.988	1											
TH	–0.49	0.44	0.845	0.77	1										
Alkalinity	–0.54	–0.48	–0.07	–0.03	–0.12	1									
PO_4	–0.22	–0.63	0.857	0.84	0.648	0.04	1								
SO_4	0.286	0.263	–0.89	–0.94	–0.604	0.014	–0.64	1							
NO_3	–0.1	–0.12	–0.12	–0.04	–0.249	–0.373	–0.249	–0.112	1						
Nitrite	–0.07	0.146	–0.37	–0.46	0.017	–0.415	–0.245	0.628	0.243	1					
Cl	–0.29	–0.18	–0.32	–0.37	–0.095	0.738	–0.07	0.515	–0.682	0.092	1				
CO_2	0.156	0.22	–0.56	–0.62	–0.242	–0.497	–0.41	0.723	0.355	0.954	–0.013	1			
Na	–0.23	–0.22	0.955	0.98	0.713	–0.092	0.731	–0.977	–0.019	–0.56	–0.475	–0.682	1		
K	–0.24	–0.18	0.96	0.95	0.791	0.009	0.779	–0.899	–0.286	–0.55	–0.256	–0.717	0.961	1	
TC	–0.52	–0.12	0.089	0.85	0.95	–0.211	0.646	–0.74	0.037	–0.02	–0.341	–0.244	0.806	0.799	1

**Correlation is significant at the 0.01 level (2- tailed)

*Correlation is significant at the 0.05 level (2-tailed)

TABLE 8.4 Correlation Coefficient Values Among the Physico-Chemical Parameters of Water Samples in Station 2 (Treatment Plant)

	Temp	**pH**	**TDS**	**EC**	**TH**	**Alkalinity**	**PO_4**	**SO_4**	**NO_3**	**Nitrite**	**Cl**	**CO_2**	**Na**	**K**	**TC**
Temp	1														
pH	0.538	1													
TDS	0.226	–0.101	1												
EC	–0.722	–0.056	0.261	1											
TH	0.297	–0.154	0.838	0.738	1										
Alkalinity	–0.765	–0.039	–0.09	0.88	–0.42	1									
PO_4	0.185	0.042	0.972	0.38	0.75	0.032	1								
SO_4	0.618	–0.061	0.788	–0.30	0.66	–0.53	0.724	1							
NO_3	0.19	0.673	–0.08	0.204	–0.31	0.411	–0.012	–0.248	1						
Nitrite	–0.371	–0.839	0.546	0.241	0.38	0.095	0.458	0.460	–0.591	1					
Cl	–0.304	0.006	–0.46	0.143	–0.8	0.412	–0.343	–0.246	0.005	0.03	1				
CO_2	0.336	0.668	0.456	0.291	0.43	0.107	0.519	0.116	0.656	–0.428	–0.567	1			
Na	–0.473	–0.917	–0.24	–0.155	–0.17	0.008	–0.389	–0.125	–0.531	0.619	0.167	–0.810	1		
K	–0.425	0.135	–0.1	0.519	–0.19	0.359	0.093	–0.23	–0.296	0.004	0.487	–0.127	–0.27	1	
TC	–0.28	–0.382	0.289	0.184	0.67	–0.184	0.235	–0.006	–0.594	0.283	–0.632	0.080	0.10	0.203	1

**Correlation is significant at the 0.01 level (2- tailed)

*Correlation is significant at the 0.05 level (2-tailed)

TABLE 8.5 Correlation Coefficient Values Among the Physico-Chemical Parameters of Water Samples in Station 3

	Temp	pH	TDS	EC	TH	Alkalinity	PO_4	SO_4	NO_3	Nitrite	Cl	CO_2	Na	K	TC
Temp	**1**														
pH	–0.071	1													
TDS	–0.32	0.918	1												
EC	–0.28	0.222	0.19	1											
TH	–0.685	0.388	0.555	0.722	1										
Alkalinity	0.36	–0.592	–0.678	–0.214	–0.722	1									
PO_4	–0.354	–0.3378	–0.16	0.16	0.516	0.5	1								
SO_4	–0.144	–0.541	–0.363	–0.775	–0.331	0.015	0.43	1							
NO_3	–0.127	–0.659	–0.55	0.469	0.099	0.542	0.227	–0.161	1						
Nitrite	0.151	–0.173	–0.106	0.592	0.185	0.349	–0.006	–0.55	0.778	1					
Cl	–0.256	0.402	0.231	–0.214	–0.093	–0.147	–0.424	0.045	–0.626	–0.782	1				
CO_2	–0.108	0.069	0.205	–0.781	–0.158	–0.415	0.256	0.793	–0.68	–0.765	0.275	1			
Na	–0.354	0.195	0.089	0.929	0.626	–0.154	0.057	–0.708	0.37	0.339	0.11	–0.746	1		
K	0.846	–0.231	–0.404	0.12	–0.45	0.51	–0.248	–0.39	0.373	0.639	–0.591	–0.532	–0.042	1	
TC	–0.063	–0.052	–0.002	–0.328	–0.46	0.677	–0.71	0.007	0.146	0.125	0.165	–0.093	–0.282	–0.008	1

**Correlation is significant at the 0.01 level (2- tailed)
*Correlation is significant at the 0.05 level (2-tailed)

TABLE 8.6 Correlation Coefficient Values Among the Physico-Chemical Parameters of Water Samples in Station 4

	Temp	**pH**	**TDS**	**EC**	**TH**	**Alkalinity**	**PO_4**	**SO_4**	**NO_3**	**Nitrite**	**Cl**	**CO_2**	**Na**	**K**	**TC**
Temp	**1**														
pH	0.142	1													
TDS	–0.036	–0.45	1												
EC	–0.07	–0.545	0.988	1											
TH	0	–0.727	0.692	0.758	1										
Alkalinity	–0.743	–0.527	–0.173	–0.052	0.148	1									
PO_4	0.324	0.355	0.631	0.552	0.171	–0.721	1								
SO_4	0.527	0.211	–0.269	–0.371	–0.411	–0.556	–0.099	1							
NO_3	0.788	0.236	0.272	0.178	0.187	–0.916	0.611	0.535	1						
Nitrite	0.333	–0.499	0.815	0.827	0.523	–0.222	0.524	–0.137	0.277	1					
Cl	0.31	0.692	–0.166	–0.252	–0.103	–0.655	0.444	0.218	0.647	–0.413	1				
CO_2	0.163	0.595	–0.668	–0.764	–0.818	–0.324	–0.266	0.794	0.148	–0.614	0.316	1			
Na	0.016	–0.576	0.983	0.988	0.797	–0.119	0.536	–0.263	0.284	0.828	–0.204	–0.721	1		
K	–0.019	–0.499	0.961	0.98	0.821	–0.086	0.598	–0.448	0.239	0.792	–0.131	–0.829	0.972	1	
TC	0	0.355	–0.632	–0.678	–0.889	0.022	–0.436	0.57	–0.274	–0.38	–0.237	0.813	0.69	–0.8	1

**Correlation is significant at the 0.01 level (2- tailed)
*Correlation is significant at the 0.05 level (2-tailed)

TABLE 8.7 Correlation Coefficient Values Among the Physico-Chemical Parameters of Water Samples in Station 5

	Temp	pH	TDS	EC	TH	Alkalinity	PO_4	SO_4	NO_3	Nitrite	Cl	CO_2	Na	K	TC
Temp	**1**														
pH	–0.55	1													
TDS	–0.69	0.77	1												
EC	–0.638	0.948	0.886	1											
TH	–0.768	0.733	0.959	0.838	1										
Alkalinity	–0.627	–0.135	0.384	0.114	0.39	1									
PO_4	–0.339	–0.126	0.217	0.033	0.045	0.765	1								
SO_4	0.953	–0.749	–0.842	–0.81	–0.863	–0.491	–0.302	1							
NO_3	–0.798	0.142	0.42	0.251	0.638	0.609	0.037	–0.645	1						
Nitrite	–0.07	–0.122	0.306	0.177	0.193	0.631	0.567	–0.063	0	1					
Cl	–0.075	0.079	0.329	0.217	0.501	0.072	–0.5	–0.063	0.461	0.254	1				
CO_2	–0.119	0.619	0.688	0.615	0.459	–0.016	0.308	–0.389	–0.312	0.208	–0.158	1			
Na	–0.846	0.079	0.393	0.178	0.549	0.742	0.325	–0.687	0.937	0.009	0.16	–0.216	1		
K	–0.414	0.921	0.734	0.824	0.611	–0.178	0.024	–0.658	–0.055	–0.188	0.157	0.818	–0.019	1	
TC	0.371	0.141	–0.119	0.033	–0.394	–0.335	0.338	0.197	–0.839	0.001	–0.765	0.568	0.640	0.392	1

**Correlation is significant at the 0.01 level (2- tailed)
*Correlation is significant at the 0.05 level (2-tailed)

TABLE 8.8 Correlation Coefficient Values Among the Physico-Chemical Parameters of Water Samples in Station 6

	Temp	**pH**	**TDS**	**EC**	**TH**	**Alkalinity**	**PO_4**	**SO_4**	**NO_3**	**Nitrite**	**Cl**	**CO_2**	**Na**	**K**	**TC**
Temp	**1**														
pH	0.344	1													
TDS	–0.429	0.544	1												
EC	–0.355	0.423	0.899	1											
TH	–0.418	0.533	0.882	0.664	1										
Alkalinity	0	0.780	0.676	0.462	0.522	1									
PO_4	0.319	0.529	0.120	0.359	–0.180	0.368	1								
SO_4	–0.243	0.722	0.938	0.800	0.776	0.884	0.306	1							
NO_3	–0.133	0.402	0.792	0.700	0.841	0.318	–0.222	0.634	1						
Nitrite	–0.125	0.230	0.705	0.621	0.735	0.222	–0.348	0.528	0.973	1					
Cl	0.804	0.695	0.156	0.227	0	0.462	0.568	0.347	0.254	0.211	1				
CO_2	0.041	0.277	0.607	0.424	0.570	0.468	–0.343	0.570	0.798	0.867	0.361	1			
Na	–0.308	0.433	0.879	0.783	0.925	0.365	–0.156	0.712	0.969	0.905	0.130	0.680	1		
K	–0.248	0.419	0.846	0.729	0.908	0.359	–0.222	0.684	0.988	0.942	0.163	0.749	0.993	1	
TC	0.476	–0.102	–0.746	–0.586	–0.850	–0.187	0.516	–0.523	–0.905	–0.913	0.159	–0.733	–0.913	–0.929	1

**Correlation is significant at the 0.01 level (2- tailed)
*Correlation is significant at the 0.05 level(2-tailed)

calculation. The average value of water quality parameters of the public distribution system are given in Table 8.9.

TABLE 8.9 Showing the Desirable Limits of Water Quality Parameters According to BIS (1991)

Serial No	Parameter	Desirable Limit
1	pH	6.5
2	Total Hardness (mg/L)	300.0
3	Alkalinity (mg/L)	200.0
4	T.D.S (mg/L)	500.0
5	Sodium (mg/L)	0.3
6	Potassium (mg/L)	10
7	Phosphate (mg/L)	0.1
8	Sulfate (mg/L)	200.0
9	Nitrate (mg/L)	45.0

8.6.16.1 UNIT WEIGHT (WI)

Unit weight implies importance of each parameter in the overall water quality. It is dependent on the desirable limits of drinking water. Parameters having low desirable limits can harm the quality of water to a large extent even by a slight increase and such parameters have high weightings. On the contrary, parameters which have high desirable limit have low weightings (Kaur et al., 2001) Wi was calculated using the formula given below-

$$\mathrm{Wi} = \frac{K}{V_i}$$

where, Vi = Desirable limits.

$$K = 1 \Big/ \sum_{i=1}^{11} \frac{1}{Vi}$$

8.6.16.2 RATING VALUE (VR)

In order to calculate WQI, parameters were assigned rating values which ranged from 0 to 100 (Table 8.10). For this, the range of their concentration in water was divided into five intervals Viz Vr = 0 indicated that the parameter exceeded the desirable limit and that the water was not clean; Vr = 100 implied that the parameter had most desirable value that water was absolutely clean. Other ratings implied intermediate status is given in Tables 8.11–8.14.

TABLE 8.10 Water Quality Status of Public Water Distribution System

Rating Value	Quality of Water
0	Not clean
0–24.9	Partially clean
25–49.9	Clean
50–74.9	Very clean
75–99.9	Very-very clean
100	Absolutely clean

TABLE 8.11 Physical and Chemical Quality of Dug Well Water (Mean Value)

Serial No.	Water Quality Parameters	Estimated Values (Mean)
1	pH	6.78
2	Total hardness (mg/L)	17.32
3	Alkalinity (mg/L)	27.71
4	T.D.S (mg/L)	0.036
5	Sodium (mg/L)	20.72
6	Potassium (mg/L)	1.74
7	Phosphate (mg/L)	2.48
8	Sulfate (mg/L)	21.17
9	Nitrate (mg/L)	7.93

TABLE 8.12 Rating Values (Vr) Assigned to Parameters on the Basis of Their Concentrations in Water

Serial No.	Parameters	Rating Values				
		100	75	50	25	0
1	pH	8.5–8.1	8.0–7.6	7.5–7.0	6.9–6.5	<6.5
2	Total hardness (mg/L)	0.0–74.0	75.0–149.0	150.0–224.0	225.0–300.0	>300
3	Alkalinity (mg/L)	0.0–49.0	50.0–99.0	100.0–149.0	150.0–200.0	>200
4	TDS (mg/L)	5.0–124.0	125.0–249.0	250.0–374.0	375.0–499.0	>500

TABLE 8.12 *(Continued)*

Serial No.	Parameters	Rating Values				
		100	75	50	25	0
5	Sodium (mg/L)	0.0–49.0	50.0–99.0	100.0–149.0	150.0–200.0	>200
6	Potassium (mg/L)	0.0–2.49	2.50–4.49	5.00–7.49	7.5–9.99	>10
7	Phosphate (mg/L)	0.0–0.024	0.025–0.049	0.050–0.074	0.075–0.099	>0.1
8	Sulfate (mg/l)	0.0–49.0	50.0–99.0	100.0–149.0	150.0–199.0	>200
9	Nitrate (mg/l)	0.0–11.3	11.4–22.5	22.6–33.7	33.8–44.9	>45

Calculation of unit weight (wi) of parameters (Table 8.12)

$$Wi = \frac{K}{Vi} \qquad K = \frac{1}{\sum_{I=1}^{12} \frac{1}{Vi}}$$

$$ie\ K = \frac{1}{\frac{1}{V_1} + \frac{1}{V_2} + \frac{1}{V_3} + \ldots\ldots \frac{1}{V_{11}}}$$

$$ie\,K = \frac{1}{\frac{1}{6.5} + \frac{1}{300} + \frac{1}{200} + \frac{1}{500} + \frac{1}{200} + \frac{1}{10} + \frac{1}{0.1} + \frac{1}{200} + \frac{1}{45}}$$

$$= \frac{1}{10.295} = 0.097$$

K of 0.097 was found to be the unit weight of parameters.

TABLE 8.13 Calculation of Unit Weights (Wi) of Different Parameters

Serial No.	Parameter	K/Vi	Unit Weight (Wi)
1	pH	0.097/6.5	0.014
2	Total Hardness (mg/l)	0.097/300	0.0003
3	Alkalinity (mg/l)	0.097/200	0.0004

Serial No.	Parameter	K/Vi	Unit Weight (Wi)
4	TDS (mg/L)	0.097/500	0.0001
5	Sodium (mg/L)	0.097/200	0.0004
6	Potassium (mg/L)	0.097/10	0.0073
7	Phosphate (mg/L)	0.097/0.1	0.97
8	Sulfate (mg/L)	0.097/200.0	0.0004
9	Nitrate (mg/L)	0.097/45.0	0.0016

TABLE 8.14 Water Quality Index Calculated from Wi and Vr

Serial No.	Parameter	Estimated Value (Mean)	Vr	Wi	Wi×Vr
1	pH	6.78	0	0.014	0
2	Total Hardness (mg/L)	17.32	100	0.0003	0.03
3	Alkalinity (mg/L)	27.71	75	0.0004	0.03
4	T.D.S (mg/L)	0.036	50	0.0001	0.005
5	Sodium (mg/L)	20.72	100	0.0004	0.04
6	Potassium (mg/L)	1.74	0	0.0073	0.00
7	Phosphate (mg/L)	2.48	0	0.97	0.00
8	Sulfate (mg/L)	21.17	100	0.0004	0.04
9	Nitrate (mg/L)	7.93	100	0.0016	0.16
					ΣWi x Vr = 0.305

The WQI of water samples of public water supply is shown in Table 8.14. The overall WQI of Japan drinking water is 0.305, indicated that the water was partially clean on the basis of the desirable limits set by BIS (1991) for physio-chemical parameters.

8.6.17 *SURVEY RESULTS*

In addition to the numerical data of water quality analysis gathered from the different sampling stations of Japan public distribution system, there is also a customer's satisfaction survey had been carried out to obtain primary data and information directly from customers using a structured questionnaire. The main aim of the survey was to accesses the customer's expectation and experiences regarding the water supply of Japan drinking water project distributing through various panchayats. The questionnaire includes different types of questions regarding water supply services, water quality and quantity, customer's level of satisfaction, health status due to poor water quality, etc.

In total, 100 households were included in the survey. The households were selected by random sampling, but included those where water samples were taken. The main drinking water source of the area are dug wells and public water supply 70% of the people are using dug well, 25% of people public water supply and 5% both dug well and public water supply. Households were asked to give their opinion about taste, color, and smell of water. These were classified from excellent to bad, very clean to very dirty and from no smell to serious smell respectively. 70% of respondents opined that the water they use were good, clean, and no smell and other 30% answered that the water is not good and clean having bad smell of chlorine.

It was observed from the survey 80% of the households boil water, 10% filter and rest 10% both boil and filter water for drinking purpose. Most of the respondents (70%) stated that their drinking water was of sufficient quality and 20% responded insufficient quality and 10% referred to bad taste and overdose of chlorination. Household had to store their water for 0–3 days before consumption as supply was not continuous. Most of the respondents (70%) had separate storage containers, and 30% of households store in plastic cans or aluminum vessels. The study showed that 60% households get water in good quantity and 30% get the water in a low quantity. This is because many areas are located at higher elevation or at the far end of the supply system due to low pressure and flow. Further, once the pumping was on, the areas that are at high elevation or at the end of the pipe would not receive water before storage reservoirs in the immediate households were filled up. In this way, many households remained deprived of water before the end of the pumping schedule. From the survey it was also found that the main reason of their dissatisfaction with the water supply services was non-equitable distribution of water.

8.7 CONCLUSION

The present study was carried out in order to assess the physicochemical and microbiological parameters of Japan drinking water project. Analysis of water samples revealed that pH is slightly acidic in the main source (Kallada River). Other parameters such as conductivity, hardness, alkalinity, TDS, nitrate, nitrite, sulfate, sodium, potassium are within the limit set by BIS. Phosphate concentration is above the limit set by BIS. All the samples were highly contaminated with coliforms. This is an indication of fecal contamination of water from the original source. Presence of coliforms in household tap water may be due to ineffectiveness of the treatment process in the treatment plant. It needs further investigation. There may be leakage of distribution pipelines with domestic waste, and it needs more emphasis, chlorination process may be made effective and treated water must be checked for residual chlorine before it is distributed for drinking purposes. Authorities may also check randomly samples at all levels to ensure the supply of safe drinking water. As a recommendation, a certain quality of chemicals used in the water for treatment and fixing up their optimum dosages. Daily or weekly sample analysis throughout the distribution systems. Frequently monitor any epidemiological in this locality. Maintenance and replacement of leaks, old age pipes, etc. whenever necessary.

KEYWORDS

- **chlorine content**
- **contamination**
- **drinking water**
- **quality monitoring**
- **water supply**

REFERENCES

APHA, (1998). *Standard Methods for the Examination of Water and Waste Water*. American Public Health Association, Washington 16th edition. Assessment 2000 Report, Washington, DC.

Besner, M. C., Gauthier, V., Servais, P., & Camper, A., (2002). Explaining the occurrence of coliforms in distribution systems. *Journal of American Water Works Association*, *94*(8), 95–109.

BIS, (1991). *Specifications for Drinking Water, IS: 10500: 1991*. Bureau of Indian Standards. New Delhi.

Chaudhary, S., Anuradha, & Sastry, K. V., (2005). Groundwater quality in Faridabad an industrial town of Haryana. *Journal of Ecotoxicology and Environmental Monitoring*, *15*(3), 263–271.

Economic Review, (2016/2017). State Planning Board, Government of Kerala, Thiruvananthapuram.

Egorov, A., Ford, T., Tereschenko, A., Drizhd, N., Segedevich, I., & Fourman, V., (2002). Deterioration of drinking water quality in the distribution system and gastrointestinal morbidity in a Russian city. *International Journal of Environment Health, 12*(3), 221–223.

EPA, (2006). *Total Coliform Rule and Distribution System.* Issue papers overview.

George, A., & Prakasam, V. R., (2010). Investigations on water resources of Edamulakal grama panchayat of Kerala. PhD Thesis, University of Kerala.

ICMR, (1975). *Manual of Standards of Quality of Drinking Water Supplies*. Indian council of medical research, New Delhi.

John, K. N., Orish, EO.,& Linus, O. E., (2008). Some physic-chemical parameter of potable water supply in Warri, Niger Delta area of Nigeria. *Scientific Research, 3*(11), 547–551.

Kaur, H., Syal, J., & Dhillon, S. S., (2001). Water quality index of river Sutlej. *Pollution Research, 20*(2), 199–204.

Kumar, A., (1998). *Technologies to Improve Efficiency in Distribution Organization* (pp. 577–579). Geneva, Switzerland.

Patel, M. R., Lautre, H. K., & Shah, P., (2016). Physicochemical analysis of groundwater quality of Dhrol. *Anveshana's International Journal of Research in Engineering and Applied Science, 1*(9), 11–12.

Trivedi, R. K., & Goel, P. K., (1986). *Chemical and Biological Methods for Water Pollution Studies*. Environmental Publications, Karad, 415110, India. *6*, 10–12.

WHO & UNICEF, (2000). *Global Water Supply and Sanitation.*

WHO, (1997). Guidelines for drinking water quality (Vol. 3): Surveillance and control of community supplies. World health system with intermittent supplies. *Wat. Suppl., 16*(1/2), pp. 238.

Yesodharan, E. P., Kokkal, K., & Harinarayanan, P., (2007). *State of Environment Report, Kerala* (Vol. 1), ISBN: 81-863666-59-8; pp. 1–244.

CHAPTER 9

Effect of Environmental Pollution on Health and Its Prevention: An Overview

PALLAVI CHATTOPADHYAY,[1] PINAKI CHATTOPADHYAY,[2] and DEBNATH PALIT[2]

[1]*Department of Microbiology, Ranigaunj Girls College, West Bengal, India, Mobile: +91 8116670112, E-mail: pallavi96chatto@gmail.com*

[2]*Department of Botany, Durgapur Government College, West Bengal, India*

ABSTRACT

Rise in pollutants in the environment is emerging as a serious matter worldwide as it has a negative outcome on public health and environmental health at a large scale with a higher magnitude. Environmental pollution affects all abiotic and biotic forms, from unicellular prokaryote to multicellular eukaryotes. Environmental pollution includes air, water, soil, noise, and radioactive contamination. Improper waste management causes pollution in the environment. Pollution can be prevented by practicing some activities and care for nature can help us to have a healthy environment in the future. As we know that prevention is better than cure, so we have to choose alternative methods in favor of sustainability. In this chapter, we have discussed the causes of pollution and its most affected area like health and few steps taken to prevent pollution. Some initiatives have also been discussed taken by organizations, and we have performed within our own capacity.

9.1 INTRODUCTION

Earth is a unique planet in our solar system, which is able to support life in all its forms, from microorganisms to human beings. After billions of years of evolution, life refines itself, and the planet earth provides all sorts of ingredients. Earth's relative position with respect to the sun that has an ambient environment which makes the possibility of life on our planet. The biosphere of Earth also includes breathable oxygen, presence of an ambient atmosphere that contains just the proportionate amounts of gases that are extremely important for the survival of living species, the temperature of the earth is also habitable, presence of water and radiant energy of the sun. In the presence of sunlight, plants are able to perform photosynthesis to synthesis food for themselves. These living creatures are well balanced to have a healthy growth of life in this world. This unique balance contributes to the life cycles of all living creatures on earth and supports the continuous sources of minerals and energies, which play a crucial role in human civilization. All living beings, including humans, have been on earth for so many years.

Nature has an ongoing cycle that has never stopped, but lots of evolution and adaptation have taken place. As the human population is increasing, the demand for food, shelter, and energy is increasing. Humans are developing with industrialization and urbanization. Pollution is at its best with development and modernization in our livelihoods and resulting in global warming and human illness. Development in science and technology, human potentials have cropped up, and it results in losing control over pollution and become a victim of his own formulations. Pollution is present in the environment from many centuries but only started to be significant from the 19th Century's industrial revolution. Industrialization is important for development, but non-sustainable rapid industrialization is a threat to nature. Pollution occurs when the nature degradation system is not able to degrade or decompose an element without creating harm or damage to itself. Those elements are not obtained from nature; destroying process can vary from days to years to decades. Any kind of pollution doesn't stay at one place, they dispersed and affect everything. Environmental Pollution also effects health of humans and other living organisms. Human beings are becoming more susceptible towards diseases.

Environmental pollution includes:

1. Air pollution;
2. Water pollution;
3. Soil pollution;
4. Noise pollution;
5. Radioactive pollution.

Each of this type of environmental pollution has its own distinctive causes and after-effect. Consequences of releasing pollutants in the environment have been discussed in the following sections. Humans are the major producers of all types of pollution and will suffer its consequences if not taken as priority. We are already experiencing its effects in form of global warming, climate change, contaminated seafood, increased cases of diseases, etc.

9.2 AIR POLLUTION

Atmosphere is the layer of air around the earth includes nitrogen, oxygen, and minute amount of other gases. Each gas has a specific proportion in the composition of atmosphere. The imbalance in proportion of gases cause hazardous environment for survival. When harmful substances including gases, particulates, and biological molecules are introduced into atmosphere then it creates air pollution. Contamination occurs of the air present in the atmosphere by addition of some harmful gases, particulates, and dust, elevation of carbon-dioxide and depletion of oxygen (Figure 9.1).

FIGURE 9.1 Smog in Durgapur, West Bengal on 27.03.2018 at 7:51 AM.

9.2.1 WHAT CAUSES AIR POLLUTION?

Air pollution comes in many forms. There are various activities which are capable of releasing and dispersing pollutants into the atmosphere. This can be divided into two groups as natural air pollution and anthropogenic air pollution. Natural air pollution is caused by volcanic eruption, dust of sand, and bushfires. Man-made air pollution or anthropogenic air pollution is caused by industries, burning fossil fuels and motor vehicles.

9.2.1.1 EMISSIONS FROM FACTORIES AND INDUSTRIES

Manufacturing industries discharge a huge amount of harmful gases and colloidal particles like hydrocarbons, organic compounds, carbon monoxide, and other chemicals into the air, which affects the air quality. Certain industries release gases like sulfur oxides, carbon oxides, Nitrogen oxides and many more harmful gases which get mix in air and form clouds, circulates in nature following the water cycle and which cause acid rain.

9.2.1.2 INDOOR AIR POLLUTION

In developing and poor countries, some population of people uses biomass for cooking which includes wood, crop waste, and dung. Burning of traditional biomass is also a source of air pollution. It creates indoor air pollution as well as outdoor air pollution. Household cleaning products, wall paint supplies emit toxic chemicals and contribute to indoor air pollution. Paint on walls creates some sort of smell which sometimes makes impossible to breath. SPM (Suspended Particulate Matter) is another cause of pollution. Emissions from paint, varnish aerosol sprays, hair spray, and other solvents was estimated to reason for pollution from volatile organic compounds (VOCs).

9.2.1.3 BURNING OF FOSSIL FUELS

Under continuous heat and pressure in the earth's crust over hundreds of millions of years, fossil fuels are hydrocarbons, primarily coal, fuel oil or natural gas, formed from decayed plants and animals that have been converted to coal, natural gas and crude oil. Incomplete combustion generally emits carbon monoxide along with nitrogen oxide which produced both natural and man-made. The combustion of fossil fuels like coal, petroleum, and other factory combustibles is one the major cause of air pollution (*Pope, 2003)*and emit sulfur dioxide. Pollution gets transported vast distances by atmospheric currents and winds. One thing we notice very often is that the origin of pollution is in one place but ends up some other place.

9.2.2 EFFECTS OF AIR POLLUTION

Air pollution has short, and long term health effects depend on type and concentration of pollutant, length of exposure, and individual characteristics of the exposed person. Air pollution is a major risk factor for a number of pollution-related diseases, including chronic and acute respiratory infections, heart disease, stroke, and lung cancer. Most of the people are in constant exposed to both indoor and outdoor air pollution. Health effects of pollutants are felt in both developed and developing countries. 92% of the population of world lives in regions having air quality below the WHO standards. In the low and middle-income countries, where air pollution is emerging at a hazardous rate. In the ambient air pollution is estimated to cause 8% of chronic obstructive pulmonary disease (COPD) deaths, 17% of respiratory infection deaths, 25% of the lung cancer deaths, 15% of chronic heart diseases and stroke. Carbon dioxide is associated with global warming and ocean acidification. Increase in amount of carbon dioxide means increase in temperature of earth.

9.2.2.1 IMPACT OF AIR POLLUTION ON HUMAN BEINGS

When polluted air is inhaled by a person then the particulate matters present in air goes inside the nostrils, if the size of particulate matter is larger than $PM_{2.5}$ generally filtered in nose and throat via cilia and mucus but if size of the particulate matter is smaller than $PM_{2.5}$ then they travel deeply into the respiratory tract, sometimes they reach to lungs. Short terms effects of pollution on health are observed like eye, nose, throat, and lung irritation, coughing, sneezing, running nose and shortness of breath after exposure to fine particles. Particulate matter and ozone at ground level are most dangerous pollutants in terms of harmful effect to humans followed by benzo(a) pyrene [Polycyclic aromatic hydrocarbons (PAHs) and nitrogen dioxide indicator] peri-nated exposure to these pollutants has been found to have effect on neuropsychological development among children. Characteristic of carbon monoxide is colorless, tasteless, and odorless but highly toxic in nature. It forms carboxy-haemoglobin by making bond with hemoglobin which is the main reason for hampering the supply of oxygen to the vital tissues or organs like brain, heart because carboxy-hemoglobin occupies oxygen-carrying space in hemoglobin that results in inefficiency of delivering oxygen to the vital body organs. Some of the ailments associated with higher intake of CO are Myocardial Ischemia (reduced oxygen to the heart) and Angina pectoris (acute chest pain). At extremely high concentration levels, CO can cause death due to carbon monoxide poisoning. Carbon monoxide sometimes effects fetus of pregnant women (Blackburn, 2007).

Acid rain forms from air pollutants and causes many skin problems to humans, eye irritation, stone leprosy and contaminate water and soil. Nowadays, air pollution contributes high negative impact on gametogenesis of both female and male. This is not just at physiological level but it is also in genetic level. Pollutants in atmosphere not only affect the quantity of egg or sperm but also their quality. Increase in number of cases of infertility and abnormal embryo development is due to air pollution. New evidence suggests air pollution is linked to dementia. Air pollution particles have been found inside the brains of people living in cities.

Indoor cigarette smoking increases cadmium concentration inside room. The average of daily exposure of cigarette smoking (e.g., 20 cigarettes in a day) is 2–4 μg of cadmium (Perkins et al., 2013). Smoking increases the daily intake of cadmium and result in excess concentration of cadmium present in body and effects health by lung injury. There is a hypothesis that excess cadmium exposure in humans might be linked to hypertension. According to WHO, in Western Europe, Australia, and the USA, the average of daily oral intake of cadmium by passive smoker or non-smokers living in unpolluted areas is 10–25 μg. Cadmium nephrotoxicity may occur due to chronic inhalation or ingestion. In animals, it has been observed that cadmium crosses the placenta, and excess cadmium causes placental injury and fetal death.

9.2.2.2 SOME MECHANISMS OF ACTION MAY BE ASSOCIATED WITH THESE HEALTH IMPACTS

9.2.2.2.1 Endocrine Disruptor Activity

A simple interaction of a substance with endocrine and results in adverse health effects-these substances are usually known to as endocrine disruptors. These substances interfere with the metabolisms of the body and disturb the biochemical pathways. Diesel incomplete combustion contains PANs and Heavy metals like Cu, Pb, Zn, etc. Diesel exhaust particles contain, for example, substances with estrogenic, anti-estrogenic, and anti-androgenic activities that can affect gonadal steroidogenesis and gametogenesis and effects human endocrine system (Takeda et al., 2004).

9.2.2.2.2 Formation of Oxidative Stress

Pollutants in the air can generate reactive oxygen species (ROS) which can damage the cell structures of living organisms by switch on the redox-sensitive pathways that lead to many biological processes such as inflammation and cell death. These cause alternations in DNA, Proteins, and Membrane lipids (Nita and Grzybowski, 2016).

9.2.2.2.3 Modifications in DNA

Through the formation of DNA adduct means DNA binds to a carcinogenic molecule, which obstruct normal functions of DNA. Proper and complete replication cannot occur to make the normal expected cell and this leads to epigenetic mutation which can affect all functions including reproduction. Air pollutants cause defects during gametogenesis in both animal and human, leading to a sudden decrease in reproductive capacities among exposed populations. Air quality has a long term effect on health because the exposure frequency is more with air than any other substance (Carré et al., 2017).

9.2.2.3 *IMPACT OF AIR POLLUTION ON BIRDS*

Human beings are not only living organisms fighting the issue of poor air quality and chemically contaminated food and water but animals are also in crises, especially avian and aquatic organisms. Harmful gases including carbon monoxide, carbon dioxide (CO_2), ozone, sulfur oxides (SO_X), smoke, and heavy metals, as well as mixtures of urban and industrial emissions and particulate matter, contribute for a negative impact on bird health. Some abnormalities which are observed in avian species are respiratory diseases, and illness, stress level increases, elevated detoxification effort, behavioral changes, immune-suppression; defective reproduction success and their offspring have lower survival rates. More contact with air pollution may furthermore reduce the population, diversity in species and species richness in bird communities. Air pollutants may even penetrate into the egg shell and cause harm to offspring. Many researches show that carbon-monoxide can diffuse through eggshells and it's practicable that other pollutants in the atmosphere may be able to do the same. According to power works incorporated, birds are even more open to colloidal particles in the atmosphere and to airborne impurities because they have very high and frequent respiratory rates. Frequency of respiration in birds is higher than any other terrestrial vertebrate. Coal miners use domestic canary to detect the presence of toxic gases in mines such as detection of dangerous level of carbon monoxide and many people take birds as 'Sentinel Species' for environmental change. The future of most bird species relies on our attitude to conserve them.

9.2.2.4 *IMPACT OF AIR POLLUTION ON PLANTS*

During photosynthesis, carbon dioxide is absorbed and oxygen is released by green plants. As plants are sensitive towards pollution, injuries caused by pollution are frequently evident in plants before its effects are observed on humans and animals. Plants respond to slight changes in environment. The fly ash, dust, and particulate matters adhere on the surface of plant leaves and clogs the stomata and inhibit the gaseous exchange and photosynthesis of the plants. In polluted regions, the numbers of stomata in leaves of the plant were less as compared to the plants in non-polluted regions. Nitrogen-dioxide and ethylene, they may hamper the plant growth and cause premature aging of the plant. Ozone at ground level is notorious as it causes small irregular lesions on the surface of the leaves, called fleck. Injury can develop on both the leaf surfaces and cause leaf blight damage. Leaf veins favor to remain green until general yellowing (chlorosis) develop. Pine needles also get chlorotic lesions and tip dieback. Generally, the lower leaves of plants get injured primarily. There are various types of pollutants present in the environment which posses wide range of effects on our environment. Pollution effected plants are much more susceptible towards diseases, viroids, and insects. Exposure to glyphosate, a broad-spectrum herbicide, has been observed to capable of change the species composition of endophytic bacteria in plant hosts, which is highly vary (Rosenblueth and Martínez-Romero, 2006).

Damages from air pollutants could be serious or minor, depend on plant or crop variety. Plant responses to air contamination are helpful in the following ways:

- Early detection of presence of air-borne impurities.
- Locating the geographical distribution of the impurities.
- Analysis of the concentration of pollutants.
- Collecting pollutants in passive system for future chemical analyses.
- On the basis of plant species and variety affected, we can obtain direct identification of different air pollutants.

There are some plants, which adapted themselves to cope with the pollution means they are pollution tolerant plants, but they are still under research. On the other hand, the elevation of carbon dioxide, affect oil quality and yield from the seed of sunflower. Study found that rising atmospheric CO_2 in coming future climate can upgrade the biomass production and yield from seed in sunflower and alter their quality of oil from seed in terms of increased condition of unsaturated fatty acids compared with saturated fatty acids and minimize the proteins and mineral nutrients of seed. Under the unstable climate scenario, for global food demand and nutritional security, the rising carbon-dioxide level plays a crucial role in crop productivity and quality (OECD, 2012).

9.2.2.5 POLLUTED ENVIRONMENTS TRIGGERS ALZHEIMER'S DISEASE

A research shows that scientists have found that elevated levels of 2 abnormal proteins, hyperphosphorylated tau, and beta-amyloid, in the brains of young children who lives in urban area with continuous exposure to fine particulate matter (P.M 2.5). Particulate matter when enter inside the body, and then it interrupts with the natural biochemical pathways and injures the internal organs. Scientists also tracked the Apolipoprotein E (APOE 4), a noted genetic risk factor for Alzheimer's disease. Studies show that Alzheimer's disease starts in childhood, and the disease progress with increasing age, APOE 4 position, and exposure to particulate matter in the environment (Calderón-Garcidueñas et al., 2018).

Bhutan is very rich in its vegetation. The government of Bhutan is very concerned about their vegetation and cleanness. Bhutan's net greenhouse gas emission is zero, because the amount of pollution which it creates is get absorbed by the forests that cover most areas of the country. The whole country at an average produces approximately 2.3 million tons of carbon dioxide a year, but the massive forest covering 72.1%. Bhutan acts as a carbon sink area, absorbs more than 4 million tons of carbon dioxide every year. People of Bhutan get most of its energy from hydroelectric power, here energy production and utilization does not emit any significant greenhouse gas.

9.2.3 PREVENTION FROM AIR POLLUTION

Air pollution is an invisible but deadly problem. Air pollution can be reduced by combined efforts of government, industries, and individuals as some measures are given below:

- Use biofuels instead of fossil fuels to ensure less air pollution.
- Practice yoga on a daily basis to enhance immunity and to become less susceptible to diseases related to pollution.
- In polluted environments, people should avoid going outdoors without masks even for activities like jogging and cycling.
- N95 masks are advised to use if it become necessary to go outdoors.
- Use of nanofilters can prevent entry of particulate matter through nostrils during breathing.
- Plants can be kept at home or in the office.
- Help your body to fight the damage air pollution causes by maintain a healthy balanced diet daily. Consume fruits which contain vitamin C, magnesium, and foods rich in omega fatty acid. Eat herbal plants daily. Eat raw turmeric daily with milk to avoid Alzheimer's disease.

- Plantation of indoor plants as indoor plants improves the air quality of the room. Those potted-plants can reliably reduce total volatile organic compound (TVOC) loads, a major class of indoor pollutants, by 74.99%, to below 100 ppb. They function equally well with or without air-conditioning.
- Increase the use of solar energy.
- Green roofs create natural habitat and cools the room. Even in multi-storeyed buildings, it has been found that green roofs can attract beneficial insects, butterflies, dragonflies, bees, and birds. Green roofs also work as a green wall, filtering pollutants and carbon dioxide out of the air, helping to lower rates of diseases such as asthma. They can also filter pollutants and heavy metals out of rainwater.
- Encourage the use of clean energy sources that do not burn coal or have incomplete combustion.
- Mercury is an element that cannot be destroyed, collect the fly ash containing mercury and other pollutants and use it for construction purpose.
- Encourage the use of sustainable methods like wind technology, algal technology, etc.
- Do not burn agricultural wastes, dry leaves, and woods. Dig a small well in soil and bury the agricultural wastes, dry leaves, and woods in the soil and cover them with soil. After few weeks or months, they will convert into manure by microorganisms present in soil and can be used for agriculture purpose as manures and maintains the soil fertility along with addition of some elements require to maintain soil health.

Since many countries are using paints which contain lead and by the use of these paints they are exploring lead pollution in their houses and also in the environment. This problem is very dangerous and there is an early need of solving the problem at global level, for that motive United Nations Environment Programme along with WHO to construct a global agreement to get rid of paint contaminated with lead. The particular initiative has been taken only to save the children from lead and also to diminish the lead-induced problems among the painters or the occupations associated with exposure of paints in large quantities. Its wide objective is to inspire a phase-out of the manufacture and sale of paints containing lead and eventually remove the risks that such paints create.

Purification of polluted gas of Industries can be done by using bioscrubber, biofilter, and biotrickling filter. The three waste gas purification systems remove air pollutants from the outcoming gaseous matter. In Switzerland, the electricity-generating network is nearly CO_2 free because there 56% of electricity generates from hydroelectricity and 39% from nuclear power.

9.3 WATER POLLUTION

Water is the basic need of every living organism. When pollutants are directly or indirectly released into water bodies without any necessary treatment, and marine organisms are unable to utilize or decompose those chemicals, and this causes water pollution.

9.3.1 SOURCES OF WATER POLLUTION

Pollutants like chemicals, pathogens, and physical changes such as a rise in temperature and change in the color of water. Water pollution can be divided into two categories, point source pollution and non-point source pollution.

Community waste water, industrial waste, agricultural waste etc. Apart from synthetic pollution, freshwater is also the final stage for biological waste, domestic wastes, human sewage, animal feces, rainwater drainage flavored by nutrient-rich fertilizers from farms and yards. These nutrients find their path through river streams into seas, sometimes creating coastal ocean zones in the absence of oxygen. In developing countries, 70.32% untreated industrial debris are dumped into water bodies, hence polluting the water. Nearly 75% of water pollution is due to sewage and domestic wastes. Approximately, about 21 million tons of fertilizers and chemicals are used annually. Every year, about 300–400 metric tons of hazardous waste is dumped by in water which contains heavy metals, oil hydrocarbons, microbial contamination, etc. Some environment pollutants enter and circulate in the water cycle.

Pollutants in atmosphere make acid rain which fall on ground and water bodies like lakes, streams, ponds, and tributaries and contaminates the water. This results in unfavorable marine environment for marine creatures. It affects the water quality by change in pH and addition of pollutants. More than 500 million people are indulging in open defecation and because of that there are a lot of environmental problems which are created the sewage go to the water source where there is a mix of the sewage with the water bodies which creates the water pollution, this leads to waterborne diseases like typhoid, cholera, hepatitis, gastroenteritis, etc. People use the same water for washing clothes, bathing, cleaning cattle and cooking in unhygienic conditions using contaminated and unhygienic water. Few major sources of pollution and introduction of xenobiotics into the environment are by some large industries like pharmaceuticals, pulp, and paper bleaching and agriculture. This mixing of sewage and usable water is the prime reason behind all illness. Some natural and anthropogenic substances may cause turbidity of the water, which obstruct light and disturbs plant growth, and cause harm to the gills of some fish species.

9.3.2 EFFECT OF WATER POLLUTION

Every living being is connected to water. Contamination in water means water pollution covers a wide range of diseases and infections to living organisms. Water pollution is the main cause of the disturbance in the life cycle and food chain of the aquatic organisms. Industrial debris and litters are dumped in water bodies which pollutes the water as it contains hazardous pesticides and insecticides and other toxic chemicals which have an outcome as life-threatening for aquatic life. Decrease in dissolved oxygen content of water bodies. The fishes under the water slowly become ill or accumulate the toxins within their body. This fish when taken by humans can even affect our health. Consumptions of these fishes may harm the health of humans. When these fishes are consumed by frogs and birds, it hampers their life too and enters into the food chain. These have a negative impact on terrestrial wildlife and aerial wildlife as well. Following that, the whole environment gets disturbed. Sometimes it enters in food chain and result in bio-magnification, as the tropic level increases the accumulation of harmful chemical increases. Water pollution also equally show negative impacts on all animals and the entire environment. Annelid worm *Tubifex* and some insect larvae act as indicator species for polluted water.

9.3.2.1 WATER POLLUTION EFFECT ON AQUATIC ECOSYSTEM

Marine biodiversity is very rich and contains a variety of living aquatic creatures, encircling variation at species to subspecies level within aquatic (marine) type of ecosystem. Oceans are home for other vertebrate and invertebrate species.

Acid rain pollutes the water bodies, as a result, the lakes lose their entire fish population, and some result in defect in laying eggs. Birds and other animals that depend on fish for food are getting affected due to decline in fish population in water-bodies. Physical and chemical changes in water affect the aquatic ecosystem. Every organism requires an optimum temperature and pH for growth. Acid rain changes the pH of the water and adds pollutant in water which disturbs the aquatic life and biogeochemical cycles. Within environment, everything is connected and every organism is linked to each other. The ocean pollution also affects the keystone species of the marine ecosystems which disturbs the food chain in aquatic species.

Some research work is needed to understand the impact of bleaching on the remainder of the reef community along with long period of affects on predation, bioerosion, competition, and substrate condition of corals, factors that determine coral growth and reef recovery. By AD 2000, an observation was made that in some projected rates of ocean warming, i.e., 2° increased in high latitude coral seas, the upper thermal acceptance limits of many reef-constructing corals could be exceeded. Current reports on climate change indicate that many corals would be not able to adjust physiologically or genetically to such an unsteady and rapid rise in temperature.

There is a huge exploitation and loss contributed by plastic pollution on marine ecosystem which we have to solve in such a way that will not affect aquatic creature as well as our plastic wastes will be disposed properly. There are many casualty cases of aquatic birds, and other aquatic flora and fauna, the most shocking thing is that the cases have now reached millions due to plastic trash content in trillions. Plastics are used once, but it stays for centuries and breaks down to microplastic, poisoning all living creatures.

Due to ocean acidity (drop in pH), the marine life suffers a lot like it is evident in the case of calcifiers (the marine creatures whose outer skeleton or outer structure is composed of calcium carbonate ($CaCO_3$) who needs calcium carbonate for their life cycle, due to sudden shift in pH, especially decrease in pH value of water of the oceans badly affects the lifecycle and shell formation which is necessary for normal living of the particular group of organism that is calcifiers. Marine pollution leads to lowering of pH of ocean water which ultimately results in disturbed biological carbon pump, which is the basic process of carbon fixation and converting the carbon-dioxide into calcium carbonate as oceans are the chief carbon sink of the earth and also promotes the construction of coral reefs in the ocean which is the habitat of many organisms such as algae, phytoplankton, etc., which are primary producers of ocean life. The competence of carbon pump is affected badly as there is a decrease in algae and phytoplankton population within the ocean which ultimately decreases the biological component that utilizes carbon-dioxide and hence contributes to carbon fixation in oceans and decreasing the carbon dioxide in the atmosphere.

Eutrophication is a process where heavy growth of toxic algal blooms, dead zones, and fish mortality occurs due to increased and excess amount of nutrients in water bodies and decreases the oxygen in water. Eutrophication cause massive damage to a water body by lowering the pH of water by producing carbon dioxide this result in ocean acidification. This inhibits the growth and lifecycle of shellfishes and some other fishes, as drop in ph of water obstructs shell formation in bivalve mollusks, and also affects the fishing of some commercial and recreational fisheries, leads to less profit in fish market as small fish harvesting (pisciculture) and rise in price of seafood.

9.3.2.2 EFFECT ON HUMAN HEALTH

Health risks from pollution vary area to area. Globally, diarrhea is the leading cause of poor health and in some cases mortality due to usage of contaminated (harmful microorganisms) water. Presence of excess cadmium in water is harmful for human consumption as it causes Itai Itai (Ouch-Ouch disease). The approximate lethal dose by oral consumption for humans is 350 to 3500 mg of cadmium; a dose of 3 mg of cadmium has negligible effects on adult (Krajnc et al., 1987). The excretory system get affected when chronic oral exposure happens, the kidney suffers the most. Cadmium interferes with the resorption function of the proximal tubules, the initial symptom being an increase in the amount of low-molecular weight proteins in Urine excretion, studied as tubular Proteinuria (Krajnc et al., 1987). When water is contaminated with nitrate, and then if a human comes in contact with that contaminated water, it affects the human and causes blue baby syndrome as excess nitrate undergoes reaction with hemoglobin and makes methemoglobin inactivated that inhibits oxygen transportation. It is known as methemoglobinemia.

Skeletal fluorosis is caused by an excess amount of fluoride in drinking water. Researchers have found that there were "strong indications" that excess exposure of fluoride, especially among growing children, is highly problematic for proper cognitive development and brain development.

Fertilizers contain some chemicals like arsenic which gets embedded in soil and afterwards get mix with groundwater during rain. Arsenic contaminants the water and there are some countries which are severely suffering from Arsenic contamination in groundwater those are India, Bangladesh, Chile, Taiwan, Argentina, Romania, China, and Thailand. In India, in West Bengal arsenic contaminations in groundwater were detected. These affected areas are multiple times arsenic-contaminated higher than the maximum concentration level (MCL) (10 micro g/l). Arsenic exposure of human resulted in health problems like degenerative, inflammatory, and neoplastic changes of the skin, blood, respiratory system, nervous system, reproductive system, and lymphatic system. No effective treatment present for chronic arsenic poisoning. The rise in risk of chronic arsenic toxicity was higher among malnutrition and low socioeconomic status population. Early detection, treatment, and prevention can save the effected people. A person suffers from black foot disease when chronic arsenic exposure happens. Arsenic contamination imposes wide range of suffers. All communities of human, plants, and animals get affected some get short term effects and some get long term effects. Even arsenic easily reaches the food chain and paddy crops gets easily contaminated with arsenic.

Xenobiotics are those chemicals which do not exist in nature. These are man-made, synthesized compounds. Man-made chemicals say xenobiotics, for example, polychlorinated biphenyls (PCBs), trichloroethylene (TCE) and PAHs are nonbiodegradable in nature and accumulate in the environment because of their recalcitrant properties and counted as an environment pollutant due to their toxicity and accumulation. These have observed particularly in the subsurface environment and water bodies, as well as in biological systems, possess the potential to give a negative impact on human health. Phosphorous is a very important element for biological system but excess amount cause harmful effects. Sometimes industrial runoff contains phosphorous which get into water bodies and toxic algal blooms grows. These algal blooms are reported as poisonous as they have been linked to paralytic shellfish poisoning in humans, most of the time, leads to death.

9.3.2.3 WHEN WATER IS POLLUTED WITH HIGHER CONCENTRATION OF MERCURY

According to World Health Organization (WHO), mercury is listed in top 10 chemicals or groups of chemicals of major hazardous to public health due to its capacity to travel long distances in any medium say atmosphere or water; its endurance in environment; its potentiality to enter in food chain and accumulate in ecosystem, its negative impact on human health and the environment. The human population which consumes seafood comes in contact with mercury biomagnifications if the water body is contaminated with mercury. Chisso-Minamata disease commonly known as Minamata disease is a neurological syndrome caused by severe mercury poisoning. Some common symptoms are numbness in the feet and hands, narrowing of field of vision, ataxia, general muscle pain, and weakness and harm a person's speech and hearing ability. In 1956, a city in Japan, Minamata city in Kumannoto prefecture, Minamata disease was first discovered. Minamata disease was caused due to release of a harmful chemical named Methylmercury in the industrial wastewater from Chisso Corporation's chemical factory from the year 1932–1968. Aquatic creatures come in contact with methyl mercury and bioaccumulation occurs in fishes in Minamata Bay and the Shiranui sea. In this way, mercury enters into the food chain. When fishes were eaten by the local residents mercury poisoning is occurs. The after effect continued in dog, cat, pig, and human deaths for many years. The cats were severely affected and later they get termed as "dancing cat fever." In March 2001, approximately 2265 patients had been officially detected as having Minamata disease (1787 of who had died) and some had received financial settlement from Chisso. In 2004, Chisso Corporation had paid approximately 86 million dollars in compensation, and in the same year was ordered to clean up its contamination from the environment (Hightower, 2008).

Different forms of mercury differ in their degree of toxicity along with their effects on the Nervous system, Digestive system, Immune system and on lungs, kidneys, skin, eyes, and brain. Excess exposure can even be fatal.

Some Factors that help us to determine whether health effects occur and their severity include: The type of mercury concerned:

1. The dose.
2. The age or developmental stage of the person exposed (the fetus is more susceptible).

3. The duration of exposure.
4. The route of exposure (inhalation, ingestion or dermal contact).

9.3.2.4 MODIFICATION IN MICROBIAL GENE

Environmental pollution has reported as a reason for number of genetic effects in the microbial world, from antibiotic and antifungal resistance to rising microbial diversity. Industrial runoffs washes off, some pharmaceuticals along with them, and get mix in water systems, where microbes have to face challenge for their existence and slowly becomes resistant to a broader class of antimicrobial drugs. For example, in South Carolina, a strain of *Escherichia coli* was isolated from Shipyard creek, which was polluted by toxic metals and other industrial wastes, has been reported as resistant to nine different classes of antibiotics. Most of the time microbes become resistance to drugs by performing horizontal gene transfer. As microorganisms are get modified in the changing environment and potentially grow more virulent and pathogenic, their host range and pathogenicity also increases.

9.3.3 PREVENTION FROM WATER POLLUTION

The effects of water pollution are not always instant. They are invisible with naked eyes, but still, contaminants are present in water. Contaminants are only detectable by performing some specific experiments. Still, water pollution has a huge impact on everybody's life. With awareness, knowledge, analysis, and initiatives, water pollution can be treated and controlled. It doesn't require much effort, just a little thought, and action.

If we protect the water bodies from getting polluted, then we can prevent those effects or diseases which are caused due to water pollution. If water is already polluted, then water treatment is a must to prevent waterborne diseases.

1. How to control human exposure from mercury-contaminated sources?

There are many ways to prevent exposure from mercury, including stop use of mercury in gold mining, promote green and clean energy, minimizing the mining of mercury and do not use non-essential mercury-containing products. By organizing mass awareness programs and public education aimed at reducing and eliminating the risk of waterborne diseases and vector-borne diseases related to water. Disinfection of drinking-water supplies has been one of the most significant public health advancements of the last century with respect to reducing waterborne disease by increasing the use of water purifiers or filters. The water purification filters must able to remove both chemical and biological contaminants. Although chlorine has been used as the disinfectant of choice for public drinking-water supplies for the past century, a number of emerging or alternative compounds are used or are being considered for the disinfection of drinking water.

Ecological sanitation is very important to maintain a healthy and hygiene environment. Ecological sanitation has been observed as a sustainable system to dealing with human excreta, using dry composting toilets. EcoSan can reduce the use and need for chemical fertilizers as it is a hygienic, efficient, practical, and cost-effective solution by which human excreta will be recycled into a resource. Later human excreta also used for manufacturing microbial fuel cells which will be able to generate electricity. Waste is not waste anymore, now wastes are resources, but we have to find them. Some places have accepted the EcoSan toilets like Kerala and Sri Lanka. EcoSan toilets are safe disposal of human excreta and urine and do not need water for Flushing after use. Promote the use of toilets for human excretion process and it can be used as sustainable process by generating electricity from excretion using microorganisms. There is still on-going research on bioelectric toilets. Microorganisms are also used in wastewater treatment. Acceleration of biodegradation of specific compounds by inoculating bacterial cells is called bioaugmentation. Bacterial cells contain specific plasmid which encodes enzymes for degradation of target compounds. Several genetically engineered strains have been developed exploiting *Pseudomonas*. Variety of plasmids has been reported from *Alcaligenes, Klebsiella, Flavobacterium, and* many more microbes.

Japan has applied genetic engineering methods and improved bioreactors for wastewater and industrial wastewater treatment. Several enzymes are use in wastewater treatments such as catalase (CAT), phosphate esterases and aminopeptidases. These enzymes can be used in freshwater to enhance biodegradation of all xenobiotic compounds. An extracellular fungal lactases are obtained from *Trametes versicolor* or *Botrytis cinerea* which can be utilized in the treatments of effluents released by the pulp and paper industry because these enzymes can be utilized for dechlorination of chlorinated phenolic compounds or oxidation of aromatic compounds even at adverse environmental conditions such as low pH, high temperature, presence of organic solvents, etc. (Berry, 2014).

Microorganisms are being used to remove metals from industrial wastewater. Microorganisms produce extracellular polymer complex compounds and accumulate metals like iron, copper, cadmium, and nickel or uranium. The accumulate metals are released from biomass upon treatment with HCl. The heavy metals should are hazardous to the health of man and animals. So heavy metals should be removed before discharging the effluents into the rivers. The process of capturing of metal contaminants from wastewaters using microorganisms is called Biosorption (Table 9.1).

TABLE 9.1 List of Some Microorganisms and Metals They Accumulate from Wastewaters Released from Industries

Serial Number	Name of Microorganisms	Type of Microorganism	Metal Removed
1	*Aspergillus niger*	Fungus	Copper, cadmium, zinc
2	*Aspergillus oryzae*	Fungus	Cadmium
3	*Chlorella vulgaris*	Algae	Gold, copper, mercury, zinc
4	*Penicillium spinulosum*	Fungus	Copper, cadmium, zinc
5	*Rhizopus arrhizus*	Fungus	Uranium
6	*Saccharomyces cerevisiae*	Yeast	Uranium
7	*Trichoderma viride*	Fungus	Copper

9.4 SOIL POLLUTION

Soil is a wonderful mixture of many things such as minerals, liquids, microorganisms, organic matter (OM), gases that together support life. Soil is also known as pedosphere, which has four major functions: it is an excellent medium for plant growth; it is a means of water storage as different type of soil have their own water holding capacity; supply and purification. When some xenobiotics or chemicals mix with soil and effects the natural environment of soil then it is known as Soil pollution.

9.4.1 CAUSES OF SOIL POLLUTION

There are several reasons behind soil pollution because soil is connected to everything present in the ecosystem. Soil erosion, deforestation, shifting agriculture, use of agrochemicals like insecticides, herbicides, and pesticides, electronic waste and improper disposal of medical and radioactive wastes. Acid rain also contributes to soil pollution as it affects the soil health by turns the pH of soil acidic, soil losses fertility and result in low crop yield. Cadmium compounds have wide use in electric batteries, nuclear reactors and electronic components. When batteries and electrical components are dumped into soil along with other garbages causes soil pollution.

The dumping of hazardous wastes from the company Union Carbide India limited in and around the site leads to soil pollution and ground water pollution. Unused and left-over chemicals at the plant are still continues to leak and pollutes the environment, especially groundwater. In 1989, UCC's laboratory experiments reports that soil and water samples collected from nearby the factory were extremely toxic to fish and humans. Factory was constructed taking a huge area and after the accident, 21 places inside the plant were observed to be highly polluted. In 1991, laboratory tests were organized and the municipal authorities gave conclusion that water samples from over nearby 100 wells found hazardous for human health and not fit for drinking and cooking purpose. In 1994, a report was published that approximately 21% of the factory area were contaminated with chemicals at danger levels. In the beginning of 1999, again samples were collected from UCIL factory and nearby areas by Green peace and others from soil, wells, groundwater, and vegetables from the residential areas showed contamination with a scale of toxic heavy metals and chemical compounds. Chemicals which were detected out of that one was found 50 times above the safety limits specified by the U.S. Environmental Protection Agency (EPA) which can damage fetal development and those chemicals are: naphthol, naphthalene, sevin, tarry residues, alpha naphthol, mercury, organochlorines, chromium, copper, nickel, lead, hexachlorethane, hexachlorobutadiene, pesticide HCH (BHC), VOCs and halo-organics. Carcinogenic chemicals were also discovered, as well as TCE. Livelihoods of people are in danger because even drinking water of wells near the site contains levels of contamination 500 times higher than the lethal dose recommended by the WHO. In 2010, a British journalist undertakes a risky action to enter into the abandoned

Union Carbide factory to inspect the allegations of leftover, leaking toxins. Later he was hospitalized for a week in Bhopal at the Sambhavna clinic, after exposure to the chemicals. Doctors treated him with painkillers, anti-inflammatory, and oxygen following a critical respiratory damage to toxic dust inside the factory

9.4.2 EFFECT OF SOIL POLLUTION

Health effects because of soil contamination may vary person to person and most importantly depend on type of pollutant, susceptibility of the exposed population and pathway of attack of the pollutant. Soil pollution also affects the ground water. It contaminates the ground water by passing of some Metals or chemicals via rain from soil to ground water. Metals like arsenic, cadmium, lead are largely found in ground water in some parts of country.

A large amount of wastes are released from the procedures of health-care activities, they contain from contaminated needle to radioisotopes. Possess a major potential for causing infection and injury among common people who come in contact, than any other type of waste, and inappropriate waste management is likely to spread serious public health illness and destructive on the environment. Improper disposal of healthcare wastes on lands are hazardous for soil health. They not only contaminate the soil but also the surface and ground waters. Proper physical sterilization process must be performed before disposal on land or landfills.

Sometimes, few chemicals get released into environment because of the treatment of healthcare wastes with chemical disinfectants, if somehow those substances are not stored, handled, and disposed in an environment suitable manner.

In United States, a herbicide named glyphosate is most of the time used on crops. This would be herbicide was discovered by Monsanto chemist John E. Franz in 1970. In some regions genetically modified crops which are resistant to glyphosate are planted under over-used and application of pesticides. Again, by passing time it embedded in soil then passed to ground water. This cause serious reproductive issues and kidney failure. If agriculture is done on polluted soil or irrigation using polluted water then crops may contain increased concentrations of pollutants, meat of animals which consume grass on polluted soil of grazing land (IARC, 1976). Kidney and liver of animals also get affected due to accumulation of cadmium in their body. During 1980 to 1988, average cadmium concentrations observed in fish were 20 µg/kg wet weights. High levels were detected in shellfish (200–1000 µg/kg) (Galal-Gorchev, 1991).

Most of the pesticides are made up of chemicals, chemical pesticides are used to secure crops against, fungi, weeds, and other pests. All over the world, approximately 1000 pesticides are used to ensure crop is protected and not damaged or destroyed or injured by pests. Pesticides perform a significant role in food production sector. They protect or enhance yields and the number of times per year a crop can be grown on the same land. Countries facing food shortages will get benefited. Every pesticide has different properties, mode of action and toxicity. Pesticides are not only effect pests but also harmful for humans they possess toxic chemicals which can have both acute and chronic health consequences, depends on the quantity and mode of exposure. It can also depend on the route by which the exposure takes place (such as swallowing, inhaling, or direct contact with the skin). Some of the old, expired, cheaper pesticides can remain for long periods, e.g., years in soil and water. These types of pesticides have been banned from agricultural use in developed countries, but they are still used in agriculture in many developing countries. Those pesticides that are authorized for use on food in international trade today are genotoxic (damaging to DNA, which can cause mutations or cancer). Deleterious effects from these pesticides occur only above a certain safe level of exposure. When people come into contact with large quantities of pesticide, this may suffer from acute poisoning or long-term health effects, including cancer and can damage reproductive system. The use of polythene bags in our country leads to landfill. Mostly, chemical plastics are foreign to the environment in which they are dumped, hence persist in nature. Microorganisms present in soil are not able to decompose the plastic and this leads to infertility of the soil.

All the essential elements and trace elements for growth of plants are equally important in the fertility management of soils. The nutrients are always being removed, which must be replaced periodically so that optimum levels of nutrients can be maintained to keep a favorable equilibrium between the gains and losses.

Electronic wastes such as cathode ray tubes mostly used in manufacturing and functioning of TVs, computer monitors, ATM, video cameras, and more) after use, Breaking, and removal of yoke, then dumping in soil contaminates the soil as it contains lead, barium, and other heavy metals that leaches into the groundwater and release of toxic elements.

Chips of electronic gadgets and other gold plated parts of electronic devices after use are processed by chemical stripping using nitric and hydrochloric acid and burning of chips, and it causes potential environmental hazards like heavy metals, PAHs, brominated flame retardants released directly into rivers which acidify the water and cause harm to fish and flora. Tin and lead contaminate the surface and groundwater. These pollutants affect the human, animal, and plant health. These pollutants contaminate the soil and groundwater. When people come in contact with contaminated soil or water then heavy metals affect them like lead damages brain by affecting the memory of a person, liver, and kidney. Arsenic causes mental disturbance, liver cirrhosis, and kidney damage.

9.4.3 PREVENTION FROM SOIL POLLUTION

Promote the growth or to cultivate the genetically modified organisms or transgenic plants which are able to bio-remediate the toxic pollutants. Proper disposal of health care wastes after sterilization because health-care waste carry potentially harmful microorganisms that can infect and spread diseases hospital patients, health workers and the general public. Several potential hazards may include drug-resistant microorganisms which can transmit from health facilities into the environment. Promote the use of biofertilizers. Some biofertilizer have nitrogen fixing microbes or phosphate solubilizing microbes or spores of VAM fungi. Process of spreading of biofertilizers to the soil includes seed treatment or by spreading it over the field during cultivation. Biofertilizers are ecofriendly and available naturally. Farmers themselves can cultivate blue green algae (BGA) and *Azolla* biofertilizers in their own land. Biofertilizers do not pollute the air, water, and soil because they are soluble in soil and increase the soil fertility. Biofertilizers are cost effective. Increase organic farming for crop production.

To control pests, pathogens, and weeds by using their biological antagonists is known as biocontrol. The antagonist may be an attenuated or normal pathogenic bacterium, fungus or a virus and in some cases plant extract. Promote the use of biocontrols in agriculture by replacing chemical pesticides and insecticides. Biocontrol is a non-hazardous, ecofriendly approach.

Waste generated after biogas production in the form of slurry, used as compost for agricultural fields. The human and animal excreta are used for biogas production. The liquid effluents are added in artificially constructed ponds meant for cultivation of algae. In such systems, $NaNO_3$ and $NaHC0_3$ are mixed so as to induce growth of *Spirulina*. Several species of birds are used as pest control agents in agriculture because some birds are insectivorous and feed on a wide variety of insect pests without harming the crops. There should be some rules and regulations on safe usage of pesticides as they are potentially toxic and easily spread in the environment, the production and distribution need strict and proper control. Examination and evaluation of residues in food and the environment is also required on a regular basis. Foods must satisfy the pesticides regulations in particular with maximum residue limits before selling or donating. Those people, who have their own vegetable gardens, must follow the given instructions and guidelines for use and handling while using pesticides. Protect themselves by wearing gloves and masks on face as a must do or switch to organic farming. Removing peels of vegetables and fruits and washing them can minimize the risk of intake of pesticide residues which further reduces other foodborne diseases as well. In international trade of food, Codex standards are considered as reference, which helps consumers from all over the world to be sure that the food they are buying meets the safe level or standards, no matter where it was manufactured. Present day, there are Codex standards for more than 100 different pesticides (Figures 9.2 and 9.3).

FIGURE 9.2 (See color insert.) Cow dung and crop residues used in agriculture at horticulture centre in Barjora, West Bengal.

FIGURE 9.3 **(See color insert.)** Organic farming in horticulture centre at Barjora, West Bengal.

9.4.4 PREVENTION FROM SOIL POLLUTION (E-WASTE)

The rapid increase in global temperature and climate change ahead the world into great problems but with technology development, world is dealing with problems like electronic wastes (e-wastes). Electronic wastes contains hazardous substances such as lead, mercury, beryllium, cadmium, chromium, brominated flame retardants, etc., that can accumulate in nature and pollute the ecosystems associated with the health risk of wildlife and humans. Phase-out use of non-essential mercury containing products and carry out safe handling, use, and disposal of remaining mercury containing products. Mercury is containing products including batteries, measuring devices (e.g., Thermometer, and Barometer), electric switches and relays in equipment, lamps, dental amalgam, skin-lightening products and other cosmetics and pharmaceuticals. E-waste requires proper disposal or recycle. E-waste should not be thrown with normal garbage.

9.5 NOISE POLLUTION

The quality of our environment is judged, apart from other factors, by the amount of noise present. Noise has been defined as "unwanted or unpleasant sounds" which is being dumped into the atmosphere to disturb the unwilling ears. It adversely affects our physiological and mental health. Noise is measured by a sound level meter and is expressed in a unit called Decibel. The laws and limitation governing environmental noise monitoring varies from country to country. According to WHO, more than 50Db is noise in residential areas.

9.5.1 SOURCES OF NOISE POLLUTION

Sources of noise pollution are industrial source, traffic noise, public systems, agricultural machines and defense equipments. Noise pollution has increased as progress in technology. Contributors of noise pollution are textile mills, metal works, printing presses, engineering establishments, etc. As the number of personal vehicle increases, there is rise in traffic and traffic jams in congested areas, where the repeated using of vehicle horns for long time by impatient drivers affects the ears of all passengers. Loud speaker contributes in its own way towards noise pollution. The reason for using loud speakers may be religious functions, birth, marriage, elections, demonstration or just commercial advertising.

9.5.2 EFFECTS OF NOISE POLLUTION

What type of impact does unwanted Noise have on your health and well-being? Noise is annoying but other than be annoying to really has any real consequence on your health or other living organisms around you? Studies find that exposure to Noise Pollution results in many health effects like neurological disorder, mental problems, behavior changes, etc. When children come in contact with noise pollution, child's epinephrine levels rose 49%, their non-epinephrine more than double and their systolic blood pressure on average went up by five points. Long exposure can damage their ear drums.

In general, when people face noise pollution, it leads to boost in heart rate, breath less deeply, release fight or flight hormones like cortisol. Release of cortisol in body has other negative effects like over eating, feeling stress. Factory workers suffer chronic hearing loss after many years of occupational noise exposure. It has adverse effects on thinking and coordination of limbs.

9.5.2.1 NOISE POLLUTION ALSO EFFECT AVIAN SPECIES

A three years research was conducted by The University of Colorado at Boulder on how noise pollution impact on bird health. They successfully proved that the noise pollution actually put a negative impact on birds and their habitats. Their communication gets disturbed when they undergo any noise. Birds use to communicate at lower frequencies and are easily cover under noise pollution, it affects their ability to attract a mate and socialize with other birds in their group. There are few birds like Finches, they communicate at a higher frequency appear unaffected by the noise pollution.

9.5.3 PREVENTION FROM NOISE POLLUTION

Noise pollution can be controlled by following measures:

1. Green Muflur Scheme: Trees such as Neem and Ashoka absorb sound vibrations to a great extent. Plantation of these trees on both sides of the roads, around silent zones reduces the menace of noise pollution.
2. Use of anti-noise devices.
3. Process and technology modification for lesser noise generation.
4. Avoid burning crackers beside the residential and hospital areas.
5. Awareness generation.

9.6 RADIOACTIVE POLLUTION

Every material is composed of atoms. Protons, electrons, and neutrons together make an atom. Protons and neutrons are present in nucleus and electrons are present in outer shells in an atom. Atom contains both positive and negative charge, nucleus carries a positive charge where as electrons carries a negative charge. These forces within the atom work toward a strong, stable balance by getting rid of excess atomic energy (radioactivity). In that process, unstable nuclei may emit a quantity of energy, and this spontaneous emission is what we call radiation. Radiation is energy given off by matter in the form of rays or high-speed particle.

The existence of radioactive compounds in the environment which can cause damage to the environment is known as radioactive pollution. A substance known as radioactive when the nuclei of its atom start to decay because it is highly unstable in nature and want to obtain a stable form and releases particles like alpha particle, beta particle, gamma rays and type of electrons. Particles of radiation can easily penetrate the skin of organisms and that is one of the most dangerous thing about the particles of radiation. Even they can cause skin cancer and other type of cancers. A living organism can be affected only if the organism is placed just within few centimeters of area of starting point or source of radiation in the case of alpha radiations. The biological creatures (plants, animals, and humans) got the effect of radiation as the other forms of radiation can spread over a large area as it can travel easily over long distances even up to several miles or kilometers.

9.6.1 SOURCE OF RADIOACTIVE POLLUTION

Sources of radioactive pollution can be divided into two groups: natural source and anthropogenic radioactive pollution sources.

9.6.1.1 NATURAL SOURCE

The natural radioactive pollution sources are:

- Cosmic rays possess high energy ionizing electromagnetic radiation.
- Small amount of naturally occurring radioisotopes and radioactive radiation are obtained from soil.
- Lithosphere contains some radioactive elements like radium, thorium, uranium, isotopes of potassium and carbon.

9.6.1.2 ANTHROPOGENIC RADIOACTIVE POLLUTION SOURCES

- Nuclear tests during some experiments;
- Nuclear reactors in industries;
- Diagnostic medical applications;
- Nuclear wastes, improper disposal of radioactive materials after use;
- Nuclear explosions;
- Nuclear metal processing;
- Medical and industrial wastes;
- Nuclear wars.

Radioactive contaminant presence is a hazard because of their radioactive decay, they continuously emit rays (ionizing radiation), alpha, beta, and gamma or neutrons. The strength of hazard can be predicted by the concentration of the contaminants, potency of the radiation being emitted, the type of radiation, and the probability of the contamination to any body organ depends on exposure.

9.6.2 EFFECTS OF RADIOACTIVE POLLUTION

Many times people get exposed to natural sources of radiation, like soil, water, and vegetation, as well as with man-made sources of radiation, X-ray, and other medical devices. Radioactive pollution affects everything present in the environment. Radiations have both immediate or short-term and delayed or long-term effects. The particles emitted by radioactive substances have high penetrating power. Alpha particle have less penetrating power and it cannot penetrate a paper sheet where as the beta and gamma ionizing radiations are able to travel long distances and have high penetrating power. Different organisms show different susceptibility to ionizing radiations. Genetic mutation, cancerous tumor formation, and therefore decreased life span are common effects on all organisms including human after exposure to radiation.

Alpha particle unable to penetrate a sheet of paper, beta particle can penetrate paper but unable to penetrate aluminum sheet and gamma ray is able to penetrate both paper and aluminum but to some extent penetration can be inhibited by using lead.

Ionizing radiation can also cause a burning effect. Skin reddening along with damaged dermal tissue and pain followed by blistering effect in skin are common symptoms of a radioactive burn on skin. The risk is calculated on several parameters like duration of radiation, radiation type, and sensitivity of the body. Radioactive can also go in internal systems of body if radioactive material or radiation contaminated food is ingested.

Radioactive pollution also affects plants and animals. It causes illness over several generations and can also hamper agricultural sector via converting the fertile soil into sterile ones hence a huge loss of crops along with radiation risk. Radioactive pollution contaminates the air, water, and soil. The living organisms present in soil and water get affected when come in contact with radiation.

A different type of introduction with bigger hazard to human physical condition occurs after inhalation or ingestion of radioactive iodine is either inhaled or ingested or consumption of radioactive substances contaminated foods, suppose when the milk, water or anything which is essential contaminated with radioactive pollution and internal exposure also happens and some of the contaminants also get accumulated in the various parts of body just like the iodine accumulation in the thyroid gland. The accumulation can also happen when the radiation is present in food and if we had consumed that particular food.

For the normal metabolism of our body the thyroid gland produces hormones by utilization of iodine, hence the gland can easily assimilate radioactive isotope of iodine as the gland cannot select between radioactive and nonradioactive isotope of iodine. Thyroid cancer can become one of the consequences and is extremely dangerous. Thyroid cancer risk is more in exposure at young age.

Pregnant women if anyhow exposed to radiation then the offspring is also radiation affected individual and has less developed brain and other parts as compared to normal individuals and the affected individual also has a great risk of having cancer.

Epidemiological studies on populations exposed to radiation, such as atomic bomb survivors or radiotherapy patients, showed a significant increase of cancer risk at doses above 100 mSv. More recently, some epidemiological studies in individuals exposed to medical exposures during childhood (pediatric CT) suggested that cancer risk may increase even at lower doses. Madam Curie's research papers are still contaminated with radioactive elements that are why it is locked in lead boxes and even radiation proof clothing is must if anybody want to consult it as the papers are full of radiation and is extremely dangerous due to same reason. Even her cook book is highly contaminated with radioactive compounds.

9.6.3 PREVENTION FROM RADIOACTIVE POLLUTION

Ionizing radiation are very useful in various sectors including medicine in various complicated diseases or disorders, research for identifying the mechanisms of various important reactions or the pathway of reactions, agricultural sector and also in industrial process, when utilized with proper protection and care but like a coin has two faces it also have the positive side as well negative or dangerous side which may cause severe radiation damages within the environment by badly affecting the biotic and abiotic factors if not treated according to the necessary protection and care.

Removal of radioactive iodine can be achieved via washing with hot water along with the application of soap during external exposure of the radioactive isotope of iodine. Urinary output of the excess iodine whether radioactive or non radioactive is possible naturally. Potassium iodide (KI) can be applied for blocking of thyroid from radiation; it can also selectively stop the radioactive iodine entry into the thyroid gland at early stages of exposure.

Proper alternative radioactive waste disposal system is to be followed because there are many cases of leakage, accidental introduction of the radioactive reaction intermediates or radioactive wastes or pure radioactive material from laboratories and various nuclear power plants. For the accidents there should always a emergency safe disposal system or to convert the radioactive substance into nonradioactive substances and then can be utilized in other forms without hampering the environment.

Nuclear Shelters: It is a specialized place for accommodating many individuals with proper protection from radiation when there is any nuclear explosion, generally the shelters are made so that it can resist the radiation during nuclear explosion and are generally made protected against gamma rays as it has very high penetrating power and for the same reason it's very dangerous.

Radioactive wastes from Atomic power plants are degraded by adopting composting method. When radioactive wastes are added to and covered with compost piles, Microbes in the compost degrade the wastes within 150 Days. Explosives such as trinitrotoluene (TNT), hexahydrotrinitro-triazine (RDX), octahydrotetranitro-tetrazocine (HMX) and tetryl in liquids can be treated in this way to reduce radioactive pollutants in the soil and water. This method is superior to conventional deep burying method that requires 1000s of years to decay the radioactive wastes.

9.7 INITIATIVES BY NATIONAL AND INTERNATIONAL ORGANIZATIONS

9.7.1 SAFAR – INDIA

System of air quality and weather forecasting and research has been installed to measure air quality and also to make it better by the help of several research activities aimed for improving the air quality by ministry of earth science, Government of India and Indian institute of tropical meteorology, Pune. A research based initiative of integrating air quality with health advisories and food security. A dedicated air quality information service for Indian metropolitan cities to make India self sufficient in providing frontier research based scientific accredited robust air quality forecasting system.

9.7.2 IR-DRDO BIO-TOILET SYSTEM

Bio-toilet is a very innovative approach towards proper human waste management via eco-friendly, cost-effective, along with proper sanitation and hygiene. It actually converts the solid egested waste of humans into biogas along with some amount of pure water via some bacterial processes.

9.7.3 STOCKHOLM CONVENTION ON PESTICIDES BANNING

Earlier in 2001, under Stockholm Convention many countries has already banned the pesticides like DDT and lindane which are responsible for biomagnifications as they can remain in lithosphere and hydrosphere for a long period of time the convention was for restricting the persistent organic type of pollutants.

9.7.4 ROLE OF WHO AND UNICEF TOWARDS GLOBAL HEALTH CARE

There was an initiative at global level for providing essential health care services like water, proper sanitation and hygiene and it also includes addressing the waste from healthcare sector by WHO and UNICEF along with other partners in the year 2015.

9.7.5 CMERI DURGAPUR: A PILOT CASE STUDY

The Central Mechanical Engineering and Research Institute, Durgapur has established a first pilot plant to produce electricity by using city garbage. The plant has a capacity to use about 500 kg garbage/ha, as a result of which about 5 KW electricity can be generated. Process of electricity production from garbage is garbage is converted into biogas under anaerobic condition and then biogas and in turn into electricity.

9.7.6 INTEGRATED FARMING IN INDIA

In India, farmers, and scientists, jointly are dedicated in finding out new ways to increase the production of food and also the quality of farmland to make money by the utilization of nuclear methods to increase no of crops and quality of livestock without harming or compromising the soil health or quality. Hence, the agricultural sector will be benefited and it will come under sustainable development as profit of mankind does not affect the environment.

Scientists at the Tamil Nadu Veterinary and Animal Sciences University have used nuclear and isotopic techniques to study soil and water use and select and grow crops that thrive on local farms. They integrated their findings with effective livestock production methods involving cattle and goats to develop and easy-to-follow, crop, and livestock-based organic farming system. The project has so far resulted in an increase in organic carbon content in the soil, which gives it structure and makes it healthier and improved for growing crops. Livestock reproductive performance has also gone up, including a 15% increase in the cattle's milk production as well as significant increases in the size of the goats.

9.7.7 ECOSAN TOILETS UNICEF, INDIA

EcoSan toilets are very effective in coastal regions with high water table areas and water scarce areas. EcoSan toilets have safe disposal of human excreta and Urine. Water is not required for flushing.

9.7.8 ROLE OF UNCHE

The United Nations Conference on Human Environment was convened for studying the observed changes within the interrelationship between the environment and human beings with respect to each other in this modern century of extremely developed science and technology.

9.7.9 ROLE OF WHO TOWARDS AIR POLLUTION

The air pollution is studied and monitored at global level via networks at international level which are installed by WHO for researching on the same and to find better solution for air pollution.

9.7.10 PERC A CASE STUDY ON PHYCOREMEDIATION

PERC (Phycospectrum Environmental Research Centre, India) is focused on the Phycoremediation of Sewage water and Industrial Effluents. Recently, Mainath lake in Uttar Pradesh has been treated where the contamination level was 50% more than the contamination level of raw sewage. PERC had employed a local microalgae system to control the pollution level within short time wherein major parameters like BOD, COD, and Coliform bacteria were reduced by 90% resulting in transparent water.

9.8 INITIATIVES FROM OUR SIDE

9.8.1 ROLE OF INDIVIDUALS TOWARDS POLLUTION CONTROL

We have participated in many awareness campaigns and social works to promote use of eco-friendly things, mass awareness for pollution and global warming under green revolution action against climate change. Participated in green revolution action against climate change. In this global certification program, we have created awareness on climate change and its prevention among school and college students. Sell seed pencils to promote plant growth. Seed pencils are paper pencils contain seeds in the bottom. When pencils are planted in pot containing soil and water then after few days or weeks it grows into small plant (Figure 9.4).

FIGURE 9.4 Nature study during educational tour exploring the impact of pollution on plants.

9.8.2 *POLLUTION CONTROL IN DURGAPUR BARRAGE AREA: A CASE STUDY*

Participated in Durgapur Barrage Cleaning Campaign 2018, whose aim was to Protect the Birds in Durgapur Barrage by creating awareness among people who came there for picnic on Holiday. Distribute leaflets among them exchange thermocol plates and bowls with Sal plates and bowls to prevent soil pollution (Figure 9.5). No DJ box from 31st December 2018 to create silent picnic in silent zones because the sound of DJ box creates noise pollution and it affects the birds of Durgapur Barrage. We aware people about wildlife of Durgapur Barrage (Figures 9.5 and 9.6).

FIGURE 9.5 During Sal leaf plate distribution by replacing thermocol plates, Durgapur Barrage, West Bengal.

FIGURE 9.6 Explaining people about wildlife of Durgapur Barrage and importance of using ecofriendly things.

9.8.3 PLANTATION IN POLLUTION CONTROL

Plant a seed on every Birthday, every year and take care of it throughout the year. This is a simple and small investment for a healthy future. Every year plant different varieties of plant because you have to maintain biodiversity as well as the number of plants. Different plants have different importance.

9.9 CONCLUSION

Interdisciplinary team needed to solve the pollution and prevent the pollution related diseases. Conservation of plants and plant diversity is required because tropical rainforest store a lot of carbon; they are very good at capturing carbon and storing it mangroves forest. Promote the use of eco-friendly things. Waste from industry or household, should be treated properly or converted into energy. Pathogens present in environment get evolved and adapted themselves quickly with changing environmental conditions and humans are becoming susceptible towards diseases. Pollution must be taken seriously, as it is reducing the distance between Existing species to endangered species to extinct species. Pollution must be controlled to protect the endangered species to become extinct species. Our actions will determine our fate; nature doesn't belong to people that is people belongs to nature. Environmental pollution can be prevented it just need a thought, determination, and action. Nature should be protected from all the human activities which affect it badly, and we must think of more innovative ideas to go with nature and development. We can do one thing now that bioremediation can be applied on the onset of releasing the pollution at industrial level then the remaining wastes after treatments should be released in the common dumping area where all the industries of the particular area will dump their wastes and the wastes are to be utilized via biological or chemical leaching or degradation into various chemicals and products for money making and to so sustainable development.

KEYWORDS

- **environmental pollution**
- **health**
- **initiatives**
- **prevention**
- **waste management**

REFERENCES

Bafana, A., Krishnamurthi, K., Sivanesan, S., & Naoghare, P. K., (2018). *Mutagenecity and Genotoxicity Testing in Environmental Pollution Control*. Elsevier BV. pp. 113–132. https://doi.org/10.1016/B978-0-12-809252-1.00006-7.

Berry, T. D., Filley, T. R., & Blanchette, R. A., (2014). Oxidative enzymatic response of white-rot fungi to single-walled carbon nanotubes. *Environ. Pollut.*, *193*, 197–204.

Blackburn, S., (2007). *Maternal, Fetal, & Neonatal Physiology: A Clinical Perspective* (4th edn., p. 325). Elsevier Health Sciences. ISBN:1–4160–2944–3.

Calderón-Garcidueñas, L., Gónzalez-Maciel, A., Reynoso-Robles, R., Delgado-Chávez, R., Mukherjee, P. S., Kulesza, R. J., et al. (2018). Hallmarks of Alzheimer disease are evolving relentlessly in Metropolitan Mexico City infants, children and young adults. APOE4 carriers have higher suicide risk and higher odds of reaching NFT stage V at ≤ 40 years of age. *Environmental Research*, *164*, 475–487.

Carré, J.,Gatimel, N.,Moreau, J.,Parin, J., & Léandri,R., (2017). Does air pollution play a role in infertility?: A systematic review. *Environ Health*, *16*, 82.

Galal-Gorchev, H., (1991). Dietary intake of pesticide residues, cadmium, mercury and lead. *Food Additives and Contaminants*, *8*, 793–806.

Hightower, J. M., (2011). *Diagnosis Mercury: Money, Politics and Poison* (2nd edn., p. 77). Island Press.

IARC, (1976). Cadmium, nickel, some epoxides, miscellaneous industrial chemicals and general considerations on volatile anesthetics. *Lyon, International Agency for Research on Cancer*, Vol. 11, pp. 39–74. (IARC Monographs on the Evaluation of the Carcinogenic Risk of Chemicals to Man).

Krajnc, E. I., Van Gestel, C. A. M., Mulder, H. C. M., De Vrijer, F. L., Sinkeldam, E. J., Vink, G. J., Canton, J. H., Van Apeldoorn, M. E., & Janus, J. A., (1987). *Integrated Criteria Document Cadmium Effects* (pp. 1–200.). Appendix to Report nr. 758476004. RIVM, Bilthoven, The Netherlands.

Nita, M., &Grzybowski, A., (2016). The role of the reactive oxygen species and oxidative stress in the pathomechanism of the age-related ocular diseases and other pathologies of the anterior and posterior eye segments in adults. *Oxid. Med. Cell Longev.*, pp. 1–23.

OECD, (2012). *Policy Coherence and Food Security: The Effects of OECD Countries Agricultural Policies*. Paper prepared for OECD Global Forum on Agriculture, Paris.

Perkins, K. A., Jao, N. C., & Karelitz, J. L., (2013). Consistency of daily cigarette smoking amount in dependent adults. *Psychol. Addict. Behav.*, *27*(3), 1–13.

Pope, C. A., (2003). Cardiovascular mortality and long-term exposure to particulate air pollution: Epidemiological evidence of general pathophysiological pathways of disease. *Circulation*, *109*(1), 71–77.

Rosenblueth, M., & Martínez-Romero, E., (2006). Bacterial endophytes and their interactions with hosts." *Molecular Plant-Microbe Interactions*, *19*(8), 827–837.

Takeda, K., Tsukue, N., & Yoshida, S., (2004). Endocrine-disrupting activity of chemicals in diesel exhaust and diesel exhaust particles. *Environmental Sciences: An International Journal of Environmental Physiology and Toxicology*,*11*(1), 33–45.

CHAPTER 10

Innovative Approach to Sustainable Development and Biodiversity Conservation

PINAKI CHATTOPADHYAY,[1] PALLAVI CHATTOPADHYAY,[2] and DEBNATH PALIT[1]

[1]*Department of Botany, Durgapur Government College, Durgapur, West Bengal, India, Mobile: +918293807145, E-mail: iampinu7@gmail.com (P. Chattopadhyay)*

[2]*Department of Microbiology, Amity University, Lucknow Campus, Lucknow, Uttar Pradesh, India*

ABSTRACT

This chapter discusses some ways of sustainable development like the utilization of flyash, bioremediation, biocontrol, bioplastic, and some new innovative ideas for sustainable development and biodiversity conservation, like spiral gardening, artificial mountain, green towers, and some green strategies for future. More research is to be done on sustainable development processes for the extending the lifeline of nature and subsequently of us. Ultimately we all have to move towards sustainable development because it is the only life jacket of human beings in the ocean of destruction.

10.1 INTRODUCTION

Sustainable development is the development to satisfy the human need without harming nature and its resources. For sustainable development, we have to reduce, reuse, and recycle the harmful products (pollutants) contributed to nature by urbanization and industrialization. So, for that proper waste (pollutants) management is compulsory. Pollutants are a kind of harmful wastes to our earth, so there is a need to transform these to a non-harmful form or to store it in such a way that it will not harm to nature, natural resources and ultimately humans. Sustainable development should be the main thinking /motive for all the human population, as it's our necessity. Here are some innovative approaches (ways) towards sustainable development.

10.2 UTILIZATION OF FLYASH

Flyash is a combustion product of coal that consists of the particulates (fine particles of fuel which has been combusted) which came out of coal-fired boilers together along with the flue gases. The ash which falls in the bottom part of the boiler is called the bottom ash. Modern coal-fired power plants, generally uses electrostatic precipitator or other type of equipment for particle filtration to capture flyash. Flyash is a very big problem which is contributed to us by sponge iron industries, thermal power plants and many other large and small industries but it can also be utilized for mankind. It is used in production of concrete actually here it is an alternative to sand and Portland cement. Now there are bricks made up of flyash which are ultimately used for construction of buildings and for construction works. Flyash pellets are now used to substitute normal aggregate in concrete mixture. It is also used for constructing embankments of water bodies and other types of structural fills. It is also used for solidification and waste stabilization, production of cement clinkers (alternative or substitute material for clay), for reclamation of mine, for soft soil stabilization, for constructing road subbase. For brick production, it is also used as an aggregate in substitute material in asphaltic concrete and used as mineral filler (Zimmer, 1970).

There are some agricultural uses too like amendment of soil, as a fertilizer, as cattle feeders, stabilization of soil in the agricultural stakes and stock feed yards. Applications also include melting the ice in parking lots, roads for controlling ice. Other applications include toothpaste, cosmetics, floor, and ceiling tiles, roofing tiles, kitchen countertops, flotation devices, stucco, utensils, structural insulated panels, bowling balls, tool handles, auto bodies, picture frames, boat hulls, PVC pipe, cellular concrete, geopolymers, railway sleepers and blasting grit. A company in Bengaluru has found a way to deal with the problem of air pollution. The idea is to separate the carbon from the particulate matter and to utilize it in making ink. These types of innovative ideas are the much-needed solution for the increasing pollution scenario across the country.

Although their product Kaalink is not commercially available as yet, they plan to install it on public vehicles so that they can send well-trained persons to take away the particulate matter and to separate it from the soot. The company is still working on improving the efficiency of the capturing process. Some workers have utilized flyash as a carrier for *Bacillus Thuringiensis Israelensis* (Bti), a biopesticide which can be used for killing a range of insect larvae (Tamilselvan et al., 2017).

10.3 BIOCONTROL

The control of the enemies of crop (pest, pathogen, and weeds) by using their biological antagonist (an organism which suppresses the growth or kills another organism) is called biocontrol or biological control, it is an eco-friendly, non- hazardous approach towards sustainable development.

Biological control of pests is of five following types:

1. Predators;
2. Parasitoids;
3. Bacterial pesticides;
4. Fungal pesticides; and
5. Viral pesticides.

1. Birds as Agents for Pest Control: There are many different species of birds who feed on a wide range of insect pests as Insectivorous. These birds help to reduce the number of pests to a considerable extent at which there is no crop loss or very less crop loss as compared to earlier. For example, the most common Indian mynah birds are often used to control locusts in citrus (Table 10.1).

TABLE 10.1 Some Predatory Birds and Their Target Pests in Crop Fields

Sl. No.	Common Name and Scientific Name of the Birds	Family	Target Pests
1	Indian pond heron (*Ardeola grayii*)	Ardeidae	A wide range of pests present in pond embankments near rice fields
2	Tit *(Parus* spp.)	Paridae	*Cameraria ohridella*
3	Indian roller (*Coracias benghallensis*)	Coraciidae	Locusts and other pests of cotton, sugarcane, and groundnut.
4	Owl		Rats and mice in paddy fields
5	Ashy wren wrabler (*Prinia socialis*)	Cisticoliae	Plant hoppers and midges in rice fields.
6	Migratory red wrabler (*Acrocephalus* sps.)	Acrocephalidae	All pests in rice fields.
7	Cattle egret (*Bubulcus ibis*)	Ardeidae	Grasshoppers and other pests in rice fields.

2. Pest-Eating Insects: Many insects feed on pests and reduce their population in crop-fields. These predatory insects are known as Pest-eating insects or pestering pests. They eat larvae and adult pests. Veladia beetle eats cottony cushion scale of Citrus in large proportion. Asian Red Weaver Ants: The Asian red weaver ants (*Oecophylla smaragdina*) are utilized in biocontrol of pests in agricultural land. Since 305 AD, Chinese farmers have been using the ants to control some leaf-feeding insects of citrus. Now, these predaceous ants are used to control:

a. Nurid leaf bugs in cocao.
b. Caterpillars and large boring beetles in date palms.
c. Coconut moth in coconut.
d. Citrus shield bugs in citrus.
e. Rhinoceros beetles (Orycetes rhinoceros) in coconut. *O. smaragdina* is introduced on Thepesia tree to propagate the red ants and ant-nests are transferred to plants to be protected.

Rodolia cardinalis (commonly known as Vedalia beetle) is a natural antagonist of Icerya cardinalis (commonly called cottony cushion scale) in Australia. Now, vedalia beetle is being used in almost all countries to control cottony cushion scales. Assassin bug (*Rhynocoris marginatus*) is a polyphagous insect occurring in crop fields, wet forests, and semi-arid regions. It is mass cultured using its prey *Corcyra cephalonica* (pest of storage grains) and disseminated into crop fields. This bug successfully controls *Helicovera armigers, Spodoptera litura, Dysdercus cingulatus, Mylabris pustulata in M. indica* (Ambrose and Clave, 2002). Some species of spiders are also known for their pest control activity. A species of spider *Lyssomanes silkimensis* eliminates mango hoppers in mango trees and wolf-Spider (*Lycosa* spp.) controls many pests in paddy fields. Parasitoids are also known to have biocontrol activity. Numerous parasitoids are there who show their controlling effect upon specific hosts (Table 10.2).

TABLE 10.2 Insects Which Parasitize to Another Insect Species is Called Parasitoids

S. N.	Parasitoids	Effective Against
1	*Aphelinus mali*	*Eriosoma laniger*
2	*Trichogramma australicum*	*Chilo indicus*
3	*Leptomastix dactylopsii*	*Planococcus citri*
4	*Trichogramma minutan*	*Argyria stricticraspis*

Similarly, there are different bacterial species which works against specific host species (Table 10.3).

TABLE 10.3 The Formulation in Which a Suitable Bacteria is Adopted for Killing Pests

SL. N.	Bacteria	Effective against
1	*Bacillus thuringiensis neoleonensis*	*Heliothis armigera*
2	*Bacillus popilliae* and *B. lentimorphus*	*Popilliae japonica*
3	*Bacillus cereus*	*Amsacta albistriga*
4	*Bacillus cereus*	*Achoea janata*

When we are discussing biopesticides, we must include the most popular fungi *Beauveria bassiana.* Kudzu bug which has been killed by *Beauveria bassiana* can be seen growing on the particular insect. Bioherbicides contains microbes, pathogens, and phytotoxins used as biological control agent for controlling weeds. Bioherbicides can be secondary metabolites of microbes or compounds derived from microbes such as bacteria or protozoa, fungi, or phytotoxic plant products, single compounds or extracts obtained from other species of plant (de Souza et al., 2017).

Natural algicides -In England, barley straw is used in mesh bags and released in ponds which have fishes or in water bodies of gardens to control algal growth without harming the flora and fauna of pond. Although the United States Environmental Protection Agency (EPA) did not approved this for using as an effective pesticide as its effectiveness as an algaecide in water bodies has given mixed results during testing in the universities of the United States and England.

3. **Fungal Pesticides:** This is a microbial pesticide and made by using some specific fungi. These are host specific (Table 10.4).

TABLE 10.4 List of Some Fungal Pesticides and Their Host

Fungi	Host
Beauveria bassiana	Coffee berry borer, Rice stem borer, Colarado beetle, Mexican bean beetle
Fusarium pallidoraseum	*Cow pea aphids*
Zoophthora radicans	*Marasmia patualis*
Metarhizium anisopliae	Rhinoceros beetles, White grubs, Cutworms, and Mosquitoes
Nomuraea rileyi	Larvae of Spodoptera, Gram pod borer (*Helicoverpa armigera*), Tobacco budworm (*Heliothis virescens*)
Paecilomyces fumosoroseus	White flies, Aphids, Thrips, and Spider mites
Entomophthora muscae	Onion maggot, Cabbage maggot, Seed corn maggot

This is a microbial pesticide which is made from some specific Families of viruses. These viral pesticides are species specific. Viral spray suspension kills the pest within 15 days after treatment (Table 10.5).

TABLE 10.5 List of Viral Pesticides Along with Their Specific Host

Virus	Effective Against Host
Granulosis Virus	Codling moth
	Sugarcane shoot borer
Rhynchophorus ferrugineus CPV	Red palm weevils
Orycetes rhinoceros CPV	Rhinoceros beetle
Lymantria disper NPV	Gypsy moth
Neodiprion sertifer NPV	Saw flies

Scientists are working with transgenic strains to improve fungi's ability to kill insects, tolerate adverse conditions and, extending beyond crops, fight against the transmission of diseases such as West Nile virus, Lyme disease and malaria. St. Leger's team is currently testing a strain of *Metarhizium* that's had a spider gene inserted that selectively targets mosquitoes. As microorganisms have host of benefits but there are some microorganisms which act as pathogens for some plants and they are controlled by using some specific group of microorganisms.

Biologically, plant pathogens are controlled by:

1. Introduction to antagonists;
2. Introduction to VAM fungi.

One group of microorganism inhibits the growth of other microorganism by releasing antibiotics, toxins, etc. This relationship is known as Antagonism. These microorganisms are Antagonist to each other. *Pseudomonas fluorescens* are used to control the growth of fungus, e.g., *Sclerospora graminicola, Pythium, Rhizoctonia solani* by secreting some antibiotics like Pyoluteorin, Pyocyanin, and Phenazine. *Pseudomonas syringae* is saprophytic in nature frequently found in Apple fruits. This bacterium is used to control the growth of *Penicillium expansum, Botrytis cinerea*, etc. Some species of *Amoeba* present in the soil, feed on pathogenic fungi or engulf those pathogenic fungi and control the plant disease. This phenomenon is called Mycophagy. Examples of Mycophagous Amoebae include *Vampyrella, Geococcus*, etc. They feed on both the conidia and hyphae of fungi. VAM Fungi (Vesicular-Arbuscular Mycorrhizal fungi) are antagonistic to some soil borne pathogenic fungi. The VAM Fungus *Glomus mosseae* is used to control *Rhizoctonia solani* that infect soyabean. These Biocontrols are user friendly as well as ecofriendly. They are promising alternatives against Chemical controls in crops. According to Wojciech et al., 2010). Microbial flora of fruit outsides have been the finest source of antagonists against fungi which causes decay after harvesting of the fruit.

Mycorrhizal association is an symbiotic or mutual association of fungi with roots of higher vascular plants, the fungi help the plant by providing water and minerals and in return it receives photosynthates from the associated plant. This type of symbiotic association is also useful in establishing forest and also in heavy metal contaminated soils like *Rhizophagus intraradices* (earlier known as *Glomus intraradices* and *Robinia pseudoacacia* (black locust or false acacia) in the soil contaminated with the heavy metal lead. For removal of lead from vetiver grass the two fungi *Rhizophagus intraradices* and *Glomus versiforme* are inoculated for bioremediation purpose. In the same way *Calendula officinalis* and AMF (Arbuscular Mycorrhizal Fungi) in the soil contaminated with the heavy metals like lead and cadmium are also used as a tool for bioremediation. AMF boosts the biological degradation of organic pollutants like ammonia, methyl tert-butyl ether and benzene from groundwater after applying the inoculation of the useful organism into *Phragmites australis* in wetlands (Deshmukh et al., 2016).

10.4 BIOPLASTIC

Plastics are actually the synthetic polymers which are derived from fossil or petroleum products and are extremely resistant towards degradation by biological agents. polypropylene (PP) and Polyethylene (PE) represent around 92% of the whole global production of plastics. There is a report of efficient PE bio-degrading larvae of the *Galleria mellonella* (a wax moth) producing ethylene glycol. *Pseudomonas putida* is competent for converting styrene oil into poly hydroxyl alkanoate (PHA) which can be utilized as a biodegradable plastic or bioplastic. In the bacterial utilization of sugars or cornstarch or glucose, a polyester type of biopolymer is produced named poly(3-hydroxybutyrate) (PHB), in terms of characteristic. It is just like petroplastic PP. This property is to be utilized by doing this on large scale, and then only we can beat plastic pollution. Bioplastics which are basically made up of esters of cellulose like nitrocellulose, cellulose acetate, and other cellulose-based derivatives are often termed as cellulose bioplastics. Wheat gluten and casein are two protein-based biopolymers which can be utilized for making bioplastics (Song et al., 2009).

10.5 BIOREMEDIATION

Bioremediation is the process of decontaminating toxic substances / pollutants via biological resources (living organisms) (Figure 10.1). Bioremediation can be done in two ways either *in-situ* or *ex-situ*. *In-situ* methods are actually on-site treatment of the contaminated stuff, whereas if the treatment is off-site or done at any other place after removing the whole stuff from the original site then it is referred to as *ex situ*. It is costly process to excavate and remove the contaminated stuff. In contrast, if the decontaminated soil is treated *in-situ*, the overall expense is far less. In washing or taking out toxic materials from the contaminated soil, the contamination is simply transferred from a place to another, and is not cured fully or eradicated properly, while problems may arise in incineration, e.g., dioxin formation, it is energy-intensive too. Bioremediation is to degrade toxic organic materials, e.g., from oil spills, industrial waste, and pesticides, at the molecular level, converting them to more non-toxic/ harmless compounds.

The main target of bioremediation is to do complete mineralization of the contaminants, i.e., their conversion to HCl, N_2, H_2O, CO_2, etc. But radioactive cations and heavy metals cannot decomposed but can be converted into their less soluble form by changing the oxidation state so that they cannot do any harm or cannot show any toxic effect as much as it can do in its optimum state or active state when present in the ground as it has been converted into less harmful/less toxic, or might be physically removed by application of plants or fungi by the process of phytoremediation and mycoremediation, respectively. Some aerobic bacterias with good degradative properties are *Pseudomonas, Alcaligenes, Sphingomonas, Rhodococcus*, and *Mycobacterium*. One of the remarkable bacteria is *Pseudomonas putida* which can destroy /degrade petroleum oil and have been used earlier to clean the oil spills. These microbes have the ability to degrade hydrocarbons (both alkanes and other similar groups of compounds) and also pesticides. Actually, these bacteria utilize the contaminant as source of carbon and energy and it is believed that this unique property is due to degradative plasmids present in them (Chakrabarty et al., 1975).

Bioremediation of polychlorinated biphenyls (PCBs) in river sediments, dechlorination of the solvent trichloroethylene (TCE), and chloroform by ligninolytic fungi and anaerobic bacteria is one of the burning research topics now. Methylotrophs constitute the group of microorganisms which utilizes reduced compound of one carbon type, like methanol or methane, as source for carbon which is for their growth; and multi-carbon compounds that contain no carbon-carbon bonds, such as dimethyl ether and dimethylamine. This group of microorganisms also includes those capable of assimilating reduced one-carbon compounds by way of carbon dioxide using the ribulose bisphosphate pathway (Anthony, 1982).

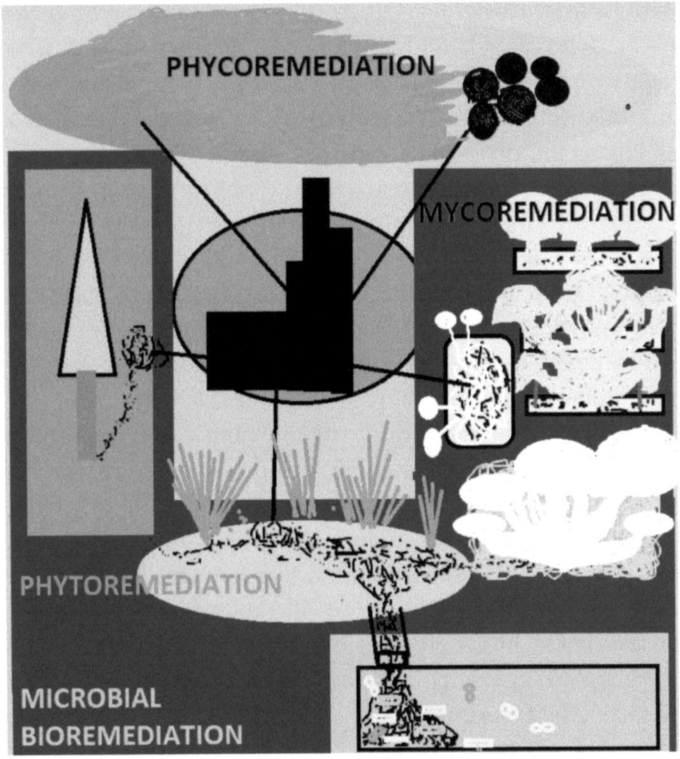

FIGURE 10.1 **(See color insert.)** Different forms of bioremediation.

10.5.1 BIOLOGICAL TREATMENT TECHNOLOGIES

10.5.1.1 BIOAUGMENTATION

This process actually deals with the increase in enzyme activities which ultimately boost the rate of degradation of the contaminants as compared to the non-augmented rates, at least at the initial level after inoculating the organism within the bioremediation technique.

10.5.1.2 BIOVENTING

This is biological degradation of organic waste from groundwater systems by utilizing suitable microorganisms in that place without transferring the whole system to another place, it is to increase the activity of the bacteria and archaea which will be utilized and stimulate the degradation of hydrocarbons biologically by enhancing airflow or oxygen flow to the unsaturated portion and, nutrients can also be added if it is necessary. In the process of bioventing, oxygen may be supplied through direct injection of air into the residual contamination in that soil. Degradation of volatile organic compounds (VOCs) and adsorbed fuel residuals are primarily assisted by the process of bioventing from the soil which is biologically active.

10.5.1.3 BIOSPARGING

It is a process of in situ biological remediation technology that utilizes the indigenous microorganisms to degrade organic constituents biologically in the potion which is saturated with the help of oxygen or air and nutrients if required to increase the activity of biological agents (microorganisms) In the process of biosparging, air (or oxygen) and nutrients (if needed) are injected into the saturated zone to increase the biological activity of the indigenous microorganisms. Concentrations of petroleum products can be reduced by the process of biosparging from contaminated groundwater or from the soil which has absorbed the contaminant.

10.5.1.4 BIOSLURPING

It is a combination of the process of vacuum-enhanced pumping and bioventing process. The aerobic breakdown of hydrocarbons via biological agents (living organisms) takes place under unsaturated zone of contaminated system when exposed to air. With the help of genetic engineering it may be possible to alter the changes in genetic level so that a particular suitable organism can produce and use a particular enzyme or enzyme system to treat wastewater by decontaminating the same.

Phytoremediation is biological remediation which performed by the biological agents like plants (angiosperms, gymnosperms, pteridophytes, and bryophytes) (Figure 10.2). Phytoextraction refers to the process by which plants can store or accumulate the contaminants majorly in the shoots and the after extracting the contaminants from the contaminated soil. Phytostabilization is the process which is accomplished first by the absorption, then accumulation and ultimately precipitation of contaminants within the area of root portion and hence protecting from the spreading of the harmful contaminant. Phytodegradation is the process by which plant enzymes are utilized in the breakdown of pollutants which are of organic type. Rhizodegradation is a process where contaminants are broken down by the group of microorganism present near root portion of plants by the stimulation from plants. That is why it is also known as phytostimulation. Phytovolatilization is a process in which contaminants are released via volatization from stomata which was taken in the plant via roots, and then it reached to leaves where this occurs. Aluminum can be phytoremediated by *Vicia faba, Solidago hispida* (*Solidago canadensis* L.), *Hydrangea* spp, *Hordeum vulgare, Agrostisca stellana, Brassica juncea, Brassica napus* can phytoremediate selenium. *Brassica oleracea, Brassica juncea, Brassica napus, Hydrilla verticillata, Vallisneria americana, Eichhornia crassipes* can phytoremediate lead *Pistia stratiotes, Hydrilla verticillata, Bacopa monnieri, Eichhornia crassipes* can *phytoremediate* mercury. Roots of plants can absorb toxic contaminants overtime from the ground water and destroyed ultimately. This process is suitable in areas where the plant roots can easily tap water present underground. There are many examples like a very common Chinese Ladder fern which is scientifically known as *Pteris vittata* L., is a good hyper bio accumulator of arsenic.

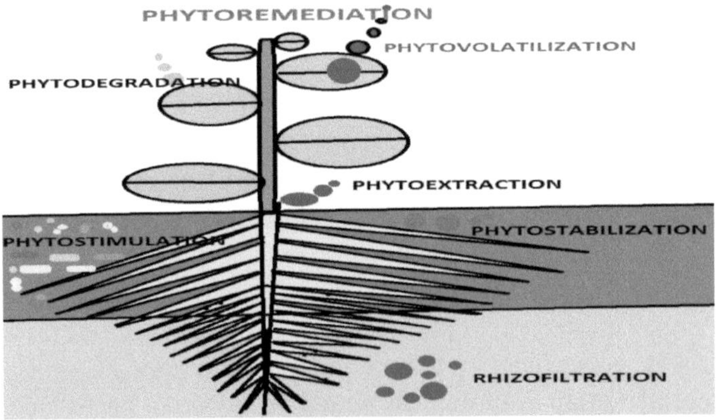

FIGURE 10.2 (See color insert.) Different forms of phytoremediation.

Agmenellum quadruplicatum, strain PR-6 can be used for biological oxidation of some particular aromatic hydrocarbons like napthalene which results in yielding 1-napthol it is one of the achievements in algal biotechnology. Organochlorine insecticide, toxic metabolite, endosulfan sulfate and endosulfan, can be degraded biologically with the help of *Klebsiella oxytoca* KE-8 (Kwon et al., 2005).

10.5.1.5 BIOREMEDIATION USING FUNGI (MYCOREMEDIATION)

This word is taken from two Latin words which means 'restoring balance.' It is a process in which fungi are used as the biological agent for biological remediation of contaminants (heavy metals, etc.) from the environment. Fungi is not so costly hence will save money, efficient, and environmental friendly as help to remove out a broad range of toxic products (like heavy metals or chemicals from leather, textile, and pharmaceutical industries) from contaminated environment without harming any natural process of environment. The process is also used for treating petroleum products, herbicides, pesticides, and some aromatic pollutants including toxins include in contaminated land or sea. The enzymes of fungi are nonspecific hence can biologically degrade a wide range of pollutants after mycoremediation the by-products can also be utilized like in the case of lactase enzyme or edible and medicinal mushrooms which can convert the remediation process into profitable investments making the remediation process even profitable. The strange but useful organism fungi is often used for degradation of some pharmaceuticals like paracetamol, the products after breakdown are harmful in earlier system of treating polluted water with *Mucor hiemalis*. Due to low cost it is used in small scale or domestic wastewater treatment by mycofiltration process which also helps in decomposition procedure of the compost toilet *Hypholoma capnoides*, an edible species of mushrooms, can be used for decontaminating Ti, Sr, and Mn and *Marasmius oreades* species can be used for Bi. In the fruiting body of *Marasmius oreades* bioconversion of Bi and Ti is possible. *Phanaerochaete chrysosporium* have unique features of biodegrading a wide range of persistent or highly toxic pollutants like polyaromatic hydrocarbons (PAH) (Singh, 2006). Common substrates of the useful fungi sawdust, straw, and corn cobs, and these are really very cheap, so now we may conclude that mycoremediation is very useful for mankind and we must utilize this in favor of nature and mankind.

Industrial methods use many of the metal-based techniques or process like electroplating of any metal over other products, or metals utilized in the manufacture of paints and leather. The agricultural utilization of the industrial wastewater is also common but very dangerous because this activity will introduce heavy metals within the food chain and hence will disturb the ecosystem and environment, in this situation mycoremediation can be applied as it is cheap and effective process as the process indulges many fungi which are either hyperaccumulator or tolerant to the particular heavy metal, hence the cytotoxicity is avoided as the intracellular uptake of the heavy metal is negligible after the biosorption of the heavy metal on surface of the fungal cell. The metals like lead, arsenic, copper, zinc, mercury, chromium, cadmium, iron, and boron can be removed easily or extracted out from a contaminated site by the help of many types of fungi like *Trichoderma, Pleurotus, Aspergillus*, etc. Property of bioaccumulation is useful for mycoremediation purpose, and also for bioindicator but can be extremely dangerous for the edible species as it can cause severe metal toxicity like in case of *Coprinus comatus* which can be used as a bio-indicator and bio-accumulator of the heavy metal mercury (Falandysz, 2016). Fungi play saprophytic, parasitic, and symbiotic roles in the ecosystem. They are extremely efficient in decaying any type of matter, due to their extracellular enzymes which breakdown cellulose and lignin, which are the major building units of the plant fibers. Organic pollutants share some structural similarity with these long-chain organic molecules. These are long-chain organic (carbon-based) compounds, structurally similar to many organic pollutants. The important enzymes which are involved in the phenomenon of degradation are ligninolytic enzymes (manganese peroxidase (MnP), versatile peroxidase, lignin peroxidase, laccase, general lipase and occasionally intracellular enzymes, specifically cytochrome P450) (Deshmukh et al., 2016; Pozdnyakova, 2012).

Fungi usually convert the harmful pollutants like phenolic substances and petroleum products into harmless or less harmful products which is very effective way for treating wastewater, PCB are removed or extracted out from contaminated soil via application of the famous edible oyster mushroom *Pleurotus ostreatus*, *Pestalotiopsis* is a fungi which can degrade polyurethane biologically as it can be used as a carbon source by the fungi in both anaerobic conditions and aerobic conditions. The mushroom may be self-sufficient, or it may also serve as a host for some microorganisms which may extract out the pollutants from the contaminated zone (Jonathan et al., 2011).

Pesticide contamination is another danger which can be degraded by the fungi, which cause white rot. These fungi are suitable due to their lignin digesting enzymes like manganese peroxidize and laccase, which efficiently degrades large quantity of insecticides like ortho-phenylphenol, imazalil, diphenylamine, endosulfan present in wastewater and the soils of clay-loamy type contains atrazine. White rot fungi digest the lignin by their extracellular enzymes and provide a white color or bleached color leaving the cellulose in undissolved state, hence the name is white rot. On the other hand in case of brown-rot fungi degrade cellulose, leaving lignin as a typical brown patch. These type of fungi generally causes shrinking and cracking in wood because they are utilizing the substrate. The *Phanerochaete chrysosporium* (a white-rot causing fungus) is an ideal role model for the process of mycoremediation because it is much more effective in the process than any other organism. It actually acts by redox reactions which represent both reducing and oxidizing mechanisms hence making it enables to use it in various conditions on different contaminants. Xenobiotics can be easily biodegraded with the application of some very common fungi which are *Lentinula edodes* (an edible and medicinal mushroom commonly known as shiitake), *Pleurotus ostreatus* (an edible mushroom, commonly called oyster mushroom), *Trametes versicolor* (commonly called turkey tail, a beautiful polypored mushroom), *Pleurotus tuber-regium, Agaricus bisporus* (an edible mushroom commonly called button mushroom), *Pleurotus pulmonarius, Bjerkandera adusta* and *Irpex lacteus*. Decontamination of soil from the crude oil which is mixed with the soil can be done easily if we can provide the proper substrate like corn cub and sawdust straw along with properly inoculating the myco-remediating agent (fungi) within the substrate with proper carbon to nitrogen ratio for the growth of the useful fungi which if utilized well can surely save the environment and ultimately humans too. Chlorophenols, polychlorinated phenolic compounds, simple phenolic compounds, pesticides, and dioxins many other harmful chemicals are released into environment daily but these chemicals can degraded by the help of mycoremediation by the proper fungi. Amazingly fungi can also biodegrade many poly aromatic hydrocarbons like coal tar, petroleum products and etc, due to this property it can be used to degrade plastic products too with the help of genetic engineering. Vanillin along with 2,2′-azinobis-(3-ethylbenzothiazoline-6-sulfonate) and heavy metal cations increases the efficiency of biological degradation of Benzo[*a*]pyrene by the common edible oyster mushroom (*Pleurotus ostreatus*). Rot fungi are actually so good at mycoremediation because of their extracellular, less specific enzyme-like MnP, lignin-peroxidase (LiP), laccase, and various enzymes which produce H_2O_2 and can degrade any of the substance resembling lignin structurally. Lignocellulosic biomass can be digested or degraded by *Lentinus squarrosulus* for helping in industrial process. For performing mycoremediation successfully we have to integrate knowledge of different aspects of fungi like fungal biochemistry which includes all the biochemical reactions and also the enzymology of how the fungus breaks down the contaminant biochemically. Further, its physiology deals with the physiological changes or adaptation or physiological stress which is to be applied or the fungi is already under some type of physiological stress or not how does it affects the efficiency of mycoremediation because many organisms are useful in bioremediation under a stressed condition. It's engineering along with its ecology and genetics are also essential to know whether the degradation of contaminant is accomplished only by the fungi or with the help of other associations with higher plants or algae or bacteria or any other organism. We can find the exact gene for mycoremediation of the particular contaminant and with the help of molecular biology and engineering we may prepare some genetically modified organisms with better efficiency.

Fungi can also contribute towards the biological degradation of extremely toxic polluting agents like some nerve gases and the famous explosive trinitrotoluene (TNT) by the inoculation of mycelia of *Pleurotus ostreatus* along with the contaminated soil sample. It is also evident that this experiment has proven as an excellent method as it can degrade almost all the contaminated matter within the time period of one month and some days. But there is a huge need of discovering some organism which can degrade these types of explosives or toxic chemicals from soil in an in-situ condition. The microorganisms are partners of fungi in decomposition process which ultimately leads to full mineralization of contaminants into carbon dioxide and water molecules. In 2007, the oil spills in the shoreline of San Francisco were cleared by using oyster mushrooms as tool for degrading the oil. Mycofiltration is the procedure responsible for this phenomenon, in which the fungal mycelia are utilized just like a filter for removal of toxic contaminants. Mycoremediation can be beneficial for converting wasteland into agricultural land as it can lead to addition of many essential substances and also the removal of toxic components which may hamper agricultural practices on that particular land.

10.5.1.6 BIOCHAR

Biochar is charcoal of very solid type, which is used for soil amendment and its carbon content is high. Biochar is produced with the help of pyrolysis of biomass. Biochar is an innovative approach towards carbon sequestration which helps us to stop climate change and its effects. It is under research because it has the potential to:

- Protect the plants from soil-borne and foliar diseases;
- Can increase soil fertility of low pH soils;
- Increase yield in agricultural land.

The agricultural sector also contributes to global warming as result of greenhouse effect by releasing a huge quantity of greenhouse gases (GHGs) like carbon dioxide via combustion and degradation of useless agricultural biomass in natural way which leads to decomposition of the same. But the most awesome thing about biochar is that it can help in storing large amounts of GHGs for a long period of time hence performing as a greenhouse gas capturing and reducing agent. It has been estimated by researchers that the sustainable use of biochar can be helpful to control emissions of carbon dioxide, carbon monoxide, methane, nitrous oxide, and other greenhouse gases. Biochar is of very porous type which increases or enhances soil health because of its typical structure it can hold water-soluble minerals, nutrients, and water very efficiently. Hence the biochar can serve as a habitat for many microbes which promote soil amendment. Biochar can improve water quality, pH, and potash content of the soil as well as it can also reduce:

- Amount of greenhouse gases which are liberated via soil;
- Acidity in soil, it actually adjusts the pH of the soil;
- Leaching of nutrients;
- Requirement of fertilizers and also irrigation.

Under some circumstances biochar can also induce systemic responses to the fungal diseases in the leaves of plants and also to pathogens present in soil.

The factors which affect the activity of biochar are:

- Type of soil;
- Nutrient quality of soil;

- Humidity in the soil;
- Temperature.

In the soils with low nutrient content biochar can be used to increase nutrient quality. Biochar along compost can be applied to increase agricultural quality, productivity, and also has positive outcomes on soil humidity. Designed biochar may be utilized for enhancing different properties of soil which can lead to incline in rate and amount of nutrient uptake by the agricultural crops for the maximum profit of agricultural sector as the crops will get more nutrition and can develop properly (Ponge et al., 2006).

The process of slash-and-char:

- can leave a large amount of carbon up to 50% in a very secure form;
- can decrease emission of GHGs like carbon-dioxide;
- can decline rate of deforestation.

So it is better to transfer from slash-and-burn as it will not give these profits. It will also act as nitrogen containing fertilizer hence can reduce the overall cost of agricultural practices which was earlier used in producing, purchasing, and transportation of the fertilizers used in agricultural sector.

As the biochar has a high surface area and is of porous nature, hence it can easily absorb and can also retain water. Biochar is an efficient soil material due to its hygroscopic property hence can control the amount of fertilizer reaching ultimately to groundwater as the growth of plants will be better than earlier because essential nutrients for plant growth will be retained for the benefit of plants and hence reducing the amount of fertilizers. Biochar can also replace coal-based power production but with little modification and some more research. Biochar can be utilized via pyrolysis to produce electrical energy in a less costly manner than the conventional methods of energy electrical production. The organic acids present in bio-oils may corrosion of the container and the high moisture content can interrupt in the burning process and fine particles of the biochar can even block the system but more research is needed to improve biochar so that we may also use it for biodiesel and also for energy production on large scale. Hence we can also use the biomass to generate electrical energy (Bridgwater et al., 2002).

The biofuels may be produced from the agricultural and forest waste by the help of pyrolysis. Biochar can be produced by the process of pyrolysis of the waste biomass and can be applied in different sectors to earn profit from the waste, as it was not used earlier hence can be said as waste earlier but now it can be said as highly important thing as it can boost agricultural productivity with less use of fertilizers and can also act as a water adsorbent and nutrient storage along with GHG capturing and it can also be applied for carbon sequestration for an extensive-time period.

10.6 BIOFUEL

Biodiesel is non-toxic, renewable, and biodegradable, and indeed that it is environment-friendly. It is very famous worldwide due to the above-mentioned reasons, and these reasons lead to its utilization as an alternative or substitute for the conventional fuel. Biofuel is always derived from or via biological agents. Biofuels can be produced by the trans-esterification of lipid-based or oil-based precursors like that of fatty acids, inedible oils, algal-based oils, waste oil from markets which use it for cooking and frying purpose, waste cooking oils from households, oils derived from algae, and waste oil from refined vegetable oils (Hilal, 2009).

10.6.1 BIOETHANOL

Ethanol which is produced from or via biological agents (living organism or dead biomass) is called bioethanol. Cellulosic biomass, sorghum, grains, sugarcane or other sugar sources like that of some fruits, and other carbohydrate sources may also be utilized for manufacturing bioethanol for better emission, and octane number ethanol is added along with gasoline to run vehicles although ethanol itself can also be used but the working of the currently working engines may not adjust properly within as they could with petroleum or gasoline and hence we have to develop more efficient engines that will adjust with bioethanol and hence we can transfer a large amount of biomass into biofuel. There are vehicles running on biofuel (bioethanol) in Brazil and USA.

10.6.2 JATROPHA BIOFUEL

The toxic plant *Jatropha curcas* is one among the good research topics for researchers working on biofuels (Divakara et al., 2010; Jhariya et al., 2011). Mostly the researchers focused on genetic manipulation of the species to increase yield of oil for maximum profit, they are also focusing on the horticultural improvements and scientific aspects of soil. Some of the researchers utilized biotechnology and molecular breeding to manufacture or produce some selected hybrid variety of seeds which has very good improvement in yield over first generation seed varieties. Some modified varieties had also shown improved resistance towards diseases and pests, synchronization of flowers also enhanced along with gaining cold-tolerance property. Many research institutes in the world are dedicated towards development of *Jatropha curcas* based biofuels to a high level (large scale) from which it can substitute petroleum fuels, for the same reason the crop has been grown in many farms.

10.6.3 FUNGAL BIOFUEL

In 2008, a paper published regarding isolation of huge quantity of lipids from just a unicellular type of fungi which can convert that lipid into biofuel via very less expensive method, by a team of researchers of Russian Academy of Sciences of Moscow. The fungus was *Cunninghamella japonica*. These particular fungi need more research investigations to get utilized on a large scale (Sergeeva et al., 2008). There was a newly discovered fungi *Ascocoryne sarcoides* (earlier named as *Gliocladium roseum*) from the rainforests northern areas of Patagonia which has the property of biochemically converting cellulose into some organic compounds generally found in biodiesel, actually the biodiesel is also called myco-diesel as it is derived from or via fungi (Strobel et al., 2008). At the same time we need to search for more options in kingdom fungi which is neither plant nor animal and something very different but possessing some similarities with the two, because there are very few fungi which has been isolated or known to us but they have a huge potential for sustainable development which can lead to a excellent future. Fungi are actually the decomposers of ecosystem along with bacteria and can convert many polymers into different products which are needed to be utilized for bio-fuel production and other vital applications so fungi are the future.

10.6.4 MICROBIAL GASTROINTESTINAL FLORA BIOFUEL

These biofuels can also be produced utilizing various microorganisms. There are many microorganisms which can be utilized for this purpose are like TU-103 strain of gram-positive bacteria *Clostridium* which was isolated from the feces of zebra. This particular strain of bacteria has the capacity of converting cellulose into the biofuel butanol. There are some researchers who are working on the wastes excreted by pandas for isolation of some useful bacteria or any other microorganism which can help in these type of important bioconversion processes, if they can succeed then we may make biofuels from bamboo sticks and their plant materials. In the same direction, many scientists are also trying to utilize the microbial enzymes for bioconversions of lignocellulosic components into biofuels after isolating bacteria and other common gut microbial flora of wood degrading insects (Prasad et al., 2017).

10.6.5 ISOBUTANOL

The research was successful in utilizing *Synechococcus elongatus* PCC7942 which was a genetically manipulated strain of cyanobacteria in producing biofuels like that of isobutanol which can act in addition with gasoline. The most interesting thing was that the cyanobacterial strain just uses carbon dioxide and utilize it for production of biofuels like 2-methyl-1-butanol, which is often termed as (2MB), which is an energy-rich molecule. It was a landmark in lipid bio-fuels production, but more research is needed to increase its efficiency. It is one of the better candidates for this job as compared to other cyanobacteria because it does not follow polyhydroxy butyrate (PHB) pathway (Shen and Liao, 2012).

The cyanobacteria *Synechococcus* is the most efficient in yield of lipids if compared to *Phormidium* and *Cyanobacterium aponinum* which are also cyanobacteria but not efficient in biofuel production as *Synechococcus* is the useful strain of cyanobacteria is also known for competing the pathways for synthesis of free fatty acid for carbon available. *Synechococcus* sp. strain PCC 7002 is also one of the good candidates for producing biofuels from carbohydrate sources. There was an experiment of production of isobutanol in a gram-negative rod-shaped bacteria *E. coli,* and it was successful. Yeasts can also be used for the same purpose as it can do fermentation and hence can be applied for lignocelluloses, they had shown tolerance to isobutanol and many type of toxic compounds. Yeast can be genetically manipulated for biofuel production from lignocellulosic precursors.

10.7 IDEA TOWARDS SUSTAINABLE DEVELOPMENT THROUGH BIODIVERSITY CONSERVATION

Nowadays there is a tremendous increase in urbanization, industrialization which directly and indirectly (via pollution leads to climate change) badly affects the biodiversity. So, for conserving biodiversity an artificial mountain can be built which should contain the following things:

a. Multiple ecosystem which consists of grassland, shrub forest and tree forest of the vegetation which is to be destroyed for industrialization and urbanization and there are other plants too for their role in building the forest. It should be done along with giving a pyramid shape to the mountain to maintain the structural stability of the artificial mountain.
b. Use flyash which is rich in nitrogen and have fewer amounts of heavy metals as it will be first treated in a vermicompost where due to formation of metallothionin complex inside the body of earth worm. Here the use of this processed flyash is only as a starting fertilizer and *Cassia siamea* will be grown there to reduce the amount of toxic heavy metals as it can phytoremediate the heavy metals in flyash hence this processed flyash will increase plant growth.
c. Introduction of mucigel and the spores of those particular fungi with which the particular plant causes mycorrhizal association in the forest hence the plants will get the necessary benefits from the mycorrhizal association.
d. The tree forest should be moist for the growth of other epiphytes like mosses, ferns, or epiphytic angiosperms which include orchids and some climbers, it should be done to increase the number of plants in that artificial mountain hence to make a larger carbon sink in the polluted city/town/district/state/country/continent/world.
e. At the humid forest there will be a controlled zone where we may grow various types of fungi on highly contaminated (polluted) soil, from which we may get the gene for turning a particular contaminant (pollutant here) into non toxic form as we cannot destroy it completely but it can be augmented to give some relief to environment. This site will be only for research purpose.

f. There will be some small ponds for forest and wildlife use and on some other ponds /water tanks we may grow algae of particular strain or of mixed type to get the benefits from algae.
g. There we may build some funnel-shaped huge structures connected to the wells which to the bottom tank of the mountain for rainwater harvesting which will ultimately be stored and may get used in deficiency of freshwater.

10.7.1 BIODIVERSITY CONSERVATION

Biodiversity conservation is a need for all of us because maximum products which we enjoy daily are of biological origin or have been produced with the help of some biological activities or organisms. For example -from the morning tea to rice or wheat-based products are of biological origin, if there were no diversity, then there will be no variety of organisms which are used for different purposes like food, medicines, fodder, etc. The biological diversity needs to be conserved along with exploration because there are many plants, animal, fungi, and microorganisms which got extinct hence without knowing them, their properties, economic importance, ecological significance and other potentials (Jhariya and Raj, 2014; Raj et al., 2018). It may be possible that from that species which got extinct before exploration or after due to unsuitable polluted environment can actually produce the medicines for the most dangerous diseases, disorders, syndromes. So there are many species in the world which are yet to be discovered, conserved, and utilized for nature and mankind. One of the reasons behind it is the habitat loss by climate change, so we have planed an innovative way to restore biodiversity at a particular place or area where we have to build and artificial mountain which will have the following:

- Multilayered ecosystem like the uppermost portion will be of grassland type, the lowermost will be of forest type, etc., types will be established there to give habitat to many organisms.
- The mountain will be made from flyash bricks which will give nutrients and hence will be a part of sustainable development.
- There will be an artificial lake surrounding the mountain which will be used to conserve aquatic biodiversity.
- There will be constructed wetlands between the artificial mountains and artificial lake.

- The kaccha houses in the whole system will became as the major source of foods for the species which depends on the human waste, non-utilized human food means the species which depends on human habitation like some mousses, some snakes, etc.
- There will be some research centers for conducting research on exploring biodiversity and its utilization for nature and mankind.

Fungi along with bacteria are the major decomposers of our ecosystems starting from leaf litter to polymers they can degrade many things as presented here in various topics. Fungi are also utilized in biocontrol (bio-insecticide), biofuel, bioremediation (mycoremediation) and also in various other applications. The edible fruiting body of fungi are called mushrooms which contains large amounts of essential amino acids, vegetable protein, minerals, vitamins (vitamin B, C, D, K) and many other nutrients which are necessary for our body, interestingly mushrooms can also decrease cholesterol from the body and it often regarded as "vegetable meat" due to its similarity in amount of protein content but like meat it does not give calories or cholesterol after consumption hence is also termed as "delight of diabetic" (de Silva et al., 2012).

Recently mushroom-based leather is also available in the market known as muskin. Which is prepared from *Phellinus ellipsoides* which can grow nearly the size of a cow within a week and the leather have the following useful properties:

- 100% vegan leather;
- Stymie bacterial proliferation;
- Animal free;
- Cruelty free;
- Good for skin as it can absorb and slowly release the moisture;
- Now muskin is also prepared from *Ganoderma lucidum.*

There is a list of antibiotics and enzymes of fungal origin which shows amensalism against a wide range of bacteria (Tables 10.6 and 10.7). Besides these, fungi are also utilized in beverages manufacturing (Table 10.8) and in bio-pesticides (Table 10.9).

TABLE 10.6 List of Antibiotics and the Fungi Utilized to Make Them

Name of Antibiotic	Name of the Fungi Utilized
Griseofulvin	*Penicillium griseofulvum*
Penicillin-G	*Penicillium chryzogenum and Aspergillus niger*
Uridin	*Trichiderma viridi*
Clavicin	*Aspergillus clavatus*

TABLE 10.7 List of Some Fungi Along With the Enzymes They Produce

Fungi	Enzyme Produced
Aspergillus niger	Amylase, lipase, and glucose oxidase
Aspergillus oryzae	Protease, amylase, lipase
Aspergillus flavus	Urate oxidase
Mucor micheli	Rennet
Penicillium notatum	Glucose oxidase
	Now glucose oxidase is used in biosensor for measuring blood sugar

TABLE 10.8 Listing of Fungi Utilized for Manufacturing of Beverages

Fungi	Product
Saccharomyces cerevisiae	Whiskey
Saccharomyces ellipsoideus	Wine
Saccharomyces cerevisiae	Brandy and malt beverage

TABLE 10.9 List of Fungal Bio-Pesticides (Entomopathogenic Fungi) Along With the Name of the Pest

Name of the Fungi	Name of the Pest
1) *Fusarium pallidoroseum*	*Aphis craccivora, A. unguiculata*
2) *Zoophthora radicans*	*Marasmia patnalis*
3) *Verticilium lecanii*	*Myzus persicae*
4) *Nomuraea rileyi*	1. *Spodoptera litura*
	2. *Helicoverpa armigera*
	3. *Heliothis virescens*

Recently an arsenic methyltransferase gene (*WaarsM*) has been reported from a soil fungi *Westerdykella aurantiaca* when exposed in rice able to convert more toxic inorganic arsenicals to methylated or volatile, less toxic arsenic species, therefore, reduce the bioaccumulation of arsenic in the rice grains (Verma et al., 2017).

- Fungi as biofertilizers;
- *Aspergillus awamorii;*
- *Aspergillus niger;*
- *Aspergillus flavus;*
- *Aspergillus fumigates;*
- *Penicillium digitatum.*

Fungi are a part of mystery as we have not discovered most of the species but only few species as compared to overall population of the organism in the world because it comes under the kingdom fungi which are not a plant, nor an animal. It has macroscopic as well as a microscopic members which are sometimes pathogenic or symbiotic and till date it has given so many products to us, and the most interesting thing is that we just need to pick up the fruiting body and no need to pick up the whole mycelium for manufacturing the products. Hence, we are not killing the organism but we can earn profit from it because in favorable conditions the mycelium will again produce fruiting body. Earlier in the chapter, we have discussed the mycoremediation and many other utilizable properties of fungi. We all have to do more research on biological diversity of this planet because there are many species which are not discovered yet and needs exploration at multilevel; starting from morphology to genetics of the organism everything is useful for us. Biological agents are very much efficient in removing or substituting our chemical products in an environment-friendly way which makes them suitable for every use in sustainable development. We have an idea which can be applied for sustainable development, biodiversity conservation and also for profit-making because without profit no one will invest in the idea and the investment will be one time, and that will also give us the maintenance cost of the project if applied. There is production of many products which may be utilized to earn the profit and may also to overcome the overall maintenance money to some extent but it is to be calculated and then can be applied on a large scale (Figure 10.3).

Here in this picture, it has been shown that there will be an artificial circular lake for conserving freshwater flora and fauna and there will be a constructed wetland for conservation of wetland species (Figure 10.3). There will be some "kaccha houses" which are made from mud and paddy straw for low cost factor and also environmental friendly nature of the materials used to make it. These houses are for giving shelter with food to the species in and around artificial mountain and constructed wetland within the boundary of artificial lake in the controlled environment. Here we need to provide some of the food from our plate to the nature as a "tax" or donation to nature. Actually it should be a "tax" not a donation because we are getting all the food stuff or its precursors basically from nature which is very little like if we consume 1 kg of rice per day then we can easily donate 5 grams of rice in daily basis to the nature because all the organisms have equal "rights" on natural resources and as humans have developed it to a large scale and as we are the most successful organism on earth it is our duty to conserve nature and its resources with developing ourselves. Just imagine if nature has not produced fruits, vegetables, plants, and other organisms which we consume then we could have not existing still now and have never survived on earth so nature should be respected and we must pay the "tax" for nature. The houses are connected to the main building from which food will be discharged and to be received by kaccha houses where there are rats, snakes, etc. species which are generally killed via a man-animal conflict. For conserving those species, we need to transfer them to these kaccha houses and the animals will remain peacefully as same as humans. So very selective species of trees and plants are to be planted which needs conservation and also those which are the habitat of other species. The black squares with red circles represent the research centers which will pursue research on the particular area for development of biodiversity conservation and sustainable development, and the red lines represent the pipes by which the food (so-called "natures tax") is to be transferred to the "kaccha houses" from human habitations. The food will be beneficial to all the organisms directly or indirectly in the "kaccha houses" of this artificial mountain and also in the constructed wetland (Figure 10.4).

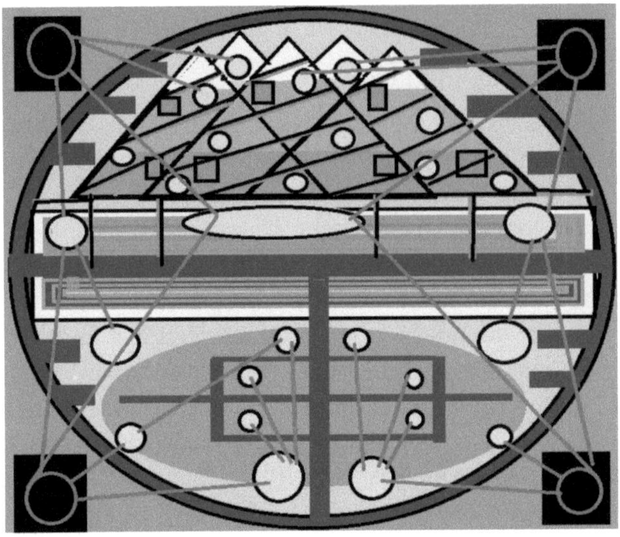

FIGURE 10.3 (See color insert.) Idea for sustainable development, biodiversity conservation, and profit making.

FIGURE 10.4 (See color insert.) Inter-linkage and dependency of organism in ecosystem.

As shown in the picture, there are many species which are directly or indirectly dependent on a tree (Figure 10.4). The birds depend for nesting, food, and also for hiding from its predators. The insects lay eggs on the leaf lamina and also on bark. The spiders are also present for shelter and food. The ants are present for utilizing the fruits or they may have some other type of relation with the tree, they may utilize the leaf or just they may have some ant domatium on the tree, when the leaves, branches, and fruits of the tree fall on the ground may be utilized by ants to grow some fungus which can act as nutrient source to ants or they can also utilize the natural the biomass of fungal flora, not only ants but many insects can also do the same. The tree may be a host plant or a nectar plant for insects and butterflies. The tree may be a host plant for some snakes. The tree may have mycorrhizal association. The tree may have some symbiotic or non-symbiotic bacterial, actinomytes, cyanobacterial colonies associated with the roots of the tree for nitrogen-fixing purpose. Some insects also take shelter near roots. Some insects like honeybee may make take shelter and food (nector) on the same or different tree. The tree may be host to various crustose, foliose, and fructicose lichens. Lichen is actually a mutualistic relation between a photosynthetic alga and a shelter providing fungi. The tree may give shelter to a number of climbers, epiphytic plants, ferns, mosses, and algae. The dead fallen woods, leaves, and fruits are ultimately degraded by the decomposers (fungi and bacteria). The dead woods can be utilized by termites and can build termite hill and snakes will obviously enter inside and would take shelter and food there. The over-ripened fruits or immature fruits may contain many worms too. The tree may also have bracket fungi which can act as a habitat of many insects in dead conditions, and some gilled mushrooms can also help frogs by giving shelter. Some other organisms like squirrel, monkey, etc., are dependent on tree. The tree may have different microbial flora on leaves and in and outside the roots. There may be question that why we have also included pathogens and infectious agents, because they have a significance of declining some population of certain species by causing diseases because any species if gets excess in environment will get dominant and will disturb the environment, so some destruction is also needed with a lot of creation but actually nothing is destroyed in nature as nature can use anything and everything weather created by man or by nature itself.

Anything excess is poison and so a species may also follow the same. As ultimately all the dead species will contribute to manure in the soil by decomposing in soil. Nature has its own systems for maintaining balance. Nature does not create or evolve anything which has no role in it, and everything has a specific role in nature. There was a saying that one tree means one life but we think that one tree is directly or indirectly equivalent to many lives because without the tree all these organisms would suffer habitat loss. So we have to conserve plants in such a way so that we can go parallel with nature in the same direction but not in the opposite direction, we mean sustainable development.

10.7.2 RAIN WATER HARVESTING IN THE ARTIFICIAL MOUNTAIN

This will be the setup for collecting the rainwater in these funnel-like structures and storing it at the bottom tank of mountain, so these will act like freshwater reservoirs and will be utilized for forest growth and a part of this water will be supplied to human habitats for agricultural and various other purposes (Figure 10.5).

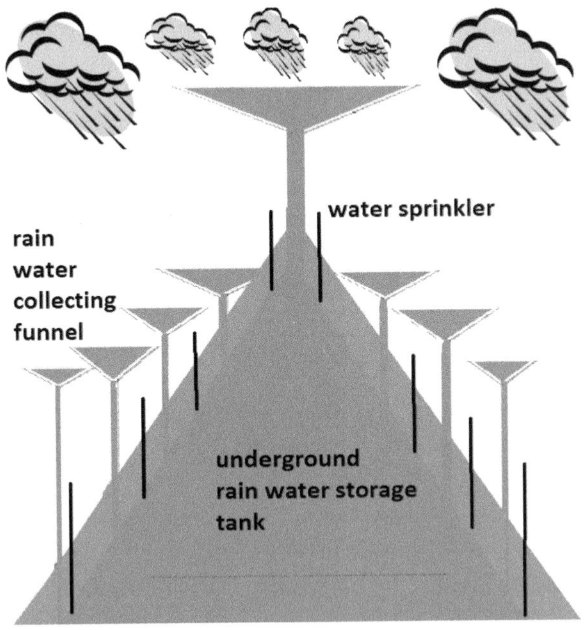

FIGURE 10.5 Rainwater harvesting in artificial mountain.

All the wastewater which is already polluted will be collected in the wetland (Kidneys of Earth), and there will be a decontamination of pollutants from water by the help of bioremediation, which will include phytoremediation, phytoremediation, mycoremediation, and bacterial remediation. Along with these if we can utilize the biomass left for biofuel and biochar production then it is profitable and subsequently if we can utilize microbial fuel cells electrical energy by setting one in the constructed wetland, other in the artificial mountain, and also near artificial lake then also we can use the biomass which is the waste (dead plant and animal material) coming from forest, lake, and wetland.

Advantages/profit from this idea if applied on a large scale are:

1. Piscicultural profit;
2. Biodiversity conservation;
3. Carbon credit;
4. Profit of electricity from the microbial fuel cell;
5. Sustainable use of flyash;
6. Profit from selling methane gas;
7. Profit from selling the biofertilizers;
8. Profit from selling biofuels;
9. Conservation of wetland bioresources;
10. Bioremediation will be eco-friendly too;
11. Microalgae in bioelectrical system;
12. Other different metabolites from algae;
13. Biochar production;
14. Valuable biomass;
15. Biohydrogen production will lead to profit;
16. Increase in number of plants, hence a larger carbon sink if applied on large scale;
17. Subsequently decrease in total air pollution, climate change, global warming, and greenhouse effect;
18. Decrease the total soil, water pollution;
19. Less amount of money investment for bioremediation purpose;
20. Increase in employment, hence increase in national income;
21. Water conservation will lead to earn some profit too by selling the freshwater collected in the bottom tank;
22. Treatment of wastewater within a low cost;
23. Papermaking from algae will also give profit;

24. More research will also give more profit and will be the most important part of this.

10.7.3 SPIRAL GARDENING

Spiral gardening is a new idea to increase number of plants on multi-storeyed buildings. This spiral garden will be an extra added structure on the building along with varnish gardening, balcony shelf gardening, green roofing, terrace farming and vertical garden (on the side which has no balcony or window). Spiral garden basically consists of shrubs, bonsai, and climbers, thereby increase the aesthetic value of the surroundings (Figure 10.6).

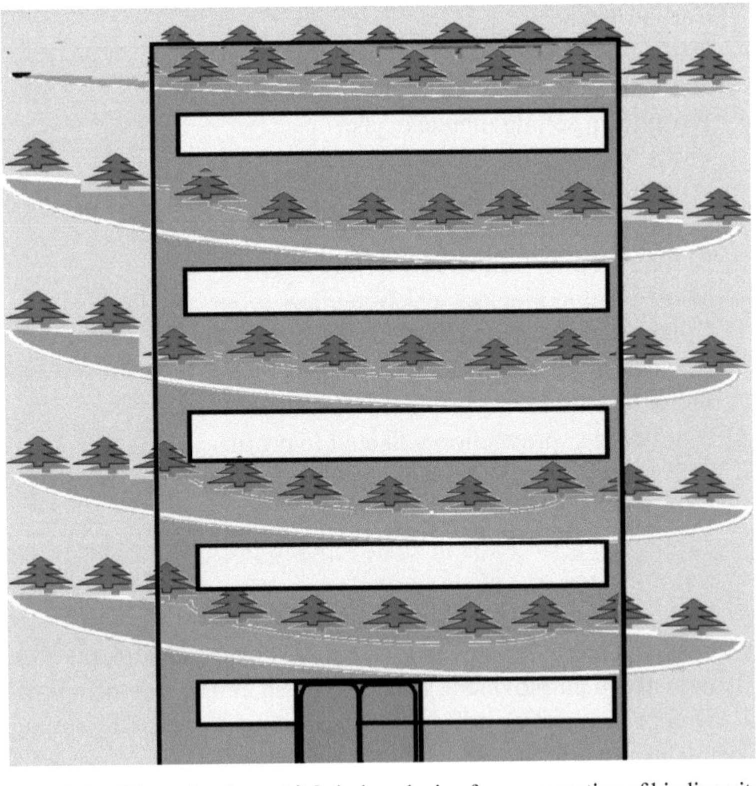

FIGURE 10.6 (See color insert.) Spiral gardening foe conservation of biodiversity and to increase green plants in a polluted city with utilizing very less amount of space but can fight with air pollution in the polluted cities, where there is no land for forest building or another green belt or green cover.

10.7.4 GREEN TOWERS

Green towers are for an increasing number of plants in a polluted city. We can grow the plants in a tower and can connect all the towers with one another by making corridors between them, on which we will grow plants to increase the carbon in the city.

10.7.5 GREEN STRATEGIES FOR FUTURE

For maintaining the sustainable development cost is an essential term which can be minimized by introducing environmental tax which is like this (these are some of my ideas for economically supporting sustainable development.

1. Every person with a motor vehicle which runs on petroleum fuels has to pay a minimum amount of tax which will be utilized for sustainable development.
2. Any person if having more than one light motor vehicle, let us say if two motor vehicle and two persons have the authority to run that vehicle will have to pay four times of the amount which can be easily calculated with a formula where total environmental tax is equal to tNnv1 or tNnv2.
3. Where(t) stands for the amount of tax to be given if one person have authority to run one light motor vehicle and if he/she have only one light motor vehicle, multiplied by the number of motor vehicle (N) multiplied by number of persons (n) who have the authority to run the vehicle, multiplied by type of vehicle whether light or heavy and if there is heavy motor vehicle (v2) then the whole value has to be multiplied by two and one for light motor vehicle (v1).
4. For industries who release the huge amount of pollutants they should be forced by the government of that country to help in sustainable development by providing initial investments on the different projects sustainable development. And that amount of money will be refunded without any interest to the industry back after getting profit from the project. The payback scheme will not give interest of money to the industry because they are supposed to use various controlling methods for preventing pollution, but they

are not doing their work properly hence something should be done to rectify this system.

5. If the government of the country does it mandatory for the builders to build the spiral garden, carnies garden, vertical garden, balcony shelf garden.
6. If the government of that country amends a rule that to build green towers are mandatory after completion of road-making.

10.8 CONCLUSION

The biological agents which help in bioremediation are the tools of nature by which we can rectify our biggest mistake (pollutants) to the nature. There is a huge need of more research and ideas for sustainable development because it is the only option for continuing our existence on earth. If we destroy nature then surely we will suffer the most because we cannot imagine a day without nature as we totally dependent on natural resources with some modifications. We have to earn development without losing nature because have created us and nature has the ability to destroy us also so let us handshake with nature from one hand and with development with the other hand. Sustainable development is our necessity because we are dependent of nature but nature is independent of us, from morning cup of tea to rice in dinner, everything is from nature with little modification. Nature is the most essential asset of human beings so we must try to utilize it with a lot of care for its protection. We all have succeeded to make some products which are a part of sustainable development. Genetic manipulation will be the next step of researchers for utilizing fungi to make products which will be environment-friendly and may contribute to sustainable development. If the above-mentioned ideas and green strategies are properly utilized, then it would be our next step towards sustainable development. The green towers, spiral gardening is very good for highly polluted cities where there is no land or very less land present for forest, gardens, and other green covers, the artificial mountain, constructed wetland, artificial lake, and the kaccha houses if built in the particular area then can surely conserve biodiversity and also treat the waste from industries by the application of all types of bioremediation processes in an integrative way hence protecting environment from the hazardous chemical waste from industrial origin. So, all the existing and new methods for bioremediation are to be utilized properly at every city level, district level, then state level, then ultimately at country level to ensure the maximum conversion or degradation of hazardous waste into nontoxic form or less toxic form. The least environmental tax can be implemented to collect funds for sustainable development, protection of nature, biodiversity conservation, and for solving other nature-related or environmental related issues. Human is the most successful organism that have evolved so well and it is our responsibility to take care of our nature because we believe that if wealth is lost nothing is lost, if health is lost something is lost but if nature or environment is lost or damaged then we can lose everything we have because without nature we do not have any existence. For protecting nature, we have to search out as many options as possible because the more number of options we have the less have to suffer because everything in the world has good and bad side but we need to utilize the good side along with suppressing the bad side so for development also our focus should be sustainable development and utilization of biodiversity but without harming the same. Nature is the creator of all the organisms, including humans; it can create as well as destroy too; hence nature, should be respected.

KEYWORDS

- **air pollution**
- **biodiversity**
- **forestland**
- **green belt**
- **photosynthetic plant**

REFERENCES

Ambrose, D. P., & Claver, M. A., (2002). Suppression of cotton leafworm *Spodoptera litura*, flower beetle *Mylabris pustulata* and red cotton bug *Dysdercus cingulatus* by *Rhynocoris marginatus* (Fabr.) (Het., Reduviidae) in cotton field cages. *Journal of Applied Entomology*, *123*(4), 225–230.

Anthony, C., (1982). *The Biochemistry of Methylotrophs* (pp. 2, 3). Academic Press.

Bridgwater, A. V., Toft, A. J., & Brammer, J. G., (2002). A techno-economic comparison of power production by biomass fast pyrolysis with gasification and combustion. *Renewable & Sustainable Energy Rev.*, *6*, 181–248.

Chakrabarty, A. M., Mylroie, J. R., Friello, D. A., & Vacca, J. G., (1975). Transformation of *Pseudomonas putida* and *Escherichia coli* with plasmid-linked drug-resistance factor

DNA." *Proceedings of the National Academy of Sciences of the United States of America, 72*(9), 3647–3651. doi: 10.1073/pnas.72.9.3647.

De Silva, A. G., Dilani, R., Sylvie, H., Kevin, B., & Ali, (2012). Medicinal mushrooms in preventing and control of diabetes mellitus. *Fungal Diversity, 56*, 1–29.

De Souza, A. R. C., Baldoni, D. B., Lima, J., Porto, V., Marcuz, C., Machado, C., Ferraz, R. C., Kuhn, R. C., Jacques, R. J. S., Guedes, J. V. C., Marcio, A., & Mazutti, M., (2017). A selection, isolation, and identification of fungi for bioherbicide production. *Brazilian Journal of Microbiology, 48*(1), 101–108.

Demirbas, H., & Ayse, (2009). Inexpensive oil and fats feedstocks for production of biodiesel. *Energy Education Science and Technology Part A: Energy Science and Research, 23*, 1–13.

Deshmukh, R., Khardenavis, A. A., & Purohit, H. J., (2016). *Indian J. Microbiol., Diverse Metabolic Capacities of Fungi for Bioremediation, 56*(3), 247–264. https://doi.org/10.1007/s12088–016–0584–6 (Accessed on 4 August 2019).

Divakara, B. N., Upadhyaya, H. D., Wani, S. P., & Gowda, C. L. L., (2010). Biology and genetic improvement of *Jatropha curcas L.*: A review." *Applied Energy, 87*(3), 732–742. doi: 10.1016/j.apenergy.2009.07.013.

Falandysz, J., (2015). Mercury bio-extraction by fungus *Coprinus comatus*: A bioindicator and mycoremediator of polluted soils. *Environ. Sci. Pollut. Res., 23*(8), 7444–7741.

Jhariya, M. K., & Raj, A., (2014). Human welfare from biodiversity. *Agrobios Newsletter, XIII*(9), 89–91.

Jhariya, M. K., Kittur, B., & Yadaw, K. N., (2011). Jatropha: An eco-friendly sustainable fuel source. *International Research Journal Lab to Land, 3*(12), 566–569.

Jonathan, R., Russell, J. R., Huang, J., Anand, P., Kucera, K., Sandoval, A. G., et al. (2011). Biodegradation of polyester polyurethane by endophytic fungi. *Appl. Environ. Microbiol., 77*(17), 6076–6084.

Kwon, G. S., Sohn, H. Y., Shin, K. S., Kim, E., & Seo, B. I., (2005). Biodegradation of the organochlorine insecticide, endosulfan, and the toxic metabolite, endosulfan sulfate, by Klebsiella oxytoca KE-8. *Appl. Microbiol. Biotechnol., 67*, 845–850.

Ponge, J. F., Topoliantz, S., Ballof, S., Rossi, J. P., Lavelle, P., Betsch, J. M., & Gaucher, P., (2006). Ingestion of charcoal by the Amazonian earthworm Pontoscolex corethrurus: A potential for tropical soil fertility. *Soil Biology and Biochemistry, 38*(7), 2008–2009.

Pozdnyakova, N. N., (2012). Involvement of the ligninolytic system of white-rot and litter-decomposing fungi in the degradation of polycyclic aromatic hydrocarbons. *Biotechnology Research International*, pp. 1–20.

Prasad, R. K., Chatterjee, S., Sharma, S., Mazumder, P. B., Vairale, M. G., & Raju, P. S., (2017). Insect gut bacteria and their potential application in degradation of lignocellulosic biomass: A review. *Bioremediation: Applications for Environmental Protection and Management* (pp. 277–299). 10.1007/978–981–10–7485–1_14.

Raj, A., Jhariya, M. K., & Harne, S. S., (2018). Threats to biodiversity and conservation strategies. In: Sood, K. K., & Mahajan, V., (eds.), *Forests, Climate Change and Biodiversity* (pp. 304–320, 381). Kalyani Publisher, India.

Sergeeva, Y. E., Galanina, L. A., Andrianova, D. A., & Feofilova, E. P., (2008). Lipids of filamentous fungi as a material for producing biodiesel fuel. *Applied Biochemistry and Microbiology, 44*(5), 523–527.

Shen, C. R., & Liao, J. C., (2012). Photosynthetic production of 2-methyl-1-butanol from CO_2 in cyanobacterium *Synechococcus elongatus* PCC7942 and characterization of the native acetohydroxyacid synthase energy. *Environ. Sci.*, *5*, 9574–9583.

Singh, H., (2006). *Mycoremediation: Fungal Bioremediation* (pp. 1–592). Wiley Interscience.

Song, J. H., Murphy, R. J., Narayan, R., & Davies, G. B. H., (2009). Biodegradable and compostable alternatives to conventional plastics. *Philosophical Transactions of the Royal Society B: Biological Sciences*, *364*(1526), 2127–2139. doi: 10.1098/rstb.2008.0289. ISSN 0962–8436.

Strobel, G., Knighton, B., Kluck, K., Ren, Y., Livinghouse, T., Griffin, M., Spakowicz, D., & Sears, J., (2008). The production of myco-diesel hydrocarbons and their derivatives by the endophytic fungus *Gliocladium roseum* (NRRL 50072). *Microbiology*, *154*(11), 3319–3328.

Tamilselvan, S., Manonmani, A. M., & Jambulingam, P., (2017). Fly ash-based water dispersible powder formulation of *Bacillus thuringiensis var. israelensis:* Development & laboratory evaluation against mosquito immature. *Indian J. Med. Res.*, *146*, 714–721.

Verma, S., Verma, P. K., Meher, A., & Chakrabarty, D., (2017). A novel fungal arsenic methyltransferase, WaarsM reduces grain arsenic accumulation in transgenic rice (*Oryza sativa* L.). *Journal of Hazardous Materials*, *344*, 626–634.

Wojciech, J., Janisiewicz, W. J., & Buyer, J. S., (2010). Culturable bacterial microflora associated with nectarine fruit and their potential for control of brown rot. *Canadian Journal of Microbiology*, *56*(6), 480–486.

Zimmer, F. V., (1970). *Fly Ash as a Bituminous Filler.* The United States, Department of Interior, Bureau of Mines, Washington DC. Information Circular 8488, pp. 49–76.

CHAPTER 11

Prospects of Biotechnology for a Sustainable Environment

SAIKAT MONDAL[1] and DEBNATH PALIT[2]

[1]*Department of Zoology, Raghunathpur College, Purulia, West Bengal, Pincode – 723133, India, E-mail: sairaniganj@gmail.com*

[2]*Department of Botany, Durgapur Government College, West Bengal, Pincode – 713214, India*

ABSTRACT

Biotechnology is the modern tool towards attaining sustainability towards the environment. Wider applicability of biotechnology has made it a promising tool towards attaining sustainable environment. Bioremediation in the form of phytoremediation, biosensors, biomining, biosurfactants, and molecular markers plays a leading role towards forming pollution-free environment. Biotechnology is also playing a key role in the field of ecological toxicity assay. Biotechnology is being effectively utilized in the production of first, second, and third-generation biofuel. This works towards pollution abatement on one hand and conservation of fossil fuel reserves on the other. Bioplastic production is another sustainable approach of biotechnology. Biopesticide formation through the application of biotechnological principles is playing a significant role in combating pesticidal pollution problem as well as the eco-friendly environment.

11.1 INTRODUCTION

Our environment is rich with full of resources and varieties of organisms, and most of the characteristics exhibit by different biotic component is also control by the environment. The basic need of a human being is sufficient food, shelter, and cloth. The utmost need of today's world is an adequate supply of food, health services, smooth communication, and industrialization to mitigate the basis needs of the human. But this leads to degrade the quality of life, which are unfavorable to the environment in which we live. Hence the urgent need is to sustain the environment. To sustain the environment according to James (2006), "we need to responsible and effective managers of resources that allow one generation to pass on to the next generation a legacy that is as good or better than the one inherited from the previous generation." As the excellence of life is reciprocally related with all-around environmental standard and quality, preservation, and sustainability of the environment is becoming the global issue (Banerjee et al., 2018).

In this respect, the role of biotechnology is now transferring from promising to practice. Biotechnology is the practice of using a variety of organisms and microorganism, both prokaryotic and eukaryotic to ameliorate our living standards in which, our health is improved, our resources are maximized, and the cost is effective and make the environment to be friendlier. Biotechnology can be used effectively for the transformation of different hazardous and toxic pollutants into less toxic or innocuous material, produce perishable compounds, develops environmentally safe and cost-effective production and destruction procedures and overall assessment of well being of the ecosystems. Biotechnology contributes a significant role towards environmental stability and sustainability. Biotechnology helps to adapt plants and animals to the environment through genetic engineering, which is the future of environmental management by means of microorganisms. Genetically modified organisms, another gift of modern biotechnology have proven to be of much benefit to man and overall environment. Biotechnology has been proven in contributing to the stability and sustainability of the environment in various ways such as, production of alternative eco-friendly fuel to replace fossil fuels, recycling of biodegradable waste to produce biogas, organic fertilizer and electricity, treatment of contaminated water, giving an alternative and cost-effecting technology to extract minerals from ores etc. biotechnology has also a promising role in agricultural sustainability which is very much needed.

Biotechnology have been used to convert areas with poor fertility soil to areas capable of growing plants through planting of genetic modified crops which are drought resistant. Biotechnology also ensures sufficient amount of food source to feed rapidly growing population of the modern globe (Wu et al., 2009). The application of genetically modified and engineered food materials have been applied to mitigate the need of food supply in a sustainable way through the production of pest resistance, herbicide tolerance, heat, cold, and salt tolerance crops. Therefore, to look for zero environmental impacts, creating greenly environment and moving towards sustainability, biotechnology has provided much of its benefits in the recent scenario and future also through continuous modification and upgradation.

11.2 ENVIRONMENTAL SUSTAINABILITY

Sustainability is the capacity of endurance. Ecologically, sustainability focuses on maintain continuous nature of diversification and productivity of the biological systems. Environmental sustainability deals with practices and maintenance which provide quality environment in a consistent and prolong manner. When natural capital is used up faster than it can be replenished, it gives rise to unsustainable situation.

11.3 SOCIO-ECONOMIC BENEFITS OF SUSTAINABLE ENVIRONMENT

- Good health of the people;
- Food security;
- Growth in the country's GDP;
- Attraction of tourist;
- Increase in local and foreign investors.

11.4 BIOTECHNOLOGY

The term "biotechnology" originated from a mixture of biology and technology. It deals with the utilization of various bio-agents or their derivatives to produce or create needful products (Olatunji, 2007). In another word biotechnology is any technological approach where live or dead biological entity has been used to generate and modification of products or procedure for specific purposes (UNCBD, 1992). Disilva (2004) has classified biotechnology in various types which are listed in Table 11.1.

TABLE 11.1 Classification of Biotechnologies

Biotechnology Types	Application
Green	Agriculture
Red	Medical
Blue	Aquatic
White	Industrial
Grey	Environment management
Brown	Arid lands
Gold	Nanotechnology, bioinformatics, and computer science
Purple	Patents
Dark	Warfare, biological weapons
Yellow	Food biotechnology

11.5 BIOTECHNOLOGY TOWARDS ENVIRONMENTAL SUSTAINABILITY

We human beings continuously put our various activities in our environment. Most of the activities are involved in chemical conversion of different natural compounds into many other forms for utilization. We also create hazardous wastes and by-products either consciously or unconsciously in the process of creating products. The best option to minimize the generated wastes is to recycling and proper management. In this regard, biotechnology can be used as a safeguard towards sustainability. With the diverse application of biotechnology, it is now possible to solve many hurdles regarding wastes and environmental pollution. The greener way of using biotechnology is the use of different microorganisms, mainly bacteria and fungi and their enzyme product in the field of environmental management and sustainability.

11.6 BIOTECHNOLOGICAL AND SUSTAINABLE ENVIRONMENT

Biotechnological tools for environmental sustainability are different approaches which have been used for reduction of pollution, production of eco-friendly products as well as maintenance of the pristine environment by using various prokaryotic and eukaryotic organisms, microorganism, and there by-product. Broadly it is an area of environmental biotechnology deals with the maintenance of sustainable environment for man and eco-components. Some of the bio tools are as enumerated in the following.

11.6.1 BIOREMEDIATION

Bioremediation is a remediation process with the help of biological organisms, mainly plants and microorganisms. Bioremediation can be used as *in-situ* or *ex-situ*. Bioremediation can happen itself or can be induced by biostimulation. Bioremediation opens a path which provides methods to degrade various pollutants using biological activity. Bioremediation provides low-cost technology with high social acceptance. Bioremediation depends on proper a environmental condition which allows microbial growth and more importantly microbial enzymes must available for degradation of pollutants. Application of bioremediation procedure in different environmental process is shown in Table 11.2.

TABLE 11.2 Application of Bioremediation in Different Environmental Process

Environmental Hazard	Microbes Used	Bioremediation Benefit
Solid waste	Bacteria, fungi	Composting and biodegradation of solid waste.
Air and waste gases	Bacteria, fungi	Application of bioreactor to filter air and gases.
Wastewater and industrial effluents	Sulfur-metabolizing bacteria	Biogas production. Removal of heavy metal and sulfur from industrially contaminated water prior to discharge in the water body.
Soil and land treatment	Bacteria and fungi	Application of bioaugmentation and phytoremediation to cleanup of soil and groundwater

11.6.2 BIOREMEDIATION TYPES

Basically, bioremediation is of two types (EPA, 2001, 2002): *In-situ* bioremediation and *ex-situ* bioremediation.

1. **In-Situ Bioremediation:** Here, removal of water or soil or excavation is not needed. Here stimulation of naturally occurring bacteria in the contaminated soil is done by supplying oxygen and

nutrients through aqueous solution circulation to degrade organic compounds.

2. **Ex-Situ Bioremediation:** In this method, contaminated substances are removed physically for treatment.

11.6.3 BIOREMEDIATION METHODS

On the basis of involved of microorganism, bioremediation is of three types: fungi associated bioremediation or mycoremediation, plant-associated bioremediation or phytoremediation, and bacterial bioremediation.

11.6.3.1 MYCOREMEDIATION

In this method, the fungus is used for the remediation process. This technology is more modern and advance type. Different types of secreted enzymes of fungus have been used to degrade toxic compounds (Barry et al., 1994). List of different fungi for bioremediation process are shown in Table 11.3.

TABLE 11.3 Different Fungi Used in Bioremediation Process

Fungi	Application	References
Phanerochaete chrysosporium	Degradation of polyaromatic hydrocarbons (PAH)	Singh, 2006
Pleurotus spp., Trametes versicolor, Agaricus bisporus, Bjerkandera adusta, Irpex lacteus and *Lentinula edodes*	Degrade persistent xenobiotic compounds,	Singh, 2006; Adenipekun et al., 2012
Pleurotus ostreatus	Degradation of Benzo[*a*] pyrene	Bhattacharya et al., 2014;
L. squarrosulus	Pretreatment and biodelignification of lignocellulosic biomass	Isikhuemhen et al., 2011

11.6.3.2 PHYTOREMEDIATION

Phytoremediation is an excellent green biotechnology mechanism for degradation of toxic compounds from the environment. In this technique plants' natural ability has been used to remove toxic contaminants from the environment (soil, water, and air). Plants involved in phytoremediation absorb, assimilate, tolerate, transfer, stabilize, and degrade toxic substances like pesticides, chlorinated compounds and polyaromatic hydrocarbons (PAH), etc. from the contaminated soil and water body.

Phytoremediation technologies uses both bioengineering & environmental biotechnology where living plant can be used to extract, concentrate, degrade, and vaporize toxic substances from the soil, water, or air. It is also cost-effective, eco-friendly, and safer methods for environmental management in comparison to the physico-chemical techniques (Raskin et al., 1997, 2000).

11.6.3.2.1 Phytoextraction

Phytoextraction is a process of up-taking and delivering of substances by the vascular plants. Vascular plants have instinct ability to absorb elements from the earth via root. This technique has been effectively used for removal of lead (Pb) and cadmium (Cd) and excessive selenium (Se) from contaminated soil (Eapen et al., 2006).

11.6.3.2.2 Phytostabilization (Immobilization of Contaminants)

Phytostabilization exploits plant's natural ability to immobilize of contaminants. This process hinders toxic compounds migration and helps to stabilize contaminated soils in place.

11.6.3.2.3 Phytovolatilization

In this process, toxic contaminants are transformed into less toxic volatile vapor form and transpiration of these less toxic vapors with the help of transpiration organs of a plant. So the transpiration rate is a key factor. With the help of phytovolatization tritium, mercury, selenium can be released into the atmosphere through the leaves (Cunningham et al., 1996). It has been reported that *Arabidopsis thaliana* is very effective for the conversion of convert mercury salts into volatile form (Watanabe, 1997).

11.6.3.2.4 Rhizofiltration

Rhizofiltration is a mechanism by which plant roots adsorb and precipitate metallic contaminants from water. Rhizofiltration is used mainly to remove different impurities from the wastewater.

11.6.3.2.5 Phytostimulation

Phytostimulation is the process of degradation of organic pollutants of soil with the help of microorganism present in rhizosphere of root. Phytostimulation is very important as rhizosphere gives particular habitat to the microorganisms and also serves as unique soil habitats; presence of organic compounds provides energy to the microorganisms. The organic compounds can be the inducer for microbial degradation.

11.6.3.2.6 Mycorrhiza Assisted Phytoremediation

Symbiotic relationships are often observed naturally between plants and fungus. This type of association has two benefits: increment of root surface area and rate of nutrients absorption. *Glomus* sp. and *Entrophospora* sp. are the two mycorrhizal species which are closely associated many plant species found in heavy metal contaminated soils stipulating their clear function in tolerating and removal of metals by phytoremediation in association with the plants (Khan et al., 2000).

11.6.3.2.7 Plant Species Involved in Phytoremediation

Varieties of plant species are effective for bioremediation of polluted environment. Prominent families for phytoremediation of contaminants are given in Table 11.4 and phytoremediating abilities of plant species are listed in Table 11.5 (Warrier and Saroja, 2008).

TABLE 11.4 Plants Families for Phytoremediation of Contaminants

Plant Families	Elements for Remediation
Brassicaceae	Cadmium, Zinc, Thallium, Nickel
Lamiaceae	Cobalt, Copper
Fabaceae	Selenium
Apocynaceae	Manganese

TABLE 11.5 Phytoremediating Abilities of Plant Species

Plant Species	Elements for Remediation
Arabidopsis halleri, Rumex acetosa, Thlaspi calaminare, T. caerulescens	Zn
Acacia cana, Astragalus bisulcatus, A. racemosus, Atriplex confertifolia, Machaeranthera glabriuscula, Neptunia amplexicaulis	Se
Agrostis tenuis, Minuartia verna, Polycarpaea synandra, Thlaspi alpestre, T. rotundifolium	Pb
Berkheya coddii, Cochlearia aucheri, Dicoma niccolifera, Geissois pruinosa, Phidiasia lindavii, Phyllanthus sp.	Ni
Lemna minor L.	Hg
Thlaspi caerulescens	Cd
Pityrogramma calomelanos, Pteris vittata	As
Aeollanthus subacaulis, Eragrostis racemosa, Haumaniastrum katangese, Ipomea alpina, Pandiaka metallorum, Vigna dolomitica	Cu

11.6.3.2.8 The Hyper-Accumulator Plant Species

Hyper-accumulator plants are able to accumulate metals in stem and leaves. Varieties of plants family are good examples of hyper-accumulators; among them Rubiaceae, Asteraceae, and Brassicaceae are significant (Prasad et al., 2003). A special type of gene regulates metals uptake by the root and deposits them to the other part within the plants.

11.6.3.3 BACTERIA BIOREMEDIATION

In this kind of bioremediation bacteria is used for removing environmental pollutants with its instinct metabolic procedure. In recent years bacteria are modified by genetic engineering and recombinant biotechnology to increase the remediation process. A wide variety of toxic compounds have been degraded effectively by different types of bacteria.

Bioremediation success depends on several factors (Vallero, 2010):

1. The growth rate of microorganism.
2. Intimacy of contact between the organisms and the target toxic compounds.
3. Microorganisms availability.
4. The environment should be habitable for the microbes to thrive.

11.6.3.3.1 Biomining

Biomining is a process of mineral extraction and metallic compound from their parent ores using biological methods. Biomining is an economical, environment-friendly alternative to those traditional expensive such as roasting and smelting for extraction and treating specific mineral ores. Biomining is relies entirely on the natural interaction of biological organisms mainly bacteria and contributes to sustainable development. A wide variety of mineral oxidizing bacteria like *Sulfolobus acidocaldarius, Acidithiobacillus ferrooxidans, A. caldus, Leptospirillum ferrooxidans, Leptospirillum ferriphilum*, and *Thermothrix thiopara* has been reported to oxidize iron and sulfur-containing minerals (Clark et al., 1996; Nagpal, 1993; Leduc et al., 1994). Fungi and plant also have the potentiality in biomining. As for example, *Aspergillus niger* and *Penicillium simplicissimum* are effective mobilizer of Cu, Sn, Al, Zn, and Pb (Brauer et al., 1991).

Microorganisms depend on basically two methods: Oxidation and Reduction for minerals recovery from the ore. The microorganisms such as *Thiobacillus ferroxidans* and *T. thioxidans* uses oxidation and liberate Fe and S, respectively. *Desulfovibro desulfuricans* involved in the reduction of sulfates and form hydrogen sulfides. Biotechnological tools such as genetic engineering and conjugation have been used now a day to generate more effective and desired microorganisms and genetically modified organism to increase bio-oxidation rate hence elevating the efficiency of biomining. Biotechnology has also played a role to produce bacterial strains which are resistant to heavy metals. These heavy metals are responsible to slow down the biomining process. So heavy metal resistance bacteria could be a promising candidate for biomining.

11.6.3.3.2 Biomonitoring

Biomonitoring is a breakthrough promising way to quantify the negative adverse effect of a contaminant. Broadly, biomonitoring is the use of living organisms to detect any adverse alteration occurs in environment. A wide variety of substances involved in biomonitoring process like biomarkers, biosensors, etc.

11.6.3.3.3 Biological Markers

A biomarker is used to detect and measure the harmful consequences of a toxic compound present in the environment (Onwurah et al., 2007). Biomarker which may be a whole organism or a part of it includes all such detection that exhibits an interaction between the contaminants and life system (NRC, 1989). If a contaminant and organism react with each other, a response is generated in form of enzyme. So analyzing such responses in the biological system gives an indication or acts as a marker of those particular contaminants. Variety of organisms especially microorganisms are used as an indicator for toxicity assay. Bacteria associated tests are considered to be most simple, cost-effective and fast (Matthews, 1980) (Table 11.6).

TABLE 11.6 Biomarkers Types and Applications

Biomarker Type	Application	References
Chlorophyll in *Zea mays* L.	Detection of hydrocarbon contamination of agricultural soil.	Ezeonu et al., 2009
Azotobacter sp.	Analyzing the consequences of oil spill in aquatic environment.	Onwurah et al., 1998
Steranes and hopanes	Steranes formed as crude oil constituents and hopanes are used to identify the rock source that produces a crude oil.	Peters et al., 1993
Sensitivity of *Nitrobacter* sp.	Evaluating the effect of crude oil on nitrite oxidation.	Okpokwasili and Odokuma, 1994
Ethoxyresorufin-O-deethylase in fish	Indicates exposure of fish to planar-halogenated hydrocarbons (PAHs)	Bucheli et al., 1995; Stegeman et al., 1994

11.6.3.3.4 Biosensor

Biosensors are the future tools for biomonitoring. Practically biosensors are usually used for continuous observation of environmental hazards and also act as biocontrol tools which are very much effective to analyze informative data on which types of contaminant are present, their location and also accurate evaluation of their bioavailability. A biosensor is a device containing of two main parts: a biocatalyst (whole cell, tissue or enzyme) and a transducer used for conversion of biological signal into a detectable signal (Wilson et al., 1994). Biosensors mainly depend on biorecognition and signal transduction. Bio-recognition depends on antibodies, enzymes, whole cells, etc., and signal transduction is performed different interfaces (optical, thermal, electrochemical magnetic, etc.) (Ripp et al., 2010). There are three types of biosensors, such as biosensors-based on enzyme, biosensors-based on antibody and biosensors based on DNA.

1. **Biosensors Based on Enzyme:** An important example of biosensors depends on enzyme is the bacterial cytochrome C3 reductases which reduce sulfur and are used to reduce heavy metals. When these biosensors are applied under imitating groundwater constrain, it shows reactivity with several other metals. Perchlorate reductase is another example of this kind for detecting groundwater contamination (Okeke et al., 2007).
2. **Biosensors Based on Antibody:** In this type, antibody is utilized as immunosensors. Antibodies are very specific to the target antigens and also show stability (Ripp et al., 2010).
3. **Biosensors Based on DNA:** DNA biosensors are used to monitor structural changes in the nucleic acids structure upon reaction to a particular chemical. The structural alteration is due to the mutagenic nature of that chemical or binding ability to the nucleic acids (Ripp et al., 2010). This kind of biosensors is significant as they give an indication of existence of harmful chemicals in the environment.

11.6.4 BIOMIMETICS AND BIOMEMS TECHNOLOGIES

The fate of biosensors is shifted towards the dawning techniques of biomimetics and BioMEMs. BioMEMs abbreviated as biological micro-electromechanical systems are based on biomicro and bio-nano-technological interfaces that make lab-on-a-chip and biochip type biosensors (Ripp et al., 2010). Their main focuses are portability, low-cost, redundancy, miniaturization, sample size, and response time. Though majority of these biosensors are contributing biomedical application, but gradually and unavoidably these techniques are shifting towards monitoring and management of environment. BioMEMs basically depends on optical transducers interconnected with receptors which may be a whole-cell, nucleic acid or enzymes. Biomimetics enhances the properties of the organic material synthetically by mimicking their characteristics trait (Ripp et al., 2010).

11.6.5 APPLICATION OF BIOTECHNOLOGY IN ENVIRONMENT

The undeniable fact that we have to admit is that the capacity of the biosphere is limited. An immediate action is required to sustain the environment and minimize the anthropological impact on environment (Cunningham, 1993). In this context, there is an urgent need of superior machinery, eco-friendly technology, and sustainable developments (Fierăscu, 2009). Biotechnology can be applied in different environmental prospect.

11.6.6 BIOTECHNOLOGY FOR HEAVY METAL REMOVAL

Heavy metals are natural component of the environment but their balance and geochemical cycles has been altered due to indiscriminate use, lead to excessive heavy metal contamination in the environment (soil and water). Such a high heavy metal accumulation brings in catastrophic effects on human and aquatic ecosystem. In this regard, biotechnological tools like biosorption and bioaccumulation could be a promising alternative to different physico-chemical technology for heavy metal removal (Kapoor et al., 1995). Bioremediation is an effective technology for elimination and recovery of the heavy metals in polluted water and lands. Microorganism and plants can be exploited for bioremediation either *ex -situ* or *in- situ* as they naturally developed certain strategies to survive in heavy metal contaminated sites or habitat (Gadd, 2000; Lim et al., 2003; Malik, 2004; Lin et al., 2005).

Examples of plant along with microorganism for heavy metal remediation are listed Table 11.7 and Table 11.8, respectively.

TABLE 11.7 Examples of Plant Studied and Treated for Heavy Metals Remediation

Plant Species	Heavy Metal	References
Salix viminalis, S. fragilis	Cd, Cu, Pb, Zn	Pulford et al., 2003; Volk et al., 2006
Ricinus communis	Cd	Huang et al., 2011
Zea mays	Cd, Pb, Zn	Meers et al., 2010
Populus deltoides, P. nigra, P. trichocarpa	Cd, Cu, Pb, Zn	Ruttens et al., 2011
Jatropha curcas L.)	Cd, Pb	Abhilash et al., 2009; Jamil et al., 2009
Populus deltoides	Hg	Che et al., 2003
Brassica juncea, Astragalus bisulcatus	Se	Bitther et al., 2012
Populus canescens	Zn	Bittsanszkya et al., 2005

TABLE 11.8 Microorganisms Serve for Utilizing Heavy Metals

Microorganism Species	Compounds	References
Saccharomyces cerevisiae Cunninghamella elegans	Lead, mercury, and nickel	Talos et al., 2009; Infante et al., 2014
Cunninghamella elegans	Heavy metals	Tigini et al., 2010
Pseudomonas fluorescens and *P. aeruginosa*	Fe2+, Pb2+and Cu2	Paranthaman et al., 2015
Lysinibacillus sphaericus	Cobalt, copper, chromium, and lead	Peña-Montenegro et al., 2015
Microbacterium profundi	Fe	Wu et al., 2015
Aspergillus versicolor, Paecilomyces sp., Terichoderma sp., Cladosporium sp	Cadmium	Soleimani et al., 2015
Geobacter spp.	Fe (III), U (VI	Mirlahiji and Eisazadech, 2014
Bacillus safensis	Cadmium	Priyalaxmi et al., 2014
Pseudomonas aeruginosa	U, Ni, and Cr	Sinha et al., 2011
Aerococcus sp., *Rhodopseudomonas palustris*	Pb, Cr, Cd	Sinha et al., 2014

11.6.7 CONTROL OVER OIL SPILLAGE THROUGH BIOTECHNOLOGY

When oil or any oil products are somehow contaminated with the environments, it creates an adverse effect. So there are of need some innovative approach and techniques to overcome this situation. In this regard, biological methods have got much attention in comparison to physico-chemical methods for complete degradation or removal and also minimize the chance of toxic byproducts production (Malik and Ahmed, 2012; Lin et al., 2014).

Several breakthrough researches have already been carried out on bacteria which able to digest or degrade oil. Microorganisms are able to produce several compounds which have surfactant characteristics. These compounds are known as bio-surfactants (Souza et al., 2014; Aparna et al., 2011).

Bio-surfactants has many advantages as these are environment-friendly, degradable, shows specificity to the pollutants and more importantly active under extreme environmental conditions (Kapadia et al., 2013; Santos et al., 2013).

Microorganisms, such as fungus, yeasts, and bacteria are reported for producing biosurfactants (Banat et al., 2010; Sobrinho et al., 2013). Microorganisms which are the good producer of biosurfactants are listed in Table 11.9 (Sobrinho et al., 2013; Pacwa–Plociniczak et al., 2011) and groups of microorganism important for oil remediation are shown in Table 11.10. Biosurfactants assists in biodegradation process of the toxic contaminants because it makes the substrates more available to the micro-organism (Aparna et al., 2011; Olkowska et al., 2012).

TABLE 11.9 List of Microorganism in the Production of Biosurfactants

Biosurfactant	Microorganism Species
Glycolipids	*Pseudomonas aeruginosa, Mycobacterium* sp.
Lipopeptides	*Bacillus subtilis, Pseudomonas fluorescens, Bacillus polymyxia*
Fatty acids and phospholipids	*Corynebacterium lepus, Thiobacillus thiooxidans*
Polymeric type surfactants	*Acinetobacter calcoaceticus, Candida tropicalis, Pseudomonas fluorescens*
Particulate type surfactant	*Acinetobacter calcoaceticus*

TABLE 11.10 Microorganisms Involved in Bioremediation of Oil

Microorganism Species	Compounds	References
Fusarium sp.	Oil	Hidayat et al., 2012
Alcaligenes odorans, Bacillus subtilis, Corynebacterium propinquum, Pseudomonas aeruginosa	Oil	Singh et al., 2013
Pseudomonas cepacia, Serratia ficaria, Bacillus cereus and *Citrobacter koseri*	Oil	Kehinde et al., 2016
Aspergillus niger Saccharomyces cerevisiae and Candida glabrata	Crude oil	Burghal et al., 2016
Pseudomonas putida and *Arthobacter* sp	Diesel oil	Sukumar et al., 2016

11.7 BIOTECHNOLOGY AND POLLUTION MONITORING

Every organism, including microbes, is very much efficient to produce a report about the health the environment where it lives. Microorganisms can be an indicator of aquatic or terrestrial ecosystem healthiness as they have the ability to detect and monitor the changes on the ecosystem where they inhibit. In comparison to other organisms, microorganisms are found in large number and easy to sample. Some microorganisms are able to produce a typical protein, stress protein when exposed to certain contaminants such as benzene and cadmium. This protein might be used as an early indication about the health status of the environment. Several microorganisms can be used as an indicator of inorganic and organic pollution. Organic pollution can be monitored by certain kind of parameters such as presence of heterotrophic bacteria, total coliforms (TC) and fecal streptococci whereas inorganic pollution can be monitored by sulfur-oxidizing – reducing bacteria, iron utilizing bacteria, nitrifying bacteria (HTC, 1993; Okpokwasili, 2007). Existence of heterotrophic microorganisms in large quantity in water body is an indicator of pollution because they get their energy by oxidizing the organic molecules (Okpokwasili, 2007). So it can be concluded that the level of organic pollution is proportionally related to microbial count (Okpokwasili, 2007).

11.8 BIOTECHNOLOGY AND ENVIRONMENTAL TOXICITY

In the past few decades, major fact of the globe is environmental pollution. A wide variety of toxic compounds and contaminants, generated from unscientific anthropogenic activities, are mixed continuously in environment and the consequences become more adverse due to bioaccumulation. Prior identification of toxic compounds in the environment and their adverse consequences on organisms is hence an important and urgent need. The traditional methodology of pollution assessment, depends on chemical methods has some drawbacks such as they do not gives information about the actual toxicity (Munawar et al., 1989). Moreover, compounds which are new to the databases and that are toxic below the detection limit cannot be detected this way. In this scenario, microorganism has an important role to play. Living organisms or cells can be used as analytical devices that effectively assess negative effects of toxic compounds found in environmental samples (Rand, 1995). When living organisms or cells are exposed to environmental sample, they give response which usually gives information about toxicity, genotoxicity, estrogenic, etc. of that particular sample (Abel, 1996; Chen et al., 2002). Toxicity tests rely on prokaryotic or eukaryotic microorganisms and genetically modified microorganisms instead of higher organisms (particularly fish and mammals) has enabled rapid and cost-effective examination of environmental samples for toxicity (Farre et al., 2003; Wadhia et al., 2007).

Bioassay is an effective technique in which living test organism is used to a sample to analyze significant biological effect in response to the potential contaminants (Wadhia et al., 2007).

Bioassays are of two types: toxicity assays and genotoxicity assays. Ecotoxicity bioassays measure the physiological changes of a test organism upon exposing to sample, while genotoxicity tests particularly measure the consequences due to alteration of the genetic material (Timbrell, 2000).

11.8.1 TOXICITY BIOASSAYS

Now a day various toxicity bioassays are standardized and available, applying bacteria and yeast. Microtox is a very effective toxicological assay tool that uses a luminescent bacteria *Vibrio fischeri* with the help of MicrotoxOmni software to measure the toxicity (Blaise et al., 2005). *Vibrio fischeri* has natural bioluminescence capabilities. *Vibrio fischeri* bioluminescence inhibition assay is applicable for many environmental samples like groundwater, municipality, and industrial waste, etc. (Parvez et al., 2006). The microbial assay for toxicity risk assessment (MARA) is a recent technique for ecotoxicological assay. It uses various luminescent bacteria cultivated from ocean and freshwater (Gabrielson et al., 2003; Fai et al., 2010). Another microbial assay named as luminous microbial array for toxicity risk assessment (LumiMARA) which is recently developed as more sensitive technique for toxicity assay of environmental samples in comparison to MARA.

Other commonly used bacterial bioassays are *Bacillus* spp. dehydrogenase activity assay and *E. coli* β-galactosidase activity assay (Wadhia et al., 2007). Unicellular algae-based bioassays is also applied in environmental biomonitoring. A well-understood technique for testing the pesticide effects relies on analyzing the rate of the growth inhibition of green microalga, *Scenedesmus supspicatus* after exposure (Stauber et al., 2002). Protozoa are eukaryotic microorganisms are also shows much sensitivity to environmental alterations, hence can be an effective test-species for toxicological assays. *Tetrahymena pyriformis* and *Tetrahymena termophila,* non-pathogenic, free-living protozoa are used in most of the toxicological assessment. There are some important parameters which are assessed in protozoa based ecotoxicity assays are growth rate, natality/mortality, ATP content and acid phosphatise activity (Du, 1993; Pauli et al., 1993; Nicolau et al., 2001).

11.8.2 GENOTOXICITY ASSAYS

Genotoxicity assay is used to measure the potentiality of a sample which is responsible to induce mutagenicity and DNA damage in genetic material of any test organisms (Timbrell, 2000). *Salmonella typhimurium* based genotoxicity assays have been used widely. Ames test is the most popular, has been performed to monitor environmental water (Maron, 1983). Another useful test is the "umu-test" depends on *S. typhimurium.* In DNA repair mechanism SOS-response genes play an important role. Genotoxicity can be identified by measuring the SOS-response genes transcription. Colorimetric detection of genotoxic compounds is performed by combining SOS-response genes with reporter gene encoded by β-galactosidase (Kenyon, 1983). SOS-chromotest, another types of genotoxicity assay which applies *Escherichia coli* as test-species follow the same principle (Quillardet, 1985) These two types of genotoxicity test is used frequently for their sensitivity to the chloride pesticides (Farre et al., 2005). Green-Screen assay (GSA), a recently developed genotoxicity assay rely on yeast is highly sensitive to variety of mutagens (VanGompel et al., 2005). Another important assay is Comet assay is used to detect and measures DNA strand-breaks in cells (Singh et al., 1988).

11.8.3 ECOSYSTEM MODELING

Ecosystem is an integrated system of living and non-living factors of the environments. It consists of abiotic physico-chemical environment and biotic assemblage of producers, consumers, decomposers, and detritivores altogether form a pristine, stable, and functional unit. Any kind of disturbances to one portion of an ecosystem can have adverse effects on another.

Ecosystem modeling is a technique to speculate future consequences. With this approach, it is possible to recognize important bits of statistics about various ecosystem elements and also possible to predict the consequences of the next disturbances by using computer programming. Bioinformatics and ecological informatics are significant for ecosystem modeling. Both these technique depends on computational methods, but their focus is different. Main areas which are under supervision of bioinformatics are to determine the gene interaction and protein function (Overbeck et al., 1999; Wolf et al., 2001, Henikoff et al., 1999; Lupas et al., 1991) and also application of DNA microarray or genomic data to determine the phenotypes of organisms (Lockhardt et al., 2000), whereas, ecological informatics uses phenotypic, genomic, and environmental data to determine genotypes of populations (Doney et al., 2004; Ouborg et al., 2007) and also structuring and working of ecosystems with the help of community and climate dataset (D'Angelo et al., 1995; Chon et al., 2006; Park et al., 2003; Jeong et al., 2006). With respect to the rapid global biodiversity loss, pollution, and change in climate, it is very much urgent to make an informed decision on preserving biodiversity and sustainable environment.

11.9 ROLE OF BIOTECHNOLOGY IN BIO-FUEL PRODUCTION

In recent years, in response to mitigate the energy crisis, global warming, and climate changes there is a demand of alternative, non-fossil-based fuels. In this context, biofuel has received a great deal of interest. Biofuels derived from organic biomass are renewable and sustainable energies. Biotechnological tools and applications have a vital role to produce biofuels in a sustainable way (Biofuel Production Technologies, 2008). Different aspect of biotechnological strategies has been used, such as creation of engineered or synthetic microorganisms or genetically engineered plants to improve biofuel production. Biotechnology is extremely useful to lowering the production cost and attainment a successful bio-fuel future (Wolt, 2009; Lynd et al., 2008).

11.10 BIOTECHNOLOGY AND DIFFERENT GENERATION OF BIO-FUEL

11.10.1 FIRST GENERATION BIO-FUEL

Most renowned first generation biofuel is ethanol which is made by fermentation using food crops such as sugar cane or maize.

11.10.2 SECOND GENERATION BIO-FUEL

In this generation, bio-fuel is made from non-food crops and less expensive biomass by using lignocelluloses bio-feed stock (Banse et al., 2008). There is a rapid hike in second-generation biofuel production due to its additional sustainable features. Second generation bio-fuel may be of two types such as biochemical bio-fuel and thermo-chemical bio-fuel, depending on the conversion process. Several microorganisms have a significant role in this procedure.

11.10.3 THIRD GENERATION BIO-FUELS

Third generations of biofuels are emerging as an effective alternative where microalgal biomass is used to avoid the use of food sources in the production of bio-fuel (Peralta et al., 2010). There are numerous alternatives that can be considered as third-generation bio-fuel are fatty acid methyl esters, natural triacylglycerols, and cellulose biomass (Kleinová et al., 2012). Different bioconversion technologies have been used to obtain wide varieties of biofuel such as biochemical, direct combustion methods.

11.11 ROLE OF BIOTECHNOLOGY

Biotechnology plays a promising role in biofuel production. The use of modern molecular biology and genetically modified organisms make the biofuel production very efficiently. Different molecular biotechnological techniques have been applied to enhance microbial activity and enzymes (Lee et al., 2008). Cell wall composition and infrastructure and components of lignocellulosic can be upgraded to increase ethanol production with the help of biotechnology (Carpita, 1996). Biotechnology is used to modify physiology, photosynthetic ability, and architecture of plants to influence the yield density. Rapid advancement has been made that enable crops to grow in adverse condition and with deprived soil characteristics. Researchers have successfully able to clone the genes which are responsible for the synthesis of cellulases enzymes which helps to construct effective, low-cost, and maximum biomass conversion strategy.

11.11.1 NEED FOR BIOPLASTICS AND ROLE OF BIOTECHNOLOGY

The main concern about the use of synthetic plastic is that it is not easy to degrade and generation of harmful toxins during degradation; hence, there is an urgent demand of environment-friendly plastics or bioplastic. Polymers which are being reported extensively as bioplastics and acts as a substitute for synthetic plastics: PHAs, PBS, PPDO, PLA, and PCL. The major drawback of bioplastic generation is its high generation cost, but the scientific study suggests that this problem can be solved by the application of microbial biotechnology. Several microbes naturally accumulate certain polymer in their cells under stress conditions, and this polymer can be easily extracted and modulated. PHAs have been drawing much attention among the biodegradable plastic candidates due to its similar nature to conventional synthetic plastics and also total biodegradability (Hong et al., 2003).

Poly(3-hydroxybutyrate) (PHB) is the first discovered PHA from *Bacillus megaterium* (Lemoigne, 1926). This bacterium accumulates a homopolymer inside the cell. PHA has been reported in many bacterial genera, such as *Alcaligenes* (Khanna et al., 2005), *Azotobacter* (Pozo et al., 2002), *Azospirillum* (Sun et al., 2002), *Rhodospirillum* (Berleman et al., 2004), *Herbaspirillum* (Catalan et al., 2007), *Bacillus* (Law et al., 2003), *Pseudomonas* (Jiang et al., 2008), *Rhizobium* (Encarnacion et al., 2002 and Todd et al., 2002), *Azorhizobium*, *Streptomyces* (Ramachander and Rawal, 2005), *Methylobacterium*, and *Nostoc* (Sharma et al., 2007). In spite of the bacteria, recombinant yeasts were used in PHA biosyntheses such as *Saccharomyces* (Carlson and Srienc, 2006), *Candida* (Kim et al., 1999) and *Variovorax* (Maskow et al., 2004).

Certain bacteria like *Ralstonia eutropha, Alcaligenes eutrophus, Cupriavidus necator* and *Pseudomonas oleovorans* are accumulating PHA during stationary phase (Khanna et al., 2005, 2006); whereas *Alcaligenes latus, A. beijerinckii* and recombinant *Escherichia coli* are known to good accumulator *of* PHA during growth phase. PHA accumulates in cells when the organism is under the condition of limited nutrients and excess carbon sources in the media. When the limited nutrient supply is normal, the PHA is degraded by depolymerases and produce carbon (Du et al., 2001; Borah et al., 2002). *Alcaligenes eutrophus* has been reported as a good accumulator of PHB in large quantity from carbon sources (Khanna and Srivastsva, 2005).

Although many different microorganisms has been reported to produce PHAs, but limited bacteria has been employed for bioplastic production. Each bacterium needs different criteria for growth and PHA production (Kim, 2000; Khodair, 2003; Reddy et al., 2003). The genetically modified *E. coli* has an operon which has the genes of *R. eutropha* for PHB biosynthesis enzymes. This is modified bacteria synthesize large quantities of PHB from various organic compounds (Jung et al., 2004; Lenz et al., 2005). Recent studies in the area of transgenic plants emphasizes that biotechnology has opened up new aspect for extensive production of a very good quality PHA from the transgenic oilseed-rape as well as from the leaves of transgenic *Arabidopsis* sp., both of them are having bacterial (A. *eutrophus* and R. *eutropha*) PHA biosynthesis gene (Lee, 1996; Poirier et al., 1995).

11.12 FOSSIL FUEL AND BIOTECHNOLOGY

Biotechnology has an important role in fossil fuel industry regarding energy recovery and sustainability. It has been reported that in oil industry, high percentage of oiling ecological sediments retain there because it is thought to be mechanically irremediable though the amount of energy recovery can be dramatically increased by the microbial activity, as micro-organisms effectively transform hydrocarbon to methane. Studies indicate that about 20% to 40% of methane presents in the reservoir of oil and gas comes from the microbial conversion of carbon dioxide into methane (Katz, 2011). Methane that is produced biologically in coal indicates that biotechnology could be an effective weapon to recover the energy from the coal (Cheung et al., 2010). Biotechnology could be the best choice to enhance the quality of fossil fuel and coal by eliminating sulfur. Inorganic sulfur is removed by microbial oxidation process, and the potential micro-organism is *Thiobacillus* sp. (Okpokwasili, 2007)

11.13 BIOTECHNOLOGY FOR RESOURCE MANAGEMENT AND SAVING ENERGY

Biotechnology has an eye-catching role to save energy and precious resources in various ways. Different inorganic catalysts are used to speed up the rate of different complex reactions to make new products. These reactions need high temperatures to work efficiently and also produce toxic wastes. Different microbes and genetically modified organisms in recent days could be used as a potential alternative to that catalyst. By this way, money and resources both will be saved and reduction of hazardous end products (Su, 1998).

In the paper industry, the pulp is treated and bleached with different chemical to make the final product. This method generates a huge amount of wastes (Su, 1998). It has been reported that enzymes from fungi are used as an alternatives to those chemicals (Su, 1998). Now a day the production cost at industrial level could be reduced by using highly productive microbial strains. Various enzymes from a wide variety of microorganism have been used effectively at industrial level to minimize the energy and production cost. Pectinolytic enzymes are one of the mostly used enzymes in the extraction and clarification of fruit juices and wines, pigments, and oil extraction from plant and also in jute manufacturing. Different types of microorganism has been reported to produce pectinolytic enzymes such as *Bacillus* spp., *Clostridium* spp., *Pseudomonas* spp., *Erwinia* spp., *Aspergillus* spp., *Penicillium* spp., *Sclerotinia libertiana*, *Thermomyces lanuginosus*, etc.(Rombouts et al., 1972). Pectic enzymes are also widely distributed in plants, fungi, insects, nematodes, and protozoa (Bateman et al., 1966).

11.14 THE ENVIRONMENTAL BENEFITS FROM MOLECULAR BIOTECHNOLOGY APPLICATION

Microbes are in everywhere. Some of them are beneficial whereas some are pathogenic to us. So these microorganisms need to be monitored. In this regard molecular biotechnological tools play an important role to monitor, inspect, and detect growth pattern of different microbial population in the environment. Different biotechnological tools are available now a day; some of them are discussed in subsections.

11.14.1 POLYMERASE CHAIN REACTION (PCR)

This technique is extremely powerful and is used widely in the field of biotechnology and molecular biology. PCR is used to generate many copies of a particular portion of a DNA. PCR has the potentiality to answer many questions regarding microbial activity and also has the ability to detect and monitor different pathogens. We can prepare phylogenetic tree from the sequence data which can be used in environmental samples.

11.14.2 FLUORESCENT IN SITU HYBRIDIZATION (FISH)

This technique is used to detect the specific DNA sequences on a particular chromosome. A fluorescent probe which binds only to the complementary sequences has been used in this technique to show sequence resemblance and is observed by fluorescence microscopy. FISH can be used to identify species. FISH can be used in environmental monitoring and also used to locate environmental samples accurately.

11.15 ROLE OF BIOTECHNOLOGY IN AIR POLLUTION ABATEMENT

Biotechnology offers a platform in a most cost-effective and greener way to control air pollution and removal of industrial and municipal airstreams odorous and toxic contaminants. Inorganic odorous compounds and volatile organic compounds (VOCs) could be hazards to the environment and also bring catastrophic and health hazards to human when emitted in large amount. There is an urgent need of a kind of air pollution controlling technology that gives annoyance free, healthy air with respect to rapid urbanization and industrialization.

11.15.1 BIOREACTORS TO CONTROL AIR POLLUTION

Bioreactors use bio-reaction where microbes are used to consume and degrade pollutants from a contaminated air stream. Microbes can effectively consume and decompose both organic and inorganic contaminants in a suitable environment. Bio-reaction is a "green" process and also low cost in comparison to the traditional approaches.

11.15.1.1 COMPONENTS OF BIOREACTORS

1. **Bioreactor Media:** In general, criteria for media selection is vary with the reactor types. Media must have important characteristics such as good capacity of absorption, buffering ability, sufficient pore size and having good biodegradation capacity. Nutrient-rich organic material is used to coat the media for microbe hosting and suitable binders are often used to provide product strength, buffer capacity and stability.
2. **Microorganisms:** A wide variety of microorganisms, mainly bacterial and fungal species have been used in bioreactors for biodegradation such as *Clostridium* sp., *Pseudomonas* sp., *Alcaligenes* sp., *Bacillus* sp., *Corynebacterium* sp., *Sphingomonas* sp., *Xanthomonas* sp., *Nocardia* sp., *Mycobacterium* sp., *Rhodococcus* sp., *Xanthobacter* sp., and *Enterobacter* sp.
3. **Moisture and Temperature:** Moisture is an important factor in the bioreactors. Low moisture will greatly reduce efficiency. Temperature is the most important variable affecting bioreactor operations. Most microbes can survive and grow in a temperature range of 60 to 105 F.
4. **Bioreactors Types:** Basically there are three types of bioreactors are widely used such as biofilters, biotrickling filters, and bioscrubbers.
5. **Bioscrubbers:** Biological oxidation takes place in this bioreactor. Here culture media is suspended type. Bio-scrubbers have two compartments: i) a scrubber where volatile organic substances are transferred to a water phase and, ii) a bioreactor where biological treatment is done in a liquid phase.
6. **Biofilters and Biotrickling Filters:** Biofilters and biotrickling employ immobilized organisms. Microorganisms used in biofilters are mesophilic. The media within a biofilter are generally composed of material such as wood bark, soil, compost, coated ceramic particles or a combination of these products. Absorbed contaminants are degraded into harmless and odorless products by microbial activity.

7. **Biofertilizer:** These are one of the best modern tools and gift of biotechnology and act as an essential component of modern farming. Bio-fertilizers play a vital role to maintain long term soil fertility and sustainability and also could be a better replacement to our conventional and chemical fertilizers. A wide variety of microorganisms have been utilized effectively to produce biofertilizer. The important microorganisms which are very much potential to be act as bio-fertilizer are *Pseudomonas* sp., *Azospirillum* sp., *Bacillus* sp., *Penicillium* sp., *Azotobacter* sp., *Anabaena* sp., *Nostoc* sp., *Rhizobium* sp., etc.

11.16 BIOTECHNOLOGICAL APPROACH IN BIOPESTICIDES

In the field of agriculture, the main objective is to increase the food and production values of plant for consumption. The organisms which are impeding in this system are called pests. Biotechnology has opened an exciting frontier in pest management and agriculture, replacing chemical-based technologies. The onset of pest resistance, adverse effect on various non-target animal or plant population and other severe problems related to the environment due to chemical oriented insecticides have to bring much interest in biotechnology-based technologies, or bio-pesticide. Bio-pesticide is the most promising substitute for chemical ones for controlling pests, pathogens, and weeds.

Microorganisms based bio-pesticides provides an effective and eco-friendly gateway to pest problems. Biofungicides, bioherbicides, and bioinsecticides are the most widely used biopesticides. Using biotechnology it is now quite easy to enhance and modification some of the vital characteristics that increase the utility and potentiality of many bio-control agents. It has become possible to utilize important and useful genes from unrelated animals and plants for bio-control needs. Use of recombinant DNA techniques for biological control can be performed in different ways such as incorporation of resistance gene in the crop plants, genetically modification of organisms associated with the plant to provide protection or modification of controlling agents. In principle, biotechnology can be used for biological pest control by improvisation and modification of the bio-control agents, and crop plants are engineered in such a way that the plants are potential resistant to pests. Caterpillars and beetles are the most dangerous attacker of several major crops. In order to resist the attack Bt genes have been introduced to the major crops. In India, several types of registered biopesticides have been used widely (Table 11.11). A botanical or plant product is now emerging as an effective means to be used in crop protection. Some of the plant products which have been registered as bio-pesticides are listed in Table 11.12 (Buss and Park brown, 2002).

TABLE 11.11 Types of Bio-Pesticide

Sl. No.	Name of the Biopesticide
1	*Bacillus sphaericus*
2	Neem based pesticides
3	*Trichoderma viride*
4	*Beauveria bassiana*
5	*Pseudomonas fluorescens*
6	*Trichoderma harzianum*
7	NPV of *Helicoverpa armigera*
8	*Bacillus thuringiensis spp.*
9	NPV of *Spodoptera litura*

TABLE 11.12 Botanicals Registered as Bio-Pesticide

Plant Products	Target Pests
Neem	Aphids, stink bugs, mites, etc.
Pyrethrum	Flies, ticks, ants, and aphids
Limonene and Linalol	Wasps and flies
Rotenone	Beetles, caterpillars, thrips, and aphids

11.16.1 PROS AND CONS ABOUT BIOTECHNOLOGY

- Genetically engineered organisms so are much less predictable in comparison to the man-made substances and chemicals. They can replicate, shows movement, and can mutate also (Su, 1998).
- So much care has to be taken in genetic engineering and should be done in proper laboratory circumstances. Genetically, modified organism cannot be withdrawn after development and release.

- There is much possibility in alteration of the genetically engineered organism or even mixed with others, producing new species after releasing to the field.
- Sometime specific genetic modifications of plants have been introduced to defend viruses. In some cases, these modifications of hosts bring mutation in the viruses and make them more virulent that increases the chance of attacking non-host plant species (Su, 1998).
- There is much more possibility that survival and diversity of other species of animal and plants are reduced due to the existence of genetically engineered plants and animals (Su, 1998). The genetically engineered plants may mix with the natural species which led to the contamination of the natural population and also put an enormous pressure on the local species that inhibit there.
- The long-term effects of GMO are uncertain.

11.17 CONCLUSION

The benefits of biotechnology in environment protection, management, and sustainability are well established in today's date. The rate at which the environment is degrading requires a critical attention and effective remedial measures and the sustainability of this measure becomes ultimate to make the environment save for life continuously. Biotechnology is very important in effort towards annihilate poverty, providing food security, degradation of contaminant and attainment sustainable environment. Biotechnology gives us a platform to make use of natural resources in sustainable way, bio-fuel production, soil erosion control, production of transgenic plants and animals, solution to global warming and environmental pollution. The techniques which are discussed in this chapter are somehow linked to biotechnology directly or indirectly, have been used to control and sustain our environment.

KEYWORDS

- **bioremediation**
- **biotechnology**
- **environment**
- **phytoremediation**
- **sustainability**

REFERENCES

Abel, P. D., (1996). Water Pollution Biology (2nd edn). Taylor & Francis, London. pp. 1–286.

Abhilash, P. C., Jamil, S., & Singh, N., (2009). Transgenic plants for enhanced biodegradation and phytoremediation of organic xenobiotics. *Biotechnol. Adv., 27*, 474–488.

Adenipekun, C. O., & Lawal, R., (2012). Uses of mushrooms in bioremediation: A review. *Biotechnol. Molec. Biol. Rev., 7*(3), 62–68.

Aparna, A., Srinikethan, G., & Hedge, S., (2011). Effect of addition of biosurfactant produced by *Pseudomonas* ssp. on biodegradation of crude oil. In: *International Proceedings of Chemical, Biological & Environmental Engineering. Proceedings of the 2nd International Proceedings of Chemical* (Vol. 6, pp. 71–75). Singapore.

Banat, I. M., Franzetti, A., Gandolfi, I., Bestetti, G., Martinotti, M. G., Fracchia, L., Smyth, T. J., & Marchant, R., (2010). Microbial biosurfactants production, applications. *Appl. Microbiol. Biotechnol., 87*, 427–444.

Banerjee, A., Jhariya, M. K., Yadav, D. K., & Raj, A., (2018). Micro-remediation of metals: A new frontier in bioremediation. In: Hussain, C., (eds.), *Handbook of Environmental Materials Management* (pp. 1–36). Springer, Cham. ISBN 978–3–319–58538–3. doi: https://doi.org/10.1007/978–3–319–58538–3_10–1 (Accessed on 4 August 2019).

Banse, M., Van Meijl, H., & Woltjer, G., (2008). The impact of first and second-generation biofuels on global agricultural production, trade and land use. In: *GTAP Conference Paper*. pp. 1–14.

Barry, D. P., & Austa, S. D., (1994). Pollutant degradation by white rot fungi. *Review of Environmental Contamination and Toxicology, 138*, 49–72.

Bateman, D. F., & Millar, R. L., (1966). Pectic enzymes in tissue degradation. *Ann. Rev. Phytopathol., 4*, 19.

Berleman, J. E., & Bauer, C. E., (2004). Characterization of cyst cell formation in the purple photosynthetic bacterium *Rhodospirillum centenum. Microbiol., 150*, 383–390.

Bhattacharya, S., (2014). Mycoremediation of Benzo[*a*]pyrene by *Pleurotus ostreatus* in the presence of heavy metals and mediators. *Biotech, 4*(2), 205–211.

Biofuel Production Technologies, (2008). *Status, Prospects and Implications for Trade and Development.* United Nations Conference on Trade and Development.

Bitther, O. P., Pilon-Smits, E. A. H., Meagher, R. B., & Doty, S., (2012). Biotechnological approaches for phytoremediation. In: Arie, A. A., & Hasegawa P. M., (eds.), *Plant Biotechnology and Agriculture* (pp. 309–328). Academic Press: Oxford, UK.

Bittsanszkya, A., Kömives, T., Gullner, G., Gyulai, G., Kiss, J., Heszky, L., Radimszky, L., & Rennenberg, H., (2005). Ability of transgenic poplars with elevated glutathione content to tolerate zinc (2+) stress. *Environ. Int., 31*, 251–254.

Blaise, C., & Férard, J. F., (2005). *Small-Scale Freshwater Toxicity Investigations: Toxicity Test Methods* (Vol. 1). Springer, Dordrecht, the Netherlands.

Borah, B., Thakur, P. S., & Nigam, J. N., (2002). The influence of nutritional and environmental conditions on the accumulation of poly-β- hydroxybutyrate in *Bacillus mycoides* RLJ B-017. *J. Appli. Microbiol., 92*, 776–783.

Brauer, H., (1991). Growth of fungi and bacteria in the reciprocating jet bioreactor. *Bioprocess Engineering, 6*, 1–15.

Bucheli, T. D., & Fent, K., (1995). Induction of cytochrome P450 as a biomarker for environmental contamination in aquatic ecosystems. *Critical Reviews in Environmental Science and Technology*, *25*(3), 201–268.

Burghal, A. A., Abu-Mejdad, N. M. J. A., & Tamimi, W. H. A., (2016). Mycodegradation of crude oil by fungal species isolated from petroleum contaminated soil. *International Journal of Innovative Research in Science, Engineering and Technology*, *5*(2), 1517–1524.

Buss, E. A., & Park-Brown, S. G., (2002). *Natural Products for Insect Pest Management*, ENY–350. http://edis.ifas.ufl.edu/IN197 (Accessed on 4 August 2019).

Carlson, R., & Srienc, F., (2006). Effects of recombinant precursor pathway variations on poly-R-3-hydroxybutyrate synthesis in *Saccharomyces cerevisiae*. *J. Biotechnol.*, *124*, 561–573.

Carpita, N. C., (1996). Structure and biogenesis of the cell walls of grasses. *Annual Review of Plant Biology*, *47*(1), 445–476.

Catalan, A. L., Ferreira, F., Gill, P. R., & Batista, S., (2007). Production of polyhydroxyalkanoates by *Herbaspirillum seropedicae* Grown with different sole carbon sources and on lactose when engineered to express the lac Zlac Y genes. *Enz. Microb. Technol.*, *40*(5), 1352–1357.

Che, D., Meagher, R. B., Heaton, A. C., Lima, A., Rugh, C. L., & Merkle, S. A., (2003). Expression of mercuric ion reductase in Eastern cottonwood (*Populus deltoides*) confers mercuric ion reduction and resistance. *Plant Biotechnol. J.*, *1*, 311–319.

Chen, C. Y., & Lu, C. L., (2002). An analysis of the combined effects of organic toxicants. *The Science of the Total Environment*, *289*, 123.

Cheung, K., Klassen, P., Mayer, B., Goodarzii, F., & Aravena, R., (2010). Major ion isotope geochemistry of fluids and gases from coal bed methane and shallow ground water wells in Alberta, Canada. *Appl. Geochem.*, *25*, 1307–1329.

Chon, T. S., Park, Y. S., Kwak, I. S., & Cha, E. Y., (2006). Non-linear approach to grouping, dynamics and organizational informatics of benthic macroinvertebrate communities in streams by artificial neural networks. In: Recknagel, F., (ed.), *Ecological Informatics, Scope, Techniques and Applications* (2nd edn., pp. 187–238). Springer-Verlag: Berlin.

Clark, D. A., & Norris, P. R., (1996). *Acidimicrobium ferrooxidans* gen. nov., sp. nov.: Mixed-culture ferrous iron oxidation with *Sulfobacillus* species. *Microbiology*, *142*, 785–790.

Cunningham, S. D., & Berti, W. R., (1993). Remediation of contaminated soil with green plants. *Vitro Cell. Dev. Biol.*, *29*, 207–212.

Cunningham, S. D., & Ow, D. W., (1996). Promises and prospects of phytoremediation. *Plant Physiology*, 110, 715–719.

D'Angelo, D. J., Howard, L. M., Meyer, J. L., Gregory, S. V., & Ashkenas, L. R., (1995). Ecological uses of genetic algorithms: Predicting fish distributions in complex physical habitats. *Canadian Journal of Fisheries and Aquatic Science*, *52*, 1893–1908.

Dasilva, E. J., (2004). The colors of biotechnological science, development and humankind. *Electronic Journal of Biotechnology*, *7*, 3.

Doney, S. C., Abbott, M. R., Cullen, J. J., Karl, D. M., & Rothstein, L., (2004). From genes to ecosystems: The ocean's new frontier. *Front. Ecol. Environ.*, *2*(9), 457–466.

Du, G., Chen, J., Yu, J., & Lun, S., (2001). Continuous production of poly-3-hydroxybutyrate by *Ralstonia eutropha* in a two-stage culture system. *Journal of Biotechnology*, *88*, 59–65.

Encarnacion, S., Vargas, M. C., Dunn, M. F., Davalos, G., Mendoza, Y., & Mora, J., (2002). AniA regulates reserve polymer accumulation and global protein expression in *Rhizobium etli*. *J. Microbiol., 184*(8), 2287–2295.

EPA, (2001). *Seminars, Bioremediation of Hazardous Waste Sites: Practical Approach to Implementation.*

Ezeonu, C. S., & Onwurah, I. N. E., (2009). Effect of crude oil contamination on chlorophyll content in *Zea mays L. International Journal of Biology and Biotechnology, 6*(4), 299–301.

Fai, P., & Grant, A., (2010). An assessment of the potential of the microbial assay for risk assessment (mara) for ecotoxicological testing. *Ecotoxicology, 19*, 1626–1633.

Farre, M., & Barcelo, D., (2003). Toxicity testing of waste water and sewage sludge by biosensors, bioassays and chemical analysis. *Trends in Analytical Chemistry, 22*(5), 299–310.

Farre, M., Brix, R., & Barcelo, D., (2005). Screening water for pollutant through biological techniques under European Union funding during the last 10 years. *Trends in Analytical Chemistry, 24*, 532–545.

Fierăscu, R. C., Dumitriu, I., Ion, M. L., Catangiu, A., & Ion, R. M., (2009). Surface and analytical techniques study of Romanian coins. *European Journal of Science and Theology, 5*, 17–28.

Gabrielson, J., Kühn, I., Colque-Navarro, P., Hart, M., Iversen, A., McKenzie, D., & Möllby, R., (2003). Microplate-based microbial assay for risk assessment and (eco)toxic fingerprinting of chemicals. *Anal. Chim. Acta, 485*, 121–130.

Gadd, G. M., (2000). Bioremedial potential of microbial mechanisms of metal mobilization and immobilization. *Curr. Opin. Biotechnol., 11*, 271–279.

Henikoff, S., Henikoff, J. G., & Pietrovski, S., (1999). Blocks+: A non-redundant database of protein alignment blocks derived from multiple compilations. *Bioinformatics, 15*, 471–479.

Hidayat, A., & Tachibana, S., (2012). Biodegradation of aliphatic hydrocarbon in three types of crude oil by Fusarium sp. F092 under stress with artificial sea water. *Journal of Environmental Science and Technology, 5*(1), 64–73.

Hong, S. H., Park, S. J., Moon, S. Y., Park, J. P., & Lee, S. Y., (2003). *In silico* prediction and validation of the importance of the enter-doudoroff pathway in poly (3-hydroxybutyrate) production by metabolically engineered *Escherichia coli*. *Biotechnol. Bioengin., 83*(8), 854–863.

HTC, (1993). *Environmental Management in the Oil and Petrochemical Industries*. Hybrid Technologies Limited, Lagos.

Huang, H., Yu, N., Wang, L., Gupta, D. K., He, Z., Wang, K., Zhu, Z., Yan, X., Li, T., & Yang, X. E., (2011). The phytoremediation potential of bioenergy crop *Ricinus communis* for DDTs and cadmium co-contaminated soil. *Bioresour. Technol., 102*, 11034–11038.

Infante, J. C., De Arco, R. D., & Angulo, M. E., (2014). Removal of lead, mercury and nickel using the yeast *Saccharomyces cerevisiae*. *Revista MVZ Córdoba, 19*, 4141–4149.

Isikhuemhen, O. S., (2011). The tropical white rot fungus, *Lentinus squarrosulus* Mont: Lignocellolytic enzymes activities and sugar release from cornstalks under solid state fermentation. *World J. Microbiol. Biotechnol., 28*(5), 1961–1966.

James, C., (2006). Global status of commercialized biotech/GM crops. *International Services for the Acquisition of Agro-Biotech Applications, ISAAA Brief*, p. 35.

Jamil, S., Abhilash, P. C., Singh, N., & Sharma, P. N., (2009). *Jatropha curcas*: A potential crop for phytoremediation of coal fly ash. *J. Hazard. Mater, 172*, 269–275.

Jeong, K. S., Recknagel, F., & Joo, G. J., (2006). Prediction and elucidation of population dynamics of the blue-green algae *Microcystis aeruginosa* and the diatom *Stephanodiscus hantzschii* in the Nakdong river-reservoir system (South Korea) by a recurrent artificial neural network. In: Recknagel, F., (ed.), *Ecological Informatics, Scope, Techniques and Applications* (2nd edn., pp. 255–273). Springer-Verlag: Berlin.

Jiang, Y., Song, X., Gong, L., Li, P., Dai, C., & Shao, W., (2008). High poly-β-hydroxybutyrate production by *Pseudomonas fluorescens* A2a5 from inexpensive substrates. *Enz. Microb. Technol., 42*, 167–172.

Jung, Y. M, Lee, J. N., Shin, H. D., & Lee, Y. H., (2004). Role of tktA gene in pentose phosphate pathway on odd-ball biosynthesis of poly-β-hydroxybutyrate in transformant *Escherichia coli* Harboring phb CAB operon. *Journal of Bioscience and Bioengineering, 98*(3), 224–227.

Kapadia, S. G., & Yagnik, B. N., (2013). Current trend and potential for microbial biosurfactants. *Asian J. Exp. Biol. Sci., 4*, 1–8.

Kapoor, A., & Viraraghvan, T., (1995). Fungal biosorption—an alternative treatment option for heavy metal bearing wastewater: A review. *Bioresour. Technol., 53*, 195–206.

Katz, B. J., (2011). Microbial processes and natural gas accumulation. *Open Geol. J., 5*, 75–83.

Kehinde, F. O., & Isaac, S. A., (2016). Effectiveness of augmented consortia of *Bacillus coagulans, Citrobacter koseri* and *Serratia ficaria* in the degradation of diesel polluted soil supplemented with pig dung. *Afr. J. Microbiol. Res., 10*(39), 1637–1644.

Kenyon, C. J., (1983). The bacterial response to DNA damage. *Trends in Biochemical Sciences, 8*, 84–87.

Khan, A. G., Kuek, C., Chaudhary, T. M., Khaoo, C. S., & Hayes, W. J., (2000). Role of plants, mycorrhizae and phytochelators in heavy metal contaminated land remediation. *J. of Chemosphere, 41*, 197–207.

Khanna, S., & Srivastava, A. K., (2005). Statistical media optimization studies for growth and PHB production by *Ralstonia eutropha. Proc. Biochem., 40*, 2173–2182.

Khanna, S., & Srivastava, A. K., (2006). Optimization of nutrient feed concentration and addition time for production of poly-hydroxybutyrate. *Enz. Microb. Technol., 39*, 1145–1151.

Khodair, T. A., (2003). Microbial production of degradable polymers as substitute for synthetic plastics. PhD Thesis, *Agric. Microbiol. Dept., Fac. of Agric., Ain Shams Univ.*, Egypt.

Kim, B. S., (2000). Production of poly (3-hydroxybutyrate) from inexpensive substrates. *Enz. Microb. Technol., 27*, 774–777.

Kim, C. H., Hong, W. K., Lee, I. Y., Choi, E. S., & Rhee, S. K., (1999). Enhanced production of D-β- hydroxyisobutyric acid through strain improvement. *J. Biotechnol., 69*, 75–79.

Kleinová, A., Cvengrošová, Z., Rimarčík, J., Buzetzki, E., Mikulec, J., & Cvengroš, J., (2012). Biofuels from algae. *Procedia Engineering, 42*, 231–238.

Law, K. H., Cheng, Y. C., Leung, Y. C., Lo Chua, W. H. H., & Yu, H. F., (2003). Construction of recombinant *Bacillus subtilis* strains for polyhydroxyalkanoates synthesis. *Biochem. Engin. J., 16*, 203–208.

Leduc, L. G., & Ferroni, G. D., (1994). The chemolithotrophic bacterium Thiobacillus ferrooxidans. *FEMS Microbiol. Rev., 14*, 103–120.

Lee, D., Chen, A., & Nair, R., (2008). Genetically engineered crops for biofuel production: Regulatory perspectives. *Biotechnology and Genetic Engineering Reviews, 25*(1), 331–362.

Lee, S. Y., (1996). Bacterial Polyhydroxyalkanoates. *Biotechnol. Bioeng., 49*, 1–14.

Lemoigne, M., (1926). Produits de deshydration et de polymerisation de l'acide b-oxybutyric. *Bull. Soc. Chem. Biol., 8*, 770–782.

Lenz, R. W., & Marchessault, R. H., (2005). Bacterial polyesters: Biosynthesis, biodegradable plastics and biotechnology. *Americ. Chemic. Soci., 6*(1), 1–8.

Lim, P. E., Mak, K. Y., Mohamed, N., & Noor, A. M., (2003). Removal and speciation of heavy metals along the treatment path of wastewater in subsurface-flow constructed wetlands. *Water Sci. Techno., 48*, 307–313.

Lin, C. C., & Lin, H. L., (2005). Remediation of soil contaminated with the heavy metal (Cd^{2+}). *J. Hazard. Mater, 122*, 7–15.

Lin, M., Yuhua, L., Weiwei, C., Hui, W., & Xiaoke, H., (2014). Use of bacteria-immobilized cotton fuse to absorb and degrade crude oil. *Int. Biodeterior. Biodegrad., 88*, 8–12.

Lockhardt, D., & Winzeler, E., (2000). Genomics, gene expression and DNA arrays. *Nature, 405*, 827–836.

Lupas, A., Van Dyke, M., & Stock, J., (1991). Predicting coiled coils from protein sequences. *Science, 252*, 1162–1164.

Lynd, L. R., Laser, M. S., Bransby, D., Dale, B. E., Davison, B., Hamilton, R., & Wyman, C. E., (2008). How biotech can transform biofuels. *Nature Biotechnology, 26*(2), 169–172.

Malik, A., (2004). Metal bioremediation through growing cells. *Environ. Int., 30*, 261–278.

Malik, Z. A., & Ahmed, S., (2012). Degradation of petroleum hydrocarbons by oil field isolated bacterial consortium. *Afr. J. Biotechnol., 11*, 650–658.

Maron, D. M., & Ames, B. N., (1983). Revised methods for the Salmonella mutagenicity test. *Mutation Research, 113*(3/4), 173–215.

Maskow, T., Olomolaiye, D., Breuer, U., & Kemp, R., (2004). Flow calorimetriy and dielectric spectroscopy to control the bacterial conversion of toxic substrates into polyhydroxyalcanaotes. *Biotechnol. Bioengin., 85*(5), 547–552.

Matthews, P. J., (1980). Toxicology for water scientists. *Journal of Environmental Management, 11*(1), 1–16.

Meers, E., Van Slycken, S., Adriaensen, K., Ruttens, A., Vangronsveld, J., Du Laing, G., Witters, N., Thewys, T., & Tack, F. M., (2010). The use of bio-energy crops (Zea mays) for phytoattenuation of heavy metals on moderately contaminated soils: A field experiment. *Chemosphere, 78*, 35–41.

Mirlahiji, S. G., & Eisazadeh, K., (2014). Bioremediation of Uranium by *Geobacter* spp. *Journal of Research and Development, 1*, 52–58.

Munawar, M., Munawar, I. F., Mayfield, C. I., & McCarth, L. H., (1989c). Probing ecosystem health: A multidisciplinary and multi-trophic assay strategy. *Hydrobiologia, 188/189*, 93.

Nagpal, S., (1993). Effect of carbon dioxide concentration on the bioleaching of a pyritearsenopyrite or concentrates. *Biotechnol. Bioeng., 41*, 459–464.

National Research Council (NRC), (1989). *Biologic Markers in Reproductive Toxicology*, National Academic Press, Washington, DC, USA.

Nicolau, N., Dias, M., Mota, N., & Lima, (2001). *Research in Microbiology, 152*, 621.

Okeke, B. C., Ma, G., Cheng, Q., Losi, M. E., & Frankenberger, W. T., (2007). Development of a perchlorate reductase based biosensor for real time analysis of perchlorate in water. *Journal of Microbiological Methods*, *68*(1), 69–75.

Okpokwasili, G. C., & Odokuma, L. O., (1994). Tolerance of *Nitrobacter* to toxicity of some Nigerian crude oil. *Bulletin of Environmental Contamination and Toxicology*, *52*(3), 388–395.

Okpokwasili, G. C., (2007). *Biotechnology and Clean Environment*. Proc. of the 20th Annl. Conf. of the Biotechnology Society of Nigeria (BSN), 14th–17th, November at the Ebonyi State University, Abakaliki, Nigeria.

Olatunji, O., (2007). *Biotechnology and Industries in Nigeria* (pp. 36–38). Proc. 20th Annual Conf., Biotech. Soc. of Nig. (BSN), 14th–17th Nov, 2007 at Ebonyi State University, Abakaliki, Nigeria.

Olkowska, E., Polkowska, Z., & Namiésnik, J., (2012). Analytical procedures for the determination of surfactants in environmental samples. *Talanta*, *88*, 1–13.

Onwurah, I. N. E., (1998). Biochemical oxygen demand exertion and glucose uptake kinetics of Azotobacter in crude oil polluted medium. *Bulletin of Environmental Contamination and Toxicology*, *60*(3), 464–474.

Ouborg, N. J., & Vriezen, W. H., (2007). An ecologist's guide to ecogenomics. *Journal of Ecology*, *95*, 8–16.

Overbeck, R., Fonstein, M., D'Souza, M., Pusch, G. D., & Maltsev, N., (1999). The use of gene clusters to infer functional coupling. *Proceedings of the National Academy of Sciences USA*, *96*, 2896–2901.

Pacwa–Plociniczak, M., Plaza, G. A., Piotrowska–Seget, Z., & Cameotra, S. S., (2011). Environmental applications of biosurfactants: Recent advances. *Int. J. Mol. Sci.*, *13*, 633–654.

Paranthaman, S. R., & Karthikeyan, B., (2015). Bioremediation of heavy metal in paper mill effluent using *Pseudomonas* spp. *International Journal of Microbiology*, *1*, 1–5.

Park, Y. S., Verdonschot, P. F. M., Chon, T. S., & Lek, S., (2003). Patterning and predicting aquatic macro invertebrate diversities using artificial neural networks. *Water Research*, *37*, 1749–1758.

Parvez, S., Venkataraman, C., & Mukherji, S., (2006). A review on advantages of implementing luminescence inhibition test (*Vibrio fischeri*) for acute toxicity prediction of chemicals. *Environmental International*, *32*, 265–268.

Pauli, W., Berger, S., Jaskulka, L., & Schitz, S., (1993). A case for the inclusion of a protozoan test in aquatic toxicity assessment using *Tetrahymena. Science of the Total Environment Supplem.*, *1*, 779–786.

Peña-Montenegro, T. D., Lozano, L., & Dussán, J., (2015). Genome sequence and description of the mosquitocidal and heavy metal tolerant strain *Lysinibacillus sphaericus* CBAM5. *Stand. Genomic. Sic.*, *10*(2), 1–10.

Peralta, Y., Sanchez, E., & Kafarov, V., (2010). Exergy analysis for third generation biofuel production from microalgae biomass. *Chem. Eng. Transact.*, *21*(1), 1363–1368.

Peters, K. E., & Moldown, J. M., (1993). *The Biomarker Guide, Interpreting Molecular Fossils in Petroleum and Ancient Sediments*. Prentice-Hall, Englewood Cliffs, NJ, USA. OSTI ID: 6066248. https://www.osti.gov/biblio/6066248-biomarker-guide-interpreting-molecular-fossils-petroleum-ancient-sediments.

Poirier, Y., Nawrath, C., & Somerville, C., (1995). Production of polyhydroxyalkanoates, a family of Biodegradable plastics and elastomers, in bacterial and plant. *Biotechnol.*, *13*, 142–150.

Pozo, C., Toledo, M. V. M., Rodelas, B., & Lopez, J. G., (2002). Effects of culture conditions on the production of polyhydroxyalkanoates by *Azotobacter chroococcum* H23 in media containing a high concentration of alpechin (wastewater from olive oil mills) as primary carbon source. *J. Biotechno.*, *97*, 125–131.

Prasad, M. N. V., & Frietas, H., (2003). Metal hyperaccumulation in plants- biodiversity prospecting for phytoremediation technology. *Electronic Journal of Biotechnology*, *6*, 275–321.

Priyalaxmi, R., Murugan, A., & Raja, P., (2014). Bioremediation of cadmium by *Bacillus safensis* (JX126862), a marine bacterium isolated from mangrove sediments. *International Journal of Current Microbiology and Applied Sciences*, *3*, 326–335.

Pulford, I., & Watson, C., (2003). Phytoremediation of heavy metal-contaminated land by trees-a review. *Environ. Int.*, *29*, 529–540.

Quillardet, P., & Hofnung, M., (1985). The SOS Chromotest, a colorimetric bacterial assay for genotoxins: Procedures. *Mutation Research*, *147*(3), 65–78.

Ramachander, T. V. N., & Rawal, S. K., (2005). PHB synthase from *Streptomyces aureofaciens* NRRL 2209. *FEMS Microbiol. L.*, *242*, 13–18.

Rand, G. M., (1995). *Fundamentals of Aquatic Toxicology* (2nd edn.). Taylor and Francis, Bristol, PA. ISBN:9781560320913; pp. 1–1148.

Raskin, I., & Ensely, B., (2000). *Phytoremediation of Toxic Metals Using Plants* (Vol. 471, pp. 19254–19256). Wiley & Sons, Inc., Canada; ISBN.

Raskin, I., Smith, R. D., & Salt, D. E., (1997). Phytoremediation of metals: Using plants to remove pollutants from the environment. *Current Opin. Biotechnology*, *8*, 221–226.

Reddy, C. S. K., Ghai, R., & Kalia, R. V. C., (2003). Polyhydroxyalkanoates: An overview. *Biores. Technol.*, *87*, 137–146.

Ripp, S., Diclaudio, M. L., & Sayler, G. S., (2010). Biosensors as environmental monitors. *Environmental Microbiology* (pp. 213–233). Wiley-Blackwell, NJ, USA.

Rombouts, F. M., (1972). *Occurrence Properties of Bactrial Pectate Luases* (p. 779). Agric. Res. Report (Versl. Landbouwk. Onderz). Centre for Agric. Pub. And Doc., Wageningen, The Netherlands.

Ruttens, A., Boulet, J., Weyens, N., Smeets, K., Adriaensen, K., Meers, E., Van Slycken, S., Tack, F., Meiresonne, L., & Thewys, T., (2011). Short rotation coppice culture of willows and poplars as energy crops on metal contaminated agricultural soils. *Int. J. Phytorem.*, *13*, 194–207.

Santos, D. K. F., Rufino, R. D., Luna, J. M., Santos, V. A., Salgueiro, A. A., & Sarubbo, L. A., (2013). Synthesis and evaluation of biosurfactant produced by *Candida lipolytica* using animal fat and corn steep liquor. *J. Pet. Sci. Eng.*, *105*, 43–50.

Sharma, L., Singh, A. K., Panda, B., & Mallick, N., (2007). Process optimization for poly-β- hydroxybutyrate production in a nitrogen fixing cyanobacterium, *Nostoc muscorum* using response surface methodology. *Biores. Technol.*, *98*, 987–993.

Singh, A., Kumar, V., & Srivastava, J. N., (2013). Assessment of bioremediation of oil and phenol contents in refinery waste water via bacterial consortium. *J. Pet. Environ. Biotechnol.*, *4*(3), 1–4.

Singh, H., (2006). *Mycoremediation: Fungal Bioremediation* (pp. 1–592). Wiley Interscience.

Singh, N. P., McCoy, M. T., Tice, R. R., & Schnider, E. L., (1988). A simple technique for quantitation of low levels of DNA damage in individual cells. *Experimental Cell Research*, *175*(1), 184–191.

Sinha, S. N., & Paul, D., (2014). Heavy metal tolerance and accumulation by bacterial strains isolated from waste water. *Journal of Chemical, Biological and Physical Sciences*, *4*(1), 812–817.
Sinha, S. N., Biswas, M., & Paul, D., (2011). Biodegradation potential of bacterial isolates from tannery effluent with special reference to hexavalent chromium. *Biotechnology Bioinformatics and Bioengineering*, *1*(3), 381–386.
Sobrinho, H. B., Luna, J. M., Rufino, R. D., Porto, A. L. F., & Sarubbo, L. A., (2013). Biosurfactants: Classification, properties and environmental applications. In: *Recent Developments in Biotechnology* (1st edn., Vol. 11, pp. 1–29). Studium Press LLC: Houston, TX, USA.
Soleimani, N., Fazli, M. M., & Mehrasbi, M., (2015). Highly cadmium tolerant fungi: Their tolerance and removal potential. *Journal of Environmental Health Science and Engineering*, *13*(19), 1–9.
Souza, E. C., Vessoni-Penna, T. C., & Souza, O. R. P., (2014). Biosurfactant-enhanced hydrocarbon bioremediation: An overview. *Int. Biodeterior. Biodegrad.*, *89*, 88–94.
Stauber, J. L., Franklin, N. M., & Adams, M. S., (2002). Applications of flow cytometry to ecotoxicity testing using microalgae. *Trends in Biotechnology*, *20*, 141–143.
Stegeman, J., & Hahn, M., (1994). Biochemistry and molecular biology of monooxygenases: Current perspectives on forms, functions, and regulation of cytochrome P450 in aquatic species. In: *Aquatic Toxicology: Molecular, Biochemical, and Cellular Perspectives* (Chapter 3, pp. 120.). CRC Press, Boca Raton, Fla, USA.
Su, S., (1998). *Genetic Engineering*. Evans brothers limited, 2a Portman Mansions, Chiltern Street, London W1M 1LE.
Sukumar, S., & Nirmala, P., (2016). Screening of diesel oil degrading bacteria from petroleum hydrocarbon contaminated soil. *Int. J. Adv. Res. Biol. Sci.*, *3*(8), 18–22.
Sun, J., Dommelen, A. V., Impe, J. V., & Vanderleyden, J., (2002). Involvement of *glnB*, *glnZ*, and *glnD* genes in the regulation of poly-3-hydroxybutyrate biosynthesis by ammonia in *Azospirillum brasilense* Sp7. *Appli. Environ. Microbiol.*, *68*(2), 985–988.
Susan, E., Singh, S. S., & D'Souza, S. F., (2006). Phytoremediation of metals & radionuclides. '*Environmental Bioremediation Technologies*' (pp. 189–209). Springer Publication, NY.
Talos, K., Pager, C., & Tonk, S., (2009). Cadmium biosorption on native Saccharomyces cerevisiae cells in aqueous suspension. *Acta Univ Sapientiae Agric. Environ.*, *1*, 20–30.
Tigini, V., Prigione, V., & Giansanti, P., (2010). Fungal biosorption, an innovative treatment for the decolourisation and detoxification of textile effluents. *Water*, *2*, 550–565.
Timbrell, J., (2000). *Principles of Biochemical Toxicology* (3rd edn., pp. 1–464.). *Taylor & Francis*, London.
Todd, J. D., Wexler, M., Sawers, G., Yeoman, K. H., Poole, P. S., & Johnston, A. W. B., (2002). RirA, an iron-responsive regulator in the symbiotic bacterium *Rhizobium leguminosarum*. *Microbiol.*, *148*, 4059–4071.
UNCBD, (1992). United Nations Conference on Biological Diversity (Earth Summit) held in Rio de Janeiro, Brazil.
US EPA, (1990). *Handbook on In Situ Treatment of Hazardous Waste-Contaminated Soils*. epa/540/2–90/002.
Vallero, A. D., (2010). *Environmental Biotechnology: A Biosystems Approach* (1st edn., pp. 1–750). Elsevier Academic Press, Burlington, Mass, USA.

Van Gompel, J., Woestenborghs, F., Beerens, F. D., Mackie, C., Cahill, P. A., Knight, A. W., Billinton, N., Tweats, D. J., & Walmsley, R. M., (2005). An assessment of the utility of the yeast Green Screen assay in pharmaceutical screening. *Mutagenesis, 20*, 449–454.

Volk, T. A., Abrahamson, L. P., Nowak, C. A., Smart, L. B., Tharakan, P. J., & White, E. H., (2006). The development of short-rotation willow in the northeastern United States for bioenergy and bioproducts, agroforestry and phytoremediation. *Biomass Bioener., 30*, 715–727.

Wadhia, K., & Thompson, K. C., (2007). Low cost ecotoxicity testing of environmental samples using microbial tests for potential implementation of the water framework directive. *Trends in Analytical Chemistry, 26*, 300–307.

Warrier, R. R., & Saroja, S., (2008). Impact of depolluted effluents on the growth and productivity of selected crops. *Ecology, Environment and Conservation, 6*(2), 251–253.

Watanabe, M. E., (1997). Phytoremediation on the brink of commercialization. *Journal of Environmental Science & Technology, 31*, 182–186.

Wilson, K., & Walker, J. M., (1994). *Principles and Techniques of Practical Biochemistry* (4th edn.). Cambridge University Press.

Wolf, Y. I., Rogozin, I. B., Kondrashov, A. S., & Koonin, E. V., (2001). Genome alignment, evolution of prokaryotic genome organization, and prediction of gene function using genomic context. *Genome Research, 11*, 356–372.

Wolt, J. D., (2009). Advancing environmental risk assessment for transgenic bio-feedstock crops. *Biotechnology for Biofuels, 2*, 27.

Wu, Y. H., Zhou, P., & Cheng, H., (2015). Draft genome sequence of *Microbacterium profundi* Shh49T, an Actinobacterium isolated from deep-sea sediment of a polymetallic nodule environment. *Genome Announcements, 3*, 1–2.

Wu, Z. Y., Zhang, W. X., Zhang, Q. S., Hu, C., Wang, R., & Liu, Z. H., (2009). Developing new sacchariferous staters for liquor production based on functional strains isolated from the pits of several famous Luzhou-flavor liquor brewers. *Journal of the Institute of Brewing, 115*(2), 111–115.

CHAPTER 12

Technological Assessment of Sewage Treatment Plants

HARSHITA JAIN,[1] RENU DHUPPER,[1] VISHAL GANDHI,[2] and GEETANJALI KAUSHIK[3]

[1]*Amity Institute of Environment Sciences, Noida, Uttar Pradesh, India, Mobile: +91-7838683947, E-mail: harshita13@gmail.com (H. Jain)*

[2]*Central Pollution Control Board, New Delhi, India*

[3]*MGM's Jawaharlal Nehru Engineering College, Aurangabad, Maharashtra, India*

ABSTRACT

This is a study to assess the efficiency of sewage treatment plants (STP) on the basis of different technologies. The performance of four STP's at Delhi and Noida working on different technology was evaluated. Sewage water samples were analyzed for various parameters like pH, biochemical oxygen demand (BOD), chemical oxygen demand (COD), total suspended solids (TSS), total nitrogen (TN), total phosphorus (TP), total coliforms (TC) and fecal coliforms (FC). The samples were collected from the inlet unit and outlet unit and then compared. Study of performance evaluation of STPs based on SBR, iSBR, BIOFAR, and ASP technologies shows that the BOD removal efficiencies were 97.72%, 94.21%, 94.49%, and 96.55% while the COD removal efficiencies were 89.85%, 94.53%, 86.9%, and 97.09%. The results were not significantly varying, and the performance of all the STP's was at par as the treated effluent was in permissible limits set by the Ministry of Environment, Forest, and Climate Change (MoEFCC). Based on the experience of this study, the performance also depends on three major factors which are regular maintenance, skilled manpower, and uninterrupted power supply. All the technologies are efficient in reducing the organic load, nutrients, and pathogens from the wastewater effectively.

12.1 INTRODUCTION

Environmental sanitation is the most important factor of concern in the process of development of a country. In many places, discharge of untreated wastewater is a common practice. Thus, due to the large gap between the generation and treatment of wastewater, the surface and groundwater is being polluted. The wastewater discharged from residents, institutions, and commercial properties is known as waste or sewage water. One of the major problems which the poor urban residents face is inadequate and improper sanitation in developing countries. This has led to a better understanding of process and treatment technology and the eventual development of water quality standards (Paranychianakis et al., 2015).

There is a need to treat wastewater due to growing environmental pollution to save and protect the health of humans and environment. Increasing levels of pollution has led to innovation and development of latest treatment technologies to efficiently remove pollutants present in wastewater. Because the higher level of treatment provided by a wastewater treatment plant, the cleaner will be the effluent and smaller the impact on the environment (EPA, 1999).

CPCB carries out a study on Municipal wastewater generation and treatment capacity in Metropolitan cities of India every year. There was a huge gap observed between sewage generation and the treatment capacity. Urbanization growth in India is at rapid rate (Figure 12.1; Table 12.1). Therefore, it is important to analyze the system to determine the overall pollution associated to these activities (Gallego et al., 2008).

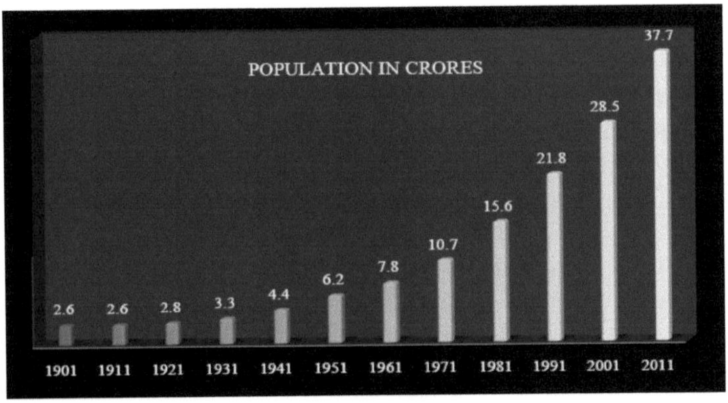

FIGURE 12.1 Urban population growth from 1901 to 2011.
(Source: CPCB, 2013).

TABLE 12.1 Decadal Urban Population Growth Since 1901

Year	Total	Rural	Urban	Urban, as % of Total	Decadal % Increase in Urban Population
1901	238,396,327	212,544,454	25,851,873	10.84	-
1911	252,093,390	226,151,757	25,941,633	10.29	0.35
1921	251,321,213	223,235,043	28,086,170	11.18	8.3
1931	278,977,238	245,521,249	33,455,989	11.99	19.1
1941	318,660,580	274,507,283	44,153,297	13.86	32.0
1951	361,088,090	298,644,381	62,443,709	17.29	41.4
1961	439,234,771	360,298,168	78,936,603	17.97	26.4
1971	548,159,652	439,045,675	109,113,977	19.91	38.2
1981	683,329,097	523,866,550	159,462,547	23.34	46.1
1991	846,302,688	628,691,676	217,611,012	25.71	36.5
2001	1,027,015,247	741,660,293	285,354,954	27.78	31.1
2011	1,210,193,422	833,087,662	377,105,760	31.16	32.2

(Source: CPCB, 2013).

Many new sewage treatment technologies have been developed in the past two decades in developing countries around the world. The technology should be a priority in environmental or public health, and both de-centralized and centralized technologies should be considered (Veenstra and Alaerts, 1996). The technology adopted should be capable of being incrementally upgraded as user demand or quality standards and treatment guidelines increase (Boller, 1997).

The technology should be low-cost which can reduce the risk of malfunction and it should require low energy output and mechanization (Frijns and Jansen, 1996; Boller, 1997). It should be simple to operate, be "local" labor intensive, maintained by the community which doesn't require additional expensive chemical inputs, like chlorine, for tertiary pathogen reductions to meet quality standards, and should be able to recover resources (Mara and Cairncross, 1989; Frijns and Jansen, 1996; Boller, 1997).

Performance evaluation of plant is extremely useful as it provides information on the treatment efficiency loading conditions of the STP. The efficiency of the STP's can be illustrated by a comparative study on of pollutants levels at the inlet and outlet units of the treatment plants.

Objective of the study is to:

- To compare the performance efficiency of sewage treatment plants (STP) based on different technologies.
- To evaluate the performance based on physiological and bacteriological parameters.
- To evaluate the performance based on working conditions of STP.

12.2 SEWAGE AND SEWAGE TREATMENT

Sewage consists of 99% water that comes from residential, industrial, and commercial establishments. Residential and municipal sewage water comprises of 99.99% water and 0.1% dissolved solids, suspended, and colloidals. Physiochemical and biological processes are used to remove physiochemical and biological contaminants (Gallego et al., 2008). In the case of sewage, the degree of treatment is considered in terms of removal of TSS, BOD, COD, nitrogen, phosphorous, and coliforms, etc. Disposal on land is preferred over water as it has an advantage of nutrient utilization for soil. BOD removal is not the only aim; other factors should also be taken care for incidental removal.

Sewage treatment plant is a structure that is designed to receive sewage water from residential, industrial, and commercial establishments to eliminate substances that are harmful to human health and environment and safely discharged into the natural environment. Its main aim is to produce a treated effluent and a sludge which is suitable for discharge or reuse back into the environment (Metcalf et al., 2003). On the basis of removal efficiency, there is a selection of treatment technologies for all parameters and to avoid nuisance conditions.

Overgrowing population and increased scarcity of water are the two main reasons for the need of a proper sewage management system. Sewage treatment is not that cheap; therefore, the main condition in this is the implementation of low-cost sewage treatment systems. If treated wastewater is to be disposed off, guidelines of CPCB (mainly $BOD5<30$ mg/l) should be followed (Table 12.2).

TABLE 12.2 Wastewater Standards for Disposal

Characteristics	General Standards	New Notified Standards (2017)
PH	5–9	6.5–9
BOD	30 mg/l	20 mg/l
COD	250 mg/l	-
TSS	100 mg/l	<50 mg/l
Coliform	<1000 MPN/100 ml	<1000 MPN/100 ml
Ammonia Nitrogen	10 mg/l	-
Phosphate	5 mg/l	-

12.3 NEED FOR SAFE SANITATION SYSTEM

Sanitation is perceived as a condition and process related to people's health, it especially deals with water supply system and human waste. It basically covers aspects like solid waste, industrial, and other hazardous wastes and stormwater drainage. The most potent pollutant of these is the sewage. When the sewage water is left untreated and becomes septic, the decomposition of its organic matter (OM) leads to production of unpleasant gases and nuisance conditions.

In addition, the untreated wastewater consists of many pathogens that dwell in the intestine tract of humans. It also has nutrients which stimulate the growth of aquatic plants, and contains many toxic components or compounds which are potentially carcinogenic or mutagenic. For these reasons, the immediate and nuisance-free removal of sewage from its sources of generation, followed by treatment, reuse, or dispersal into the environment in an eco-friendly manner is necessary to protect public health and environment (CPHEEO, 2012).

12.4 ENVIRONMENT MANAGEMENT AND SUSTAINABILITY

Water reuse and other forms of resource recovery, such as energy recovery and nutrient recycling can help in reducing the environmental impact associated with wastewater treatment facilities (Cornejo Pablo, 2015). Water scarcity, urban water demand, efficient resource utilization, and the protection of human health and ecosystem are additional factors considered towards recent movements to reclaim water and other resources (EPA, 2012b; NRC, 2012). All these factors have led to the implementation of 3,300 water reclamation systems worldwide (FAO, 2010). From a systems perspective, water reuse can offset energy and resources needed for conventional water production, and nutrient recycling can offsets chemical fertilizer usage and energy recovery can lead to energy offsets by replacing natural gas (Fine and Hadas, 2012; Mihelcic et al., 2011; Mo and Zhang, 2012). About 22% of the world's phosphorus supply could be meet through nutrient recycling through urine and feces, which ultimately leading to the reduction of anthropogenic impacts of phosphate mining and addressing phosphorus scarcity (Mihelcic et al., 2011). Integrated resource recovery, collectively through water reuse (Figure 12.2), energy recovery, and nutrient recycling can address the challenges linked with the rising environmental footprint of sewage water treatment (Cornejo Pablo, 2015).

The reuse of sewage water is becoming an important approach towards addressing water shortages. After the treated effluent meets the prescribed standards, the water can be used for irrigation, industrial, and domestic purposes like toilet flushing, garden watering and road cleaning (Xudong Chen et al., 2018).

Energy recovery can be done by anaerobic treatment of wastewater which produces biogas that can be used for various purposes. Environmental indicators require energy use as it measures utilization of resource in eliminating contaminants in sewage water. It also determines the reuse potential of the treated sewage water.

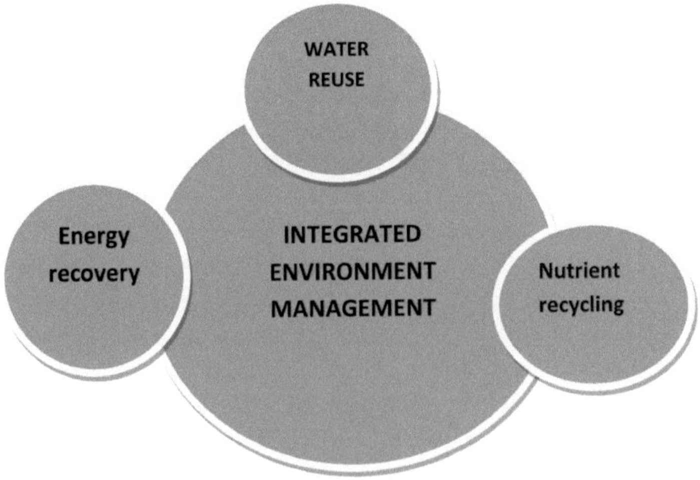

FIGURE 12.2 Integrated management towards environmental sustainability.

12.5 FUNDAMENTAL PRINCIPLE OF TREATMENT

The biological processes are used to remove contaminants, in sewage water treatment, and nowadays many biological treatment options are available and have shown encouraging results over the treatment of complex OM (Mittal, 2006; Wijetunga et al., 2010). There are mainly two types of treatment used for treating sewage water, viz., aerobic, and anaerobic treatment. Both have different uses with pros and cons.

12.5.1 AEROBIC TREATMENT

In aerobic treatment, there is the presence of oxygen and uses bacteria (aerobes), which utilize molecular air (oxygen) to breakdown organic impurity by converting them into water, CO_2, and biomass (Figure 12.3).

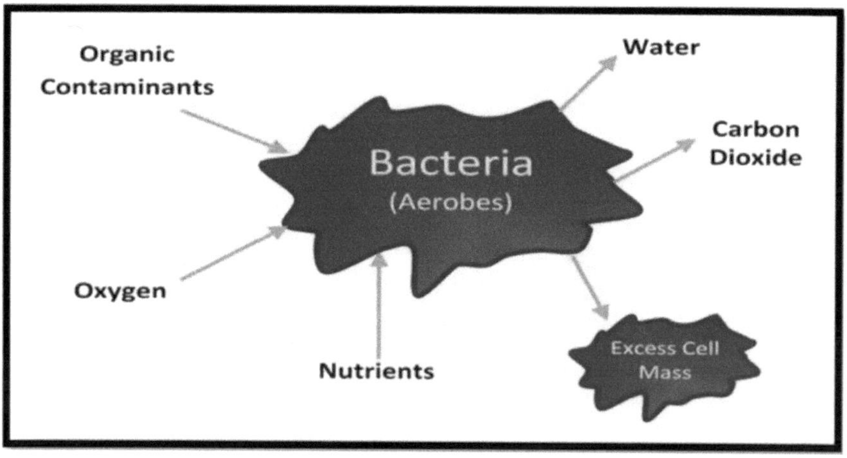

FIGURE 12.3 Aerobic treatment pathway.

In aerobic process, there are two types of treatment mechanisms: aerobic attached growth and aerobic suspended growth.

12.5.1.1 AEROBIC ATTACHED GROWTH

At growth stage, most microorganisms have the ability to colonize the surface of the substrate by means of exopolymers produced by them. So, bacteria attach themselves to the substrate initially on number of preferential sites subsequently covering the entire surface area. When the substrate is coated all over, Biofilm develops and at the same time it keeps on thickening as new cells are produced. Stratification occurs within the film with the superimposition of aerobic layer in which oxygen is diffused and a deeper anaerobic layer (which contains no oxygen). The wastewater is made diffuse through the film so that oxygen and nutrients can be treated. Depending on the reactor and substrate types, the size of these two layers will vary. Some of the treatments technologies which follow this principle are listed below:

1. Trickling filter;
2. Bio far;
3. Mix bed biofilm reactor;
4. Membrane bioreactor;
5. Rotating biological reactor.

12.5.1.2 AEROBIC SUSPENDED GROWTH

As the name suggests, suspended aerobic growth refers to the process in which there are free-floating microorganisms around which the wastewater flows. As a result, biological flocks are formed by the reaction of microbes and wastewater contaminants, which settle out of wastewater. These flocks still contain microorganism, so they are recycled for further treatment. Some of the treatments technologies which follow this principle are listed below:

1. Activated sludge process;
2. Sequential batch reactor;
3. Oxidation ponds;
4. Aerated lagoons.

12.5.2 ANAEROBIC TREATMENT

In anaerobic treatment, there is the absence of oxygen and uses bacteria that dosen't require molecular oxygen to breakdown organic impurity. They are known as Anaerobes. The end products of organic assimilation during anaerobic treatment consist of methane and carbon dioxide gas and biomass (Mittal, 2011). These process steps are important as they have positive effects like, removal of higher organic loading, less sludge production and more pathogen removal, methane gas production and low energy consumption (Nykova et al., 2002) (Figure 12.4).

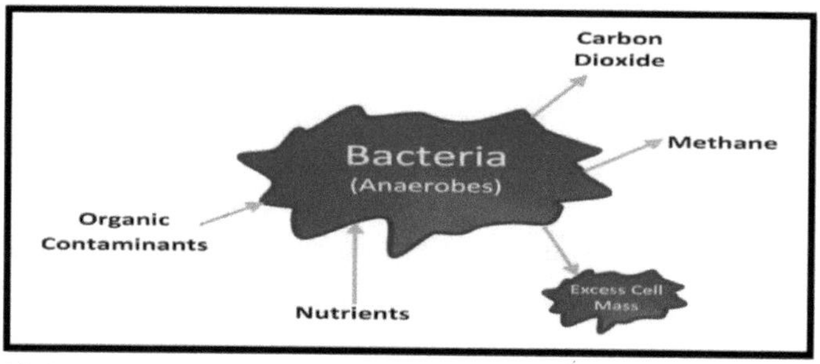

FIGURE 12.4 Anaerobic treatment pathway.

12.5.2.1 ANAEROBIC ATTACHED GROWTH

Bacteria are attached on the media and treat the OM without oxygen, i.e., in the absence of air.

Some of the anaerobic attached growth technologies used are listed below:

1. Up-flow anaerobic sludge blanket.
2. Anaerobic expanded bioreactor.
3. Fluidized bed anaerobic reactor.

12.5.2.2 ANAEROBIC SUSPENDED GROWTH

In Anaerobic suspended growth systems bacteria are freely suspended in the reactors. Microorganisms can be suspended in the wastewater, as a single cell or as their agglomerates. Some of the technologies are listed below:

1. Anaerobic lagoon;
2. Anaerobic digestor;
3. Facultative pond;
4. Claridigestor.

12.6 DESCRIPTION OF TREATMENT TECHNOLOGIES

12.6.1 ACTIVATED SLUDGE PROCESS (ASP)

The activated sludge process (ASP) uses a mass of microorganisms (usually bacteria) to treat sewage water aerobically (Nagwekar, 2014). It is a biological sewage water treatment process in which a mixture of sewage water and activated sludge is mixed and aerated. The sludge is then subsequently separated from the treated water by sedimentation and returned back to the process for further treatment. After settling of effluent clean water is discharged through the outlet of the sedimentation tank. Then after the final sampling and analysis of treated water, it is discharged into the receiving stream (Nagwekar, 2014) (Figure 12.5).

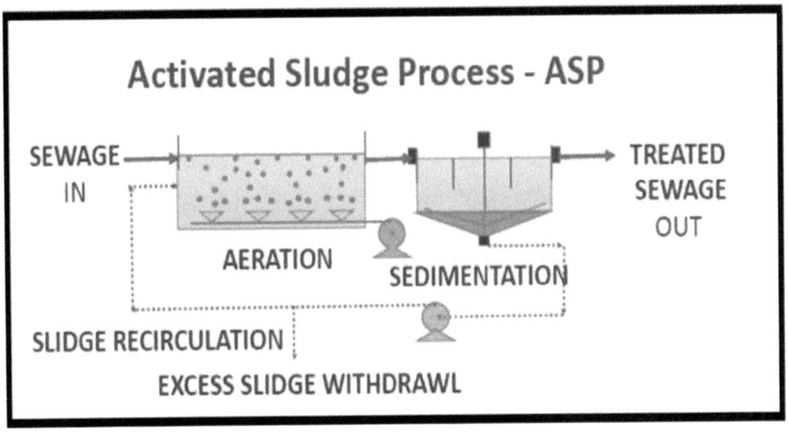

FIGURE 12.5 Activated sludge process.

Advantages:

- Since COD is uniformly low it allows good nitrification.
- It is able to handle loads.
- It is capable of removing 96% of suspended solids.
- Biological nitrification is done without addition of chemicals.

Disadvantages:

- High consumption of energy.
- Foaming observed in winter season which affects the oxygen transfer and hence, performance.

- Elaborated sludge treatment arrangement.
- Moreland requirement than SBR & MBBR.

12.6.2 SEQUENTIAL BATCH REACTOR (SBR)

Sequential Batch Reactor (SBR) is another form of the ASP. It is different from the ASP because collaborates all of the treatment steps into a single large basin whereas the ASP has separate basins for different treatment stage. SBR system has many benefits to alternative conventional process for the biological treatment of both domestic and industrial wastewater (Ying-Chih et al., 2007). Removal of nutrients in SBR process was studied about nitrification, denitrification, and biological phosphorus and monitoring and controls (Obaja et al., 2003; Akin et al., 2005).

SBR is based on a principle known as "fill and draw principle." It has five different steps which are as follows: fill, react, settle, decant, and idle. For a continuous flow process, a minimum of two SBR tanks are required so that one tank receives flow while the other completes treatment cycle (Singh and John, 2013). The process is illustrated in Figures 12.6 and 12.7.

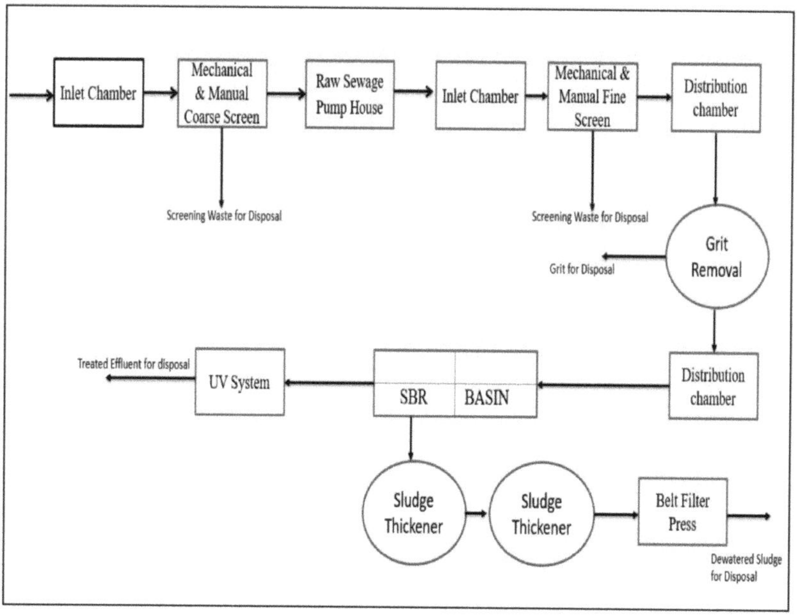

FIGURE 12.6 Sequential batch reactor process.

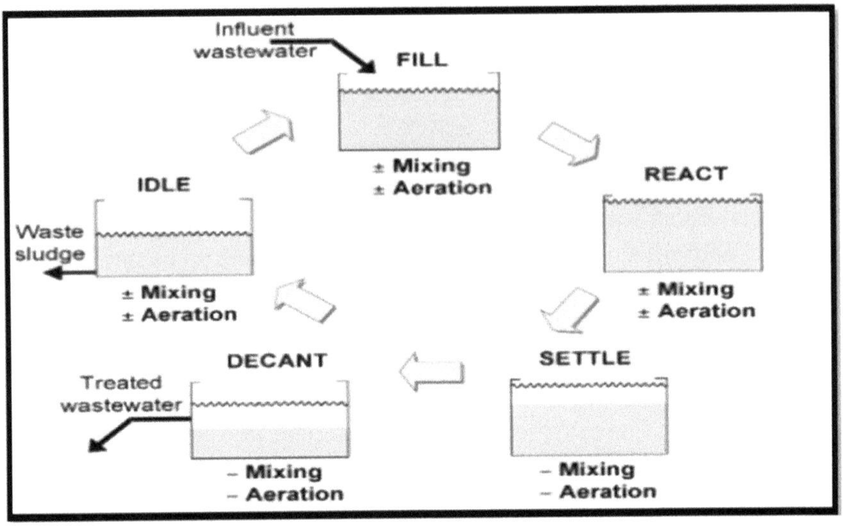

FIGURE 12.7 Schematic flow of sequential batch reactor.

The working process of SBR reactor in phase wise is described below:

1. Fill: In the fill phase, the sewage influent comes into the basin. This way the food comes for the microbes in the form of sludge by the sewage water which comes. This makes an appropriate condition for the reactions to occur by which the treatment occurs.

By varying the time and process of aeration and mixing three different styles of fills are created:

- **Static Fill:** In this fill, there is no mixing and no aeration. When sewage water comes in there is no movement that is why it is also known as an energy-saving phase.
- **Mixed Fill:** In this process, the mixing unit are active and work while the aerating unit are inactive and remain close. The constant mixing provides the fine blend of sewage water and the biomass present in the incoming influent. Because there is no aeration, as a result of anaerobic atmosphere the process of denitrification takes place.
- **Aerated Fill:** In this process both the mixing unit and aerating units work and both are active. Anaerobic atmosphere is converted to aerobic atmosphere. But for denitrification it is

necessary to switch off the oxygen supply and for nitrification aeration is required. So by adjusting the supply of oxygen by turning it on and off the blowers, both the conditions are created making both the process of nitrification and denitrification work simultaneously.

Studies recommend static fill with neither aeration nor mechanical mixing, as this helps promote high fermentation rates with allow flocculent bacteria to outcompete filamentous species, hence prevent sludge bulking (Chudoba et al., 1973; Schroeder, 1982).

2. **React:** In this phase, more reduction in levels of organics takes place. During this, the sewage does not enter the SBR unit while the mixers and aerators are active. Therefore, the reduction level increases rapidly as there is no further addition of incoming sewage water. Mostly BOD reduction occurs in this phase.

3. **Settle:** During this process phase, the sludge remains unaffected as no more sewage is added in this phase. There is no movement and the activated sludge is made to settle down on its own. No sewage flow enters and no mixing and aeration takes place. The sludge will settle down forming a flock mass at the bottom of the basin. This way there is clear upper water that is called supernatant. The sludge collected at the bottom of the basin forms a sludge blanket.

4. **Decant:** It is the withdrawal phase for the discharge of clarified effluent from the reactor. Once the previous phase is completed the decanter then works and opens its valve which is called the effluent discharge valve. There are two types of decanters viz, floating, and fixed arm. The floating decanters works at the inlet below the surface of water, they have the ability for varying the volumes during fill and draw stages. Fixed arm decanters can lower or raise the level and they are low cost also.

5. **Idle:** This phase occurs during decant and fill phases. In this process the sludge that is settled at the bottom of the tank is thrown out from the basin and it is called wasting. A small amount of activated sludge is pumped out. Time varies according to the rate of flow and strategy of operator. Alternatively, IDLE can be eliminated altogether. In instances where operation of SBR does not include an IDLE period, as noted earlier, sludge wasting may be achieved by solid wasting from the mixed liquor during the REACT phase.

Advantages:

- One single tank or basin is there for all the treatment stages.
- Enhanced phosphorous removal without chemicals.
- Controls growth of bacteria.
- Cost saving.

Disadvantages:

- More maintenance is needed.
- Depth of basin should be sufficient for an adequate clear water depth over the sludge blanket for prevention of settled solids entrainment.
- In case of small SBR units(less than 10 MLD), effluent flow balancing is needed for downstream processes such as filtration or disinfection.
- Larger capacity aeration system required.
- Higher level of sophistication is required.
- After the process, equalization is required.
- Lot of power consumption in extended aeration.

12.6.3 BIOFOR

In order to provide compact plants and to assure greater treatment efficiency and reliability, biological aerated filtration technique has been developed (Bradley et al., 2001).

BIOFOR Technology (Biological Filtration and Oxygenated Reactor) is a patented technology of degremont includes intensified aerobic treatment with Dense-Deg & BIOFER (Charu and Singh, 2013).

The main components of the treatment process of densadeg – BIOFOR plant comprise coagulation and flocculation in a specially designed clarisettler. BIOFOR technology is a patented technology of degremont includes Intensified Aerobic treatment with Dense-Deg & BIOFER (Charu and Singh, 2013).

Dosage of alum as coagulant is rather high at around 60 mg/l and then the sedimentation of flocks is enhanced by addition of polyelectrolyte. In fact a bulk of the treatment takes place at this primary clarification stage where almost 90% of suspended solids and 70% of BOD are removed. The second stage of up flow filtration is then considered more of a polishing treatment. The biological filtration is carried out in two identical successive stages. Figure 12.8 depicts schematic diagram of STP process.

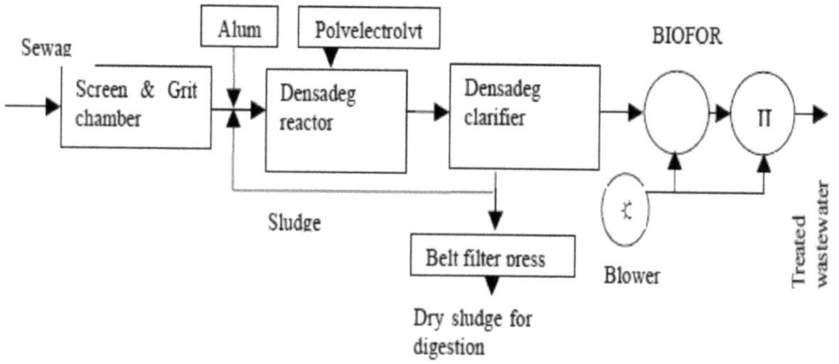

FIGURE 12.8 Schematic flow diagram of biofar process.

Advantages:

- Compact layout.
- Aeration achieved is higher by diffused aeration units,
- Less area needed as secondary sedimentation is not there.
- Capability to handle fluctuations.
- Can comply easily with less standards.
- Lesser manpower requirement.

Disadvantages:

- High and continuous chemicals added in primary treatment.
- Larger waste sludge generated due to chemicals.
- Undigested sludge requires treatment.

12.6.4 *IMPROVED SEQUENTIAL BATCH REACTOR (I-SBR)*

ISBR provides aerobic treatment and stands for improved SBR which means, that it has some improvements over the conventional SBR sewage treatment technology (Bhat, 2013).

I-SBR is based on claridigestor (anaerobically) process which is an advanced SBR (aerobically) technology. Suspended solids present in raw water expressed in terms of BOD/COD are treated anaerobically and liquid consisting soluble pollutants are treated aerobically.

Claridigester unit consists of a large tank, cylindrical shaped in which upper portion is primary clarifier and bottom portion is an anaerobic digester. The upper most part of this unit act as gas holder. Anaerobic digestion occurs in the absence of oxygen. Combitreat provides anaerobic treatment to sewage and reduces the organic load to be handled by further treatment (Bhat, 2013).

In the process, most of the OM converts into CH_4, CO_2 and water. The collected gas is scrubbed to remove and used for power generation. The schematic flow diagram of STP is depicting in Figure 12.9.

The supernatant from combitreat unit flows into I-SBR basin. I-SBR basin acts as an aeration tanks as well as settling tank. It is operated in a batch mode. The plant is equipped with 3 basins for continuous treatment and the flow is further distributed by using motorized gates located on inlet. Combitreat-ISBR requires very less electricity as compared prevalent technologies (Bhat, 2013).

When only ISBR is installed, the pumping requirement for the STP reduces because it is a constant level process and thus large saving in power is achieved. When it comes to outlet parameters, combitreat-ISBR is the frontrunner (Bhat, 2013) (Figure 12.9).

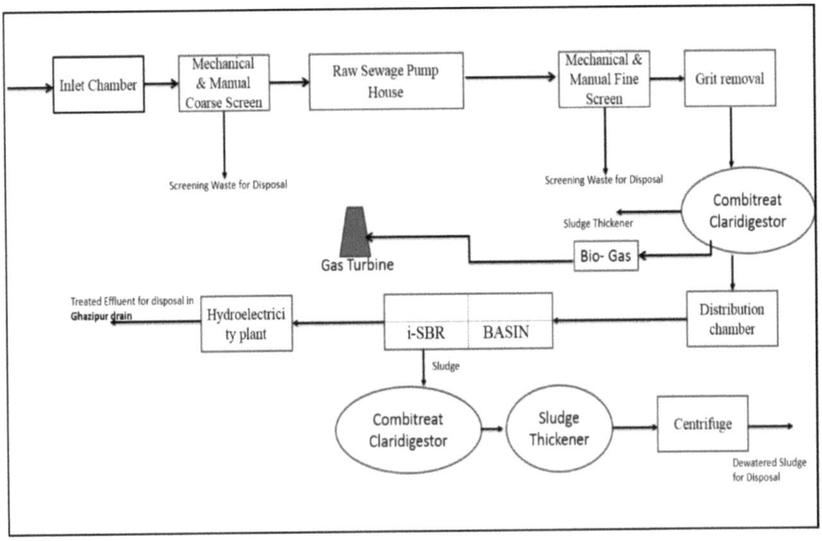

FIGURE 12.9 Schematic flow diagram of I-SBR.

12.7 PARAMETERS

Various parameters used for performance evaluation of STP are given below:

12.7.1 PH

The negative logarithm of hydrogen ion concentration is measured as the pH of a solution. On a given temperature the solution's acid or basic character is measured by the hydrogen ion concentration. If the pH values are between 0 to 7 the solution is indicated as acidic, if it is between 7 and 14, it is alkaline and if 7, it is neutral.

The value of pH in one of the most significant factor in every phase of water quality management as it is depends on pH.

Natural water pH lies between the value 4 to 9 and it is mainly basic because it has the presence of carbonates and bicarbonates of alkali and alkaline metals.

12.7.1.1 ELECTROMETRIC METHOD FOR PH ANALYSIS

Electrometric method is used for pH analysis because it is free from interferences and more accurate. The pH scale lies between 0 and 14 for an aqueous solution. In case of approximate pH value requirement, the colorimetric indicator methods are used.

The pH determination is done by measuring electromotive force of a cell which comprises of an indicator, such as glass electrode (an electrode which response to hydrogen ions) put in a test solution and a reference electrode (commonly calomel electrode). Through liquid junction, contact is made which is a part of the reference electrode. With pH meter's help the electromotive force (emf) of this cell is measured (Guide Manual, CPCB).

12.7.2 BIOCHEMICAL OXYGEN DEMAND (BOD)

The BOD is a laboratory test which is a measure of oxygen consumed for the oxidation of OM and inorganic materials present in water and sewage water at particular temperature and incubation time. Amount of oxygen required in this process is the measure of test. This test is applied for freshwater sources (lakes, rivers), polluted receiving water bodies, marine water (coastal water, estuaries), sewage water (residential, commercial), and for finding out the pollution level and performance of STP.

12.7.2.1 TITRIMETRIC METHOD FOR BOD ANALYSIS

This test is a measure of oxygen that is utilized for the oxidation of inorganic materials (sulfides and ferrous ions) and the biochemical degradation of organic material (carbonaceous demand) at a specific incubation time. This test also measures the amount of oxygen that is used for oxidation of reduced forms of nitrogen if it is not prevented by any inhibitor. The test is performed at a fixed temperature (temperature effects being held constant). The relevance of this BOD test is to compare the difference observed between initial and final dissolved oxygen levels of the samples incubation. Minimum quantity requires 1.5 liters for the testing (Guide Manual, CPCB).

12.7.3 COD

The COD test is a measure of the oxygen amount that is required during oxidation in the presence of a strong chemical oxidant. It is a significant parameter which measures the organic strength of sewage water. This procedure is useful in monitoring and evaluating the performance of STP. COD test results can be obtained in approximately 3 hrs as compared to BOD test, which takes 3 to 5 days time. Also, this method is comparatively easy and also is not affected by interferences.

Closed reflux (titrimetric and colorimetric) method using cod digester for COD analysis:

In this method, the OM gets completely oxidized on reaction with $K_2Cr_2O_7$ with Ag_2SO_4 used as a catalyst in the presence of concentrated sulfuric acid to give carbon dioxide and water. The $K_2Cr_2O_7$ which is left after the reaction is then titrated against ferrous ammonium sulfate (Fe $(NH_4)^2(SO_4)^2$). The dichromate consumed gives the amount of oxygen (O_2) that is needed for the OM oxidation. In a closed system, the volatile organic compounds (VOCs) are oxidized more completely because of longer contact time with oxidants (Guide Manual, CPCB).

12.7.4 TSS

TSS refers to the material that can be filterable or non-filterable and remains as leftover residue on evaporation and drying at a given temperature. TSS affects water quality greatly in a number of ways. High mineralized water is not suitable for many applications in industries. High suspended solids concentration in water is not very suitable for purposes such as bathing. TSS analysis is necessary for biological sewage water treatment and for the performance assessment.

12.7.4.1 METHOD FOR TSS ANALYSIS

Total solids are measured by the evaporation, and subsequent drying of a known volume of sample in the oven at a specific temperature (usually 103–105°C), the residue left behind is called total solids. Total solids are made up from both TSS and Total dissolved solids TDS (Guide Manual, CPCB).

12.7.5 PHOSPHATE

Phosphorous is present as various types of phosphates in the natural and polluted water. These may occur in the form of particles, soluble form, or in the bodies of aquatic organisms. Phosphates presence in water and sewage water testing has a significant importance. At small concentration, phosphates are being used in water supplies to reduce formation of scale, increasing carrying capacity, avoiding corrosion, removing iron and manganese in smaller quantities.

12.7.5.1 STANNOUS CHLORIDE METHOD FOR PHOSPHATE ANALYSIS

Under acidic condition, reaction of orthophosphate with ammonium molybdate yields molybdophosphoric acid. This is reduced further by adding ascorbic acid or stannous chloride to molybdenum blue. The blue color is observed after adding ammonium molybdate, measure at 690 or 880 nm within 8–10 mins after appearance of color by help of blank. From standard graph, the concentration is observed. Then the intensity of the blue colored complex is measured. This is proportional to the phosphate concentration in the sample (Guide Manual, CPCB).

12.7.6 AMMONIA

The production of Ammonia takes place by the degradation of nitrogenous matter. Ammonia concentration above a particular value in sewage water is considered toxic animals like fish.

12.7.6.1 NESSLERISATION METHOD FOR AMMONIA ANALYSIS

Ammonia yields a yellow color compound on reaction with alkaline Nessler reagent. Pre-treatment of zinc sulfate ($ZnSO_4$) and sodium hydroxide (NaOH) will precipitate calcium, ferrous, magnesium) and sulfide and removes turbidity and color. Rochelle salt solution or EDTA (Before Nessler reagent) addition will prevent precipitation of calcium and magnesium in the presence of Nessler reagent. The chemical equation (Guide Manual, CPCB) is given below:

$$2Kr_2Hgl_4 + NH_3 + 3KOH ® (NH_2Hg_2IO) + 7Kl + 2H_2O$$

12.7.7 TOTAL COLIFORM

Pathogens in water can enter the body of humans by skin, ingestion, aspiration, and inhalation or by direct contact with any part of body and can cause severe effects and diseases. Coliform bacteria (aerobic/facultative) are gram negative bacteria that release gas on fermentation in culture media within 46–48 hrs at 35°C. From historic times coliforms are used to mark the quality of drinking water. Test that is used to indicate and detect TC is Most Probable Number (MPN).

12.7.7.1 MULTIPLE TUBE FERMENTATION TECHNIQUE

For analysis of total coliform count, the MPN method is used. In this procedure a series of tubes are inoculated which contains test portion of water sample and a suitable broth culture medium such as lactose broth or MacConkey broth. This is performed at a specific incubation period and at a particular temperature. The presence of coliforms is indicated by gas formation in the tubes. This is referred to as "Presumptive positive test." Since the gas can also be produced by some other organisms, so it is followed by a confirmatory test. In a confirmatory test material taken from the positive tubes is inoculated with brilliant green bile broth. After the incubation time is complete, the tubes are again tested for formation of gas. By this way, the MPN can be obtained with the help all the positive tubes obtained in confirmatory test and total number of tubes inoculated. This procedure is known as MPN method (Guide Manual, CPCB).

12.7.8 FECAL COLIFORM

Fecal coliform bacteria can ferment lactose with the production of acid and gas at 44.5°C in the time period of one day, and that is how they are differentiated from coliforms. Some of the limitations of FC are same as coliforms. Test for fecal coliforms are same as used for total coliform (MPN, MF, and P/A).

12.7.8.1 THERMOTOLERANT METHOD FOR FC ANALYSIS

This procedure relies on filtration of a particular volume of water sample by a filter membrane which consists of cellulose with pore diameter usually 0.44 or 0.2 μm. By this, the bacteria are trapped on the surface of this filter membrane. Then this is incubated in a sterilized container at a specific tem with selective culture. After this colony of coliforms develops, which can be counted (Guide Manual, CPCB).

12.8 MATERIALS AND METHODS

Samples were collected from the inlet and outlet of each STP. A total of eight samples were collected, representing four inlets and seven outlets of the STP. The samples were tested for the following parameters: pH, BOD, COD, TSS, TN, TP, TC, and FC. The pH was measured by the electrolytic method (Guide Manual, Water, and Waste Water Analysis, CPCB). BOD was tested by titrimetric method (Guide Manual, Water, and Waste Water Analysis, CPCB). COD of the samples was measured using closed reflux (titrimetric and colorimetric) method using cod digester for cod analysis (Guide Manual, Water, and Waste Water Analysis, CPCB). Total solids are measured by the evaporation and subsequent drying of a known volume of sample in oven at a specific temperature (usually 103–105°C). Stannous chloride method was used for phosphate analysis; intensity of the blue colored complex was measured at 690nm within 8–10 minutes after appearance of color (Guide Manual, Water, and Waste Water Analysis, CPCB). Nesslerization method was used for ammonia analysis (Guide Manual, Water, and Waste Water Analysis, CPCB). For the analysis of total coliform count, MPN method was used (Guide Manual, Water, and Waste Water Analysis, CPCB). Fecal coliform was counted by Thermotolerant method (Guide Manual, Water, and Waste Water Analysis, CPCB).

12.9 PERFORMANCE EVALUATION OF STPS

12.9.1 PERFORMANCE EVALUATION OF STPS AT NOIDA

12.9.1.1 STPS INSTALLED AT NOIDA SECTOR 54 (SBR & ISBR)

02 numbers of STP of capacity 33 MLD and 54 MLD are installed in sector -54. The STP of capacity 33 MLD is receiving 18 MLD of sewage. The STP of capacity 54 MLD is utilizing 27 MLD. STP receives sewage from catchment area namely Bishanpura, Gijhapur, Makanpur, Nithari, Morna, Harola, Atta, Film City and treated sewage disposed into Noida drain (Figures 12.10–12.16).

FIGURE 12.10 Grit chamber of 33 MLD unit.

FIGURE 12.11 Inlet chamber of 33 MLD unit.

FIGURE 12.12 Sludge disposal of 33 MLD unit.

FIGURE 12.13 Grit chamber of 54 MLD unit.

FIGURE 12.14 Clari-digestor of 54 MLD unit.

FIGURE 12.15 SBR unit of 54 MLD unit.

FIGURE 12.16 Chlorination unit of 54 MLD.

Performance of STPs with respect to physio-chemical and bacteriological parameter at different unit process is depicted in Tables 12.3 and 12.4.

TABLE 12.3 Analytical Results of Noida Sector 54, SBR Based STP

SL. No.	Source	Parameter							
		pH	TSS	COD	BOD	NH3-N	PO4-	T.C MPN/ 100 ml	F.C MPN/ 100 ml
1.	Inlet	7	166	335	132	31	3.41	40 x 10^6	14 x 10^6
2.	Outlet	7.49	13	34	3	BDL	0.78	49 x 10^3	11 x 10^3

Note: All values are measured in mg/l except pH

TABLE 12.4 Analytical Results of Noida Sector 54, iSBR Based STP

SL. No.	Source	Parameter							
		pH	TSS	COD	BOD	NH3-N	PO4-	T.C MPN/ 100 ml	F.C MPN/ 100 ml
1.	Inlet	7.1	91	256	121	27	3.95	11 x 10^6	4 x 10^6
2.	Outlet	7.3	20	14	7	8	0.38	49 x 10^3	68 x 10^2

Note: All values are measured in mg/l except pH

Observations:

The following observations are made:

1. STP of capacity 54 MLD is designed on ISBR technology and 33 MLD units is designed on SBR technology. The 54 MLD unit is equipped with Gas Engine as Biogas generated from Clari-Digestor is used for power generation, and same is used in-house.
2. Both unit of capacity 54 MLD and 33 MLD has limitation for sludge disposal. The dry sludge processed through centrifuge is lying inside the premises of STP.
3. BOD of STP of 33 MLD at the inlet was measured as 132 mg/l, and with subsequent treatment process, it reduces to 03 mg/l. Similarly, TSS level show reduction of 92.16% as TSS level reduces from 166 mg/l to 13 mg/l and COD show 89.85% removal. With respect to Fecal Coliform, 99.92% removal was observed as it reduces from 14 x 10^6 MPN/100 ml to 11 x 10^3 MPN/100 ml. Nitrogen removal with respect to Ammonical Nitrogen was observed as below detection limit. Similarly, Phosphate removal was observed as it reduces from 3.41 mg/l to 0.78 mg/l.

4. BOD of STP of 54 MLD at the inlet was measured as 121 mg/l and with subsequent treatment process it reduces to 07 mg/l. Similarly, TSS level show reduction of 78.02% as TSS level reduces from 91 mg/l to 20 mg/l and COD show 94.5% removal. With respect to Fecal Coliform, 99.83% removal was observed as it reduces from4 x 10^6 MPN/100 ml to 68 x 10^2 MPN/100 ml. Nitrogen removal with respect to Ammonical Nitrogen was observed as NH_3-N reduces from 27 mg/l to 8 mg/l. Similarly, Phosphate removal was observed as it reduces from 3.95 mg/l to 0.38 mg/l.

12.9.2 PERFORMANCE EVALUATION OF STP'S AT DELHI

12.9.2.1 STP INSTALLED AT DELHI GATE

Delhi Gate (New) STP is located near Delhi Secretariat and receives sewage from Delhi Gate Drain. The installed capacity of STP is 68 MLD and actual utilization is 69 MLD. The STP is based on Densadeg-BIOFOR technology.

The main components of the treatment process of Densadeg BIOFOR plant comprise coagulation and flocculation in a specially designed Clarisettler, followed by two-stage filtration through a special medial bed where organic degradation is facilitated by external oxygenation (Figures 12.17–12.23).

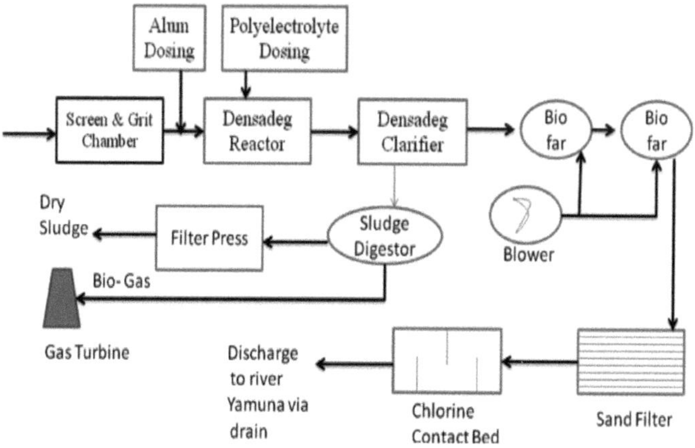

FIGURE 12.17 Schematic flow diagram of Delhi Gate (New) STP process.

FIGURE 12.18 Intake point of STP.

FIGURE 12.19 Coarse screen bar.

FIGURE 12.20 Densadeg clarifier.

FIGURE 12.21 Bio-for filter.

FIGURE 12.22 Sand filter unit.

FIGURE 12.23 Chlorination tank.

Performance of STPs with respect to physio-chemical and bacteriological parameter at the different unit process is depicted in Table 12.5.

TABLE 12.5 Analytical Results of Delhi Gate, BIOFOR Based STP

SL. No.	Source	Parameter							
		pH	TSS	COD	BOD	NH3-N	PO^{4-}	T.C MPN/ 100 ml	F.C MPN/ 100 ml
1.	Inlet	7.58	256	397	109	39	2.82	70 x 10^9	49 x 10^9
2.	Outlet	7.69	13	52	06	21	0.42	<1.8	<1.8

Note: All values are measured in mg/l except pH

Inspection and analytical results following observations are made:

1. The Plant was operated by M/s Degramont Pvt. Ltd.
2. BOD at the inlet of STP was measured as 109 mg/l and with subsequent treatment process it reduces to 06 mg/l. Similarly, TSS level show reduction of 94% as TSS level reduces from 256 mg/l to 14 mg/l and COD show 85% removal.
3. With respect to Fecal Coliform, 99.99% removal was observed as it reduces from 49 x 10^9 MPN/100 ml to 14 x 10^3 MPN/100 ml.
4. Nitrogen removal with respect to Ammonical Nitrogen was observed as NH_3-N reduces from 39 mg/l to 21 mg/l. Similarly, Phosphate reduces from 2.82 mg/l to 0.42 mg/l.
5. The sludge collected from densadeg clarifier is conveyed to sludge digester. As informed, 32 tonnes/day of sludge is generated and sludge digester has the capacity of 9000 m^3 which generates 2500 m^3/day of biogas. The biogas generated is used for production of electricity via gas turbine.
6. The plant has the power requirement of 23,000 Kwh/day and half of the requirement is met by the Gas turbine.

12.9.2.2 STP INSTALLED AT OKHLA PHASE VI

Okhla STP is located at Phase VI Okhla, Delhi Secretariat and receives sewage from surrounding areas. The installed capacity of STP is 136.2 MLD and actual utilization is 136.2 MLD. The STP is based on ASP technology (Figures 12.24–12.31).

FIGURE 12.24 Grit chamber.

FIGURE 12.25 Primary clarifier.

FIGURE 12.26 Phosphorus removal unit.

FIGURE 12.27 Anoxic zone.

FIGURE 12.28 Aeration tank.

FIGURE 12.29 Secondary clarifier.

FIGURE 12.30 Sand filtration unit.

FIGURE 12.31 Chlorination tank.

Performance of STPs with respect to Physico-Chemical and Bacteriological parameter at the different unit process is depicted below (Table12.6):

TABLE 12.6 Analytical Results of Okhla, ASP Based STP

SL. No.	Source	Parameter							
		pH	TSS	COD	BOD	NH3-N	PO^{4-}	T.C MPN/ 100 ml	F.C MPN/ 100 ml
1.	Inlet	6.99	143	172	87	25	3.44	16×10^{10}	16×10^{10}
2.	Outlet	7.17	13	5	3	19	0.62	20	20

Note: All values are measured in mg/l except pH

Observations:

Inspection and analytical results following observations are made by CPCB:

1. The plant is constructed and operated by M/s VA Tech WABAG Ltd.
2. BOD level at the inlet of STP was 87 mg/l and with subsequent treatment the BOD reduced to 03 mg/l.
3. Similarly, TSS level show reduction of 90.9%. COD level reduces from 172 mg/l to 05 mg/l.
4. With respect to fecal coliform, 100% removal was observed as it reduces from 10 x 10^{10} MPN/100 ml to 20 MPN/100 ml. Fecal coliform at outlet is meeting the proposed norms. The plant is equipped with sand filtration and chlorination facility for coliform reduction.
5. Ammonical Nitrogen and Phosphate reductions were observed as NH_3-N level at outlet of STP was 19 mg/l and Phosphate level was 0.62 mg/l.

12.10 RESULTS AND DISCUSSIONS

The team inspected STPs in Delhi and Noida on advanced treatment technology along with coliform reduction facilities. All the four STPs are designed on different treatment technology. The details of STPs are mentioned in Table12.7.

TABLE 12.7 STPs Details

SL. No.	STP Location	Technology	Capacity Installed (MLD)	Actual Treatment Capacity (MLD)
1.	Noida Sector 54 Phase I	SBR	33	18
2.	Noida Sector 54 Phase II	ISBR	54	27
3.	Delhi Gate Phase II	BIOFAR	68.1	49.94
4.	Okhla Phase VI	ASP	136.2	136.2

In order to assess the performance of the above said STPs and to efficiency removal of pollutant, in-depth analysis was carried out at inlet and outlet units of STPs. CPCB's team collected samples from inlet and outlet unit of STPs. The samples collected were analyzed for following parameters.

Sewage Treatment Plants	**Inlet**	pH, COD, BOD, SS, NH_4–N, PO^{4-}, Total, and Fecal coliform
	Outlet	pH, COD, BOD, SS, NH_4–N, PO^{4-}, Total, and Fecal coliform

Sampling of STPs is carried out at different stages/treatment process of plant with respect to physio-chemical parameters and bacteriological parameters. The summarized table indicating analytical results of all the seven STPs for above said parameters indicates analytical result of STP for Inlet and Outlet with respect to physio-chemical parameters and bacteriological parameters analyzed by CPCB.

Analytical result of STP for inlet and outlet with respect to physio-chemical parameters and bacteriological parameter (Table 12.8).

TABLE 12.8 Analytical Results of STPs for Inlet and Outlet with Respect to Physio-Chemical Parameters and Bacteriological Parameters

SL. No	STP Location	Source	Parameters							
			pH	TSS	COD	BOD	NH_3-N	PO_4-P	T.C MPN/ 100 ml	F.C MPN/ 100 ml
1.	Noida Sector 54 Phase I	Inlet	7	166	335	132	31	3.41	40×10^6	14×10^6
		Outlet	7.49	13	34	3	BDL	0.78	49×10^3	11×10^3
2.	Noida Sector 54 Phase II	Inlet	7.1	91	256	121	27	3.95	11×10^6	4×10^6
		Outlet	7.3	20	14	7	8	0.38	49×10^3	68×10^2
3.	Delhi Gate Phase II	Inlet	7.58	256	397	109	39	2.82	70×10^9	49×10^9
		Outlet	7.69	13	52	06	21	0.42	<1.8	<1.8
4.	Okhla Phase VI	Inlet	6.99	143	172	87	25	3.44	16×10^{10}	16×10^{10}
		Outlet	7.17	13	5	3	19	0.62	20	20

Note: All values are measured in mg/l except pH.

The comparative analysis of aforesaid 04 STPs were carried out to find out the removal efficiency for microbial and organic pollutant. Summary of removal efficiency is depicted in Table 12.9.

TABLE 12.9 Removal Efficiency of 04 STPs for Organic and Microbial Pollutant

SL. No.	STP Location	Treatment Technology	% BOD Removal	% COD Removal	% TSS Removal	% Total Coliform Reduction	% Fecal Coliform Reduction
1.	Noida Sector 54 Phase I	SBR	97.72	89.85	92.16	99.87	99.92
2.	Noida Sector 54 Phase II	I-SBR	94.21	94.53	78.02	99.55	99.83
3.	Delhi Gate Phase II	BIOFAR	94.49	86.90	94.92	100	100
4.	Okhla Phase VI	ASP	96.55	97.09	90.90	100	100

Note: All values are measured in mg/l except pH.

In view of the above studies and facts, observations are made with respect to achievability in reduction of organic and bacteriological pollutant. The primary goal of sewage treatment plant is to reduce organic loading. Performance efficiency is done by comparing the outlet and inlet BOD. It greatly depends on the inlet quality BOD removal efficiency also depends upon the aeration time provided which can vary according to the size of plant and quantity and incoming sewage composition. Study of performance evaluation of STPs based on SBR, iSBR, BIOFAR, and ASP technologies shows that the BOD removal efficiencies are 97.72%, 94.21%, 94.49%, and 96.55% while the COD removal efficiencies are 89.85%, 94.53%, 86.9% and 97.09% (Figures 12.32–12.34).

FIGURE 12.32 BOD reduction efficiency.

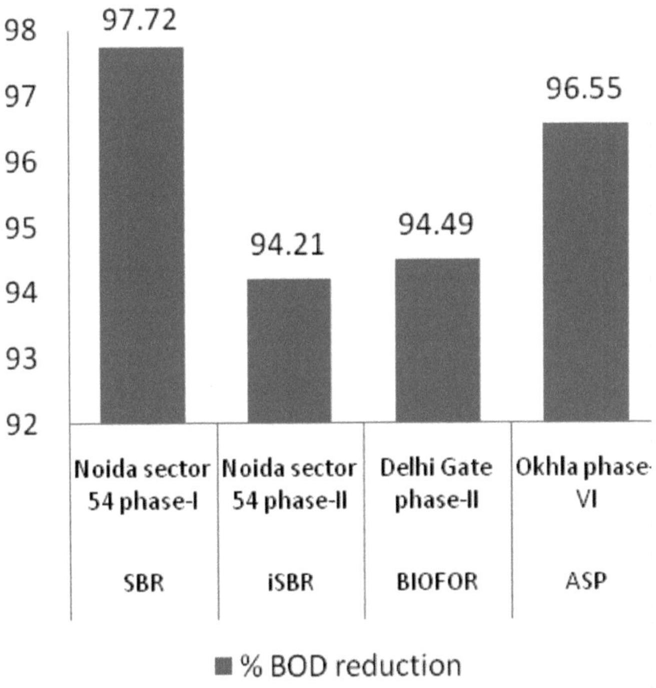

FIGURE 12.33 COD reduction efficiency.

Study of performance evaluation of STPs based on SBR, iSBR, BIOFAR, and ASP technologies shows that the TSS removal efficiencies are 92.16%, 78.02%, 94.92%, and 90.9%. The reduced TSS levels at the outlet are attributed to the STPs efficiency of TSS removal through sedimentation of settleable solids. Discharge of effluents with high levels of suspended solids into water bodies lower the quality of water and deplete dissolved oxygen availability for aquatic life (Figure 12.34).

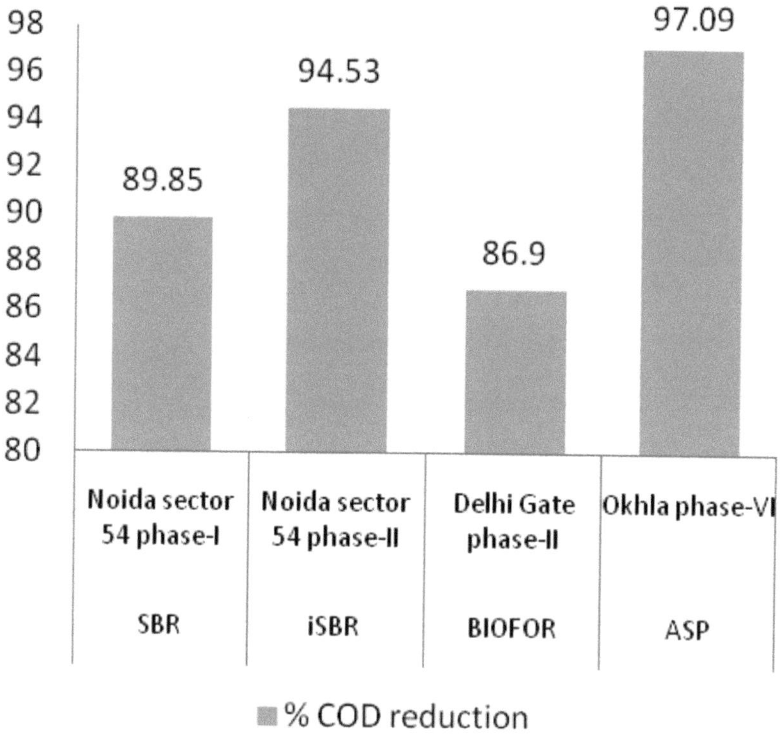

FIGURE 12.34 TSS reduction efficiency.

Study of performance evaluation of STPs based on SBR, iSBR, BIOFAR, and ASP technologies shows that the TC removal efficiencies are 99.87%, 99.55%, 100%, and 100% and FC removal efficiencies are 99.92%, 99.83%, 100% and 100%. Monitoring of biological markers in particular treated sewage means to determine the overall quality of the same. Since 1920s, TC, and FC are used as standard microbial indicators for water quality. The types and concentration of these organisms depend on various factors such as per capita water supply rate, connection of sources like hospitals, levels of disease in the community and other constituents of sewage water including chemicals which are toxic (Figures 12.35 and 12.36).

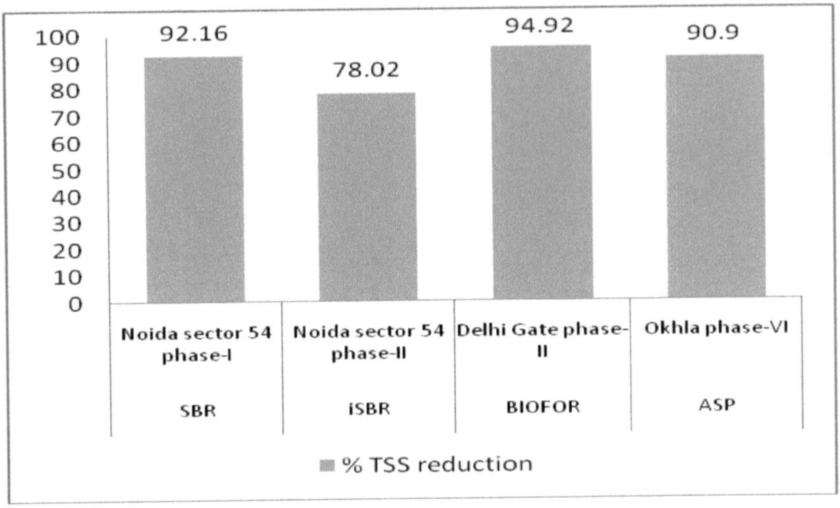

FIGURE 12.35 TC reduction efficiency.

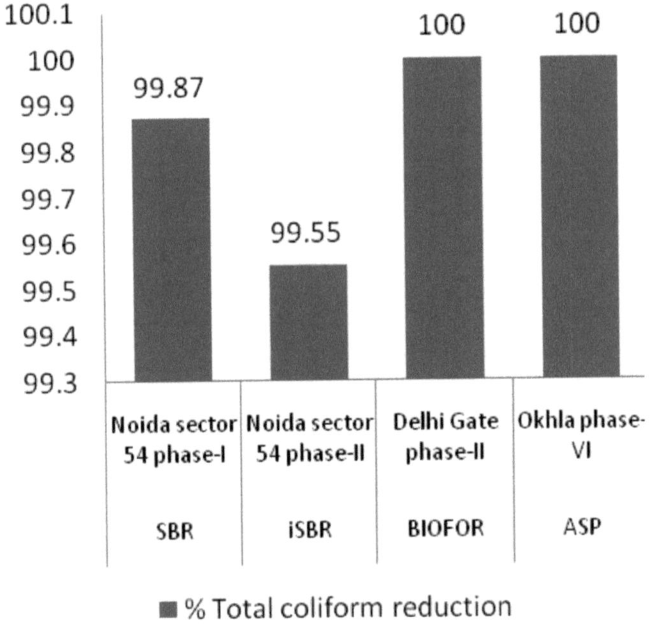

FIGURE 12.36 FC reduction efficiency.

Based on the experience of the present study, the operation and maintenance of STP's depend on three factors:

- Regular and preventive maintenance.
- Uninterrupted power supply.
- Skilled manpower.

In the case of conventional treatment technologies, more energy is required. For proper knowledge of the process and operation, skilled manpower is required for smooth functioning of the STP's. Also, maintenance is required at every stage with due diligence.

12.11 CONCLUSION

Treatment of sewage effluents is vital to prevent pollution and protect environment and public health before being discharged to aquatic bodies. In this project, an attempt is made to compare STPs based on SBR, iSBR, BIOFAR, and ASP technologies. The performance evaluation of all the STPs was done on the basis of following parameters like BOD, COD, TSS, TC, and FC. Results were almost comparable and were not varying significantly. The STPs designed on advanced technology are achieving BOD of less than 10mg/l and COD 50 mg/l. The above STPs are also achieving nutrients removal with respect to phosphate and ammonical nitrogen removal. STPs having facility of sand filtration before disinfection have achieved TSS level less than 10 mg/l and fecal coliform less than 230 MPN/100 ml. The STPs is disinfecting the treated effluent using Ultra Violet system and have a coliform reduction with 99.999% reduction. It shows that all the technologies are relatively efficient in the reduction of organic loads and nutrients from sewage water. The study depicts that there is availability of advanced technology which can reduce organic pollutant, nutrients, and bacteriological contamination at par with the standards.

KEYWORDS

- **activated sludge process (ASP)**
- **BIOFOR**
- **improved sequential batch reactor (ISBR)**
- **performance evaluation**
- **sequential batch reactor (SBR)**
- **sewage treatment plants**
- **technology**

REFERENCES

Akın, B. S., & Ugurlu, A., (2005). Monitoring and control of biological nutrient removal in a sequencing batch reactor. *Process Biochemistry*, *40*, 2873–2878.

Bhat, H. N., (2013). *Combitreat-ISBR Sewage Treatment Technology by HNB Engineers, a Part of IFAT India 2013, a Trade Fair Focusing on Water, Sewage, Refuse and Recycling Which will Premiere between October 24 and 26, 2013, at the Bombay Convention and Exhibition Center (BCE).*

Boller, M., (1997). Small wastewater treatment plants – A challenge to wastewater engineers. *Water Science and Technology*, 35, 1–12.

Bradley, B. R., Daigger, G. T., Rubin, R., & Tchobanoglous, G., (2001). Evaluation of onsite wastewater treatment technologies using sustainable development criteria. *J. of Clean Technol. and Envt.*, *4*, 87–99.

Charu, S., & Singh, S. K., (2013). Performance evaluation of sewage treatment plant Based on advanced aerobic biological filtration and oxygenated reactor (BIOFOR) technology-A case study of capital city, Delhi, India. *International Journal of Engineering Science and Innovative Technology*, *2*(4), 435–442.

Chudoba, J., Ottova, V., & Madera, V., (1973). Control of activated sludge filamentous bulking: Effect of hydraulic regime or degree of mixing in aeration tank. *Water Research*, *8*, 1163.

Cornejo, P. K., (2015). *Environmental Sustainability of Wastewater Treatment Plants Integrated with Resource Recovery: The Impact of Context and Scale*. Graduate Theses and Dissertations, University of South Florida, pp. 1–181.

CPHEEO, (2012). Central Public Health and Environmental Engineering Organization Ministry of Urban development New Delhi, Manual on Sewerage and Sewage Treatment Part A: Engineering.

EPA, (1999). *USEPA*. Wastewater, Technology Fact Sheet: Sequencing Batch Reactors, U.S. Environmental Protection Agency, Office of Water, Washington, D.C., EPA 932-F-99-073.

EPA, (2012b). *Guidelines for Water Reuse*. Washington, DC: The United States Environmental Protection Agency.

FAO, (2010). *The Wealth of Waste-the Economics of Wastewater Use in Agriculture*. Rome, Italy: Food and Agricultural Organization.

Fine, P., & Hadas, E., (2012). Options to reduce greenhouse gas emissions during wastewater treatment for agricultural use. *Science of the Total Environment, 416*, 289–299.

Frijns, J., & Jansen, M., (1996). Institutional requirements for appropriate wastewater treatment systems. In: Balkema, A., Aalbers, H., & Heijndermans, E., (eds.), *Workshop on Sustainable Municipal Waste Water Treatment Systems* (pp. 54–66). Leusdan, the Netherlands. ETC in cooperation with WASTE.

Gallego, A. Hospido, M., Moreira, T., & Feijoo, G., (2008). Environmental performance of wastewater treatment plant. *Resources, Conservation and Recycling, 52*, 931–940.

Guide Manual, Water and Waste Water Analysis, CPCB. pp. 1–183. https://www.indiawaterportal.org/sites/indiawaterportal.org/files/Water_and_wastewater_analysis_Guide_manual_Central_Pollution_Control_Board_2011.pdf.

Mara, D., & Cairncross, S., (1989). *Guidelines for the Safe Use of Wastewater and Excreta in Agriculture and Aquaculture*. Geneva: United Nations Environmental Programme/World Health Organization. ISBN: 9241542489; pp. 1–187.

Metcalf and Eddy, (2003). *Wastewater Engineering, Treatment and Reuse*. New Delhi: Tata McGraw-Hill Publishing Company Limited.

Mihelcic, J. R., Fry, L. M., & Shaw, R., (2011). Global potential of phosphorus recovery from human urine and feces. *Chemosphere, 84*, 832–839.

Mittal, A., (2011). *Biological Wastewater Treatment* (pp. 32–44). Fulltide Articles, Water Today.

Mittal, G. S., (2006). Treatment of wastewater from abattoirs before land application—a review. *Bioresource Technology, 97*(9), 1119–1135.

Mo, W., & Zhang, Q., (2012). Can municipal wastewater treatment systems be carbon neutral? *Journal of Environmental Management, 112*, 306–367.

Nagwekar, P. R., (2014). Removal of organic matter from wastewater by activated sludge process – Review. *International Journal of Science, Engineering and Technology Research (IJSETR), 3*(5), 1260–1263.

NRC, (2012). *Water Reuse: Potential for Expanding the Nation's Water Supply Through Reuse of Municipal Wastewater*. Washington, D. C.: National Research Council of the National Academies.

Nykova, N., Muller, T. G., Gyllenberg, M., & Timmer, J., (2002). Quantitative analyses of anaerobic wastewater treatment processes: Identifiability and parameter estimation. *Biotechnology and Bioengineering, 78*(1), 89–103.

Obaja, D. S., Mace, J., Costa, C., Sans, J., & Mata-Alvarez, (2003). Nitrification, denitrification and biological phosphorus removal in piggery wastewater using sequencing batch reactor. *Bioresource Technology, 87*, 103–111.

Paranychianakis, N. V., Salgot, M., Snyder, S. A., & Angelakis A. N., (2015). Quality criteria for recycled wastewater effluent in EU countries: Need for a uniform approach. *Crit. Rev. Environ. Sci. Technol., 45*, 1409–1468.

Schroeder, E. D., (1982). Design of sequencing batch reactor activated sludge processes. In: *Civil Engineering for Practicing and Design Engineers, 2*, 33–34.

Singh, D. D., & John, S., (2013). Study the different parameters of sewage treatment with UASB & SBR technologies. *LOSR Journal of Mechanical and Civil Engineering (LOSRJMCE), 6*(1), 112–116.

Veenestra, S., & Alaerts, G., (1996). Technology selection for pollution control. In: Balkema, A., Aalbers, H., & Heijndermans, E., (eds.), *Workshop on Sustainable Municipal Waste Water Treatment Systems* (pp. 17–40). Leusdan, the Netherlands. ETC in co-operation with WASTE.

Wijetunga, S., Li, X. F., & Jian, C., (2010). Effect of organic load on decolorization of textile wastewater containing acid dyes in upflow anaerobic sludge blanket reactor. *Journal of Hazardous Materials*, *177*(1–3), 792–798.

Xudong, C., Xu, Z., Yao, L., & Ma, N., (2018). Processing technology selection for municipal sewage treatment based on a multi-objective decision model under uncertainty. *International Journal of Environmental Research and Public Health, 15(3), 448.*

Ying-Chih, C., Li-Ling, L., Cheng-Nan, C., & Chao, A. C., (2007). Control of carbon and ammonium ratio for simultaneous nitrification and denitrification in a sequencing batch bioreactor. *International Biodeterioration & Biodegradation*, *59*, 1–7.

Index

B

C

P

Q

R